MATEJ ŠURC

# PATRIOTISM FOR SALE

How Independent Slovenia Fell
Foul of Crooked Arms Dealing

Translated by *Vasilij Volarič*

MATEJ ŠURC

# PATRIOTISM FOR SALE
How Independent Slovenia Fell Foul of Crooked Arms Dealing

*Translated by* Vasilij Volarič
*Editor and Introduction* Marcus Ferrar
*Foreword* Eva Joly
*Expert reviewers* Andrew Feinstein, dr. Spomenka Hribar

*Cover design* Boštjan Pavletič
*Page break* David Križ

*Dokumenta* Editions

*Published by*
Sanje Publishing, Ltd
Leskoškova 12, 1000 Ljubljana, Slovenija
Rok Zavrtanik

© Matej Šurc, 2018
All Rights Reserved.

First Edition, March 2018
Print on demand

*Hardcover* ISBN 978-961-274-547-9
*Paperback* ISBN 978-961-274-520-2
*e-Publication* ISBN 978-961-274-521-9

www.sanje.si

A CIP catalogue record for this book is available from
the National and University Library, Ljubljana

94(497.4)"1990/1992"
323(497.4)"1990/1992"

ŠURC, Matej
Patriotism for sale : how independent Slovenia fell foul of crooked arms dealing /
Matej Šurc ; translated by Vasilij Volarič ; [introduction Marcus Ferrar ; foreword
Eva Joly]. - 1st ed. - Ljubljana : Sanje, 2018. - (Dokumenta editions / Sanje)

Izv. stv. nasl.: Prevarana Slovenija

ISBN 978-961-274-547-9 (hardcover)
ISBN 978-961-274-520-2 (paperback)

293823232

MATEJ ŠURC

# PATRIOTISM FOR SALE

For my wife Marjetka, son Jernej and daughter Eva, with love.

*If an honest patriot is required to honour his homeland with his blood and life whenever necessary, he is no less required to serve his homeland with his pen to make it renowned around the world at every opportunity.*

Janez Vajkard Valvasor, "The Glory of the Dukedom of Carniola"
Bogenšperk Castle, April 15, 1689

# CONTENTS

FOREWORD ........... 13

INTRODUCTION ........... 17

- *Chapter 1* -
COMRADES AND GENTLEMEN: BETWEEN THE RED
STAR AND AN INDEPENDENT SLOVENIA ........... 19

- *Chapter 2* -
AS STORM CLOUDS GATHER, SLOVENIA RE-ARMS ........... 43

- *Chapter 3* -
SLOVENIA DECLARES INDEPENDENCE: YUGOSLAV
ARMY INVADES ........... 65

- *Chapter 4* -
SLOVENIAN IS KILLING BROTHER SLOVENIAN ........... 89

- *Chapter 5* -
"A MAN IN SHORTS WITH KALASHNIKOV DISARMS ARMY" ........... 107

- *Chapter 6* -
SURRENDER ULTIMATUM AS SLOVENIANS SEIZE
YUGOSLAV ARMS DEPOTS ........... 129

- *Chapter 7* -
SAVING LJUBLJANA FROM A BLOODBATH ........... 147

- *Chapter 8* -
CEASEFIRE AS YUGOSLAV ARMY DISINTEGRATES ........... 165

- *Chapter 9* -
SLOVENIA SETTLES AT BRIONI … BUT A MAJOR
KNOWS TOO MUCH ........... 177

- *Chapter 10* -
SHOWDOWN IN BELGRADE: SLOVENIA TRAINS
CROATIAN USTASHES ........... 199

- *Chapter 11* -
SLOVENIANS SWINDLE CROATIAN DEFENDERS AND
LINE THEIR OWN POCKETS ........... 215

- Chapter 12 -
SLOVENIA'S WEAPONS BAZAAR ... 231

- Chapter 13 -
YUGOSLAV ARMY PULLS OUT: SLOVENIA RECOGNISED ... 251

- Chapter 14 -
THE VANISHING PROCEEDS FROM ILLEGAL ARMS SALES
TO CROATIA ... 273

- Chapter 15 -
THE FOX GUARDS THE HENHOUSE ... 291

- Chapter 16 -
SIX MEN CHARGED AND A PROSECUTOR INFATUATED ... 305

- Chapter 17 -
THE DUBIOUS MIDDLEMEN OF SLOVENIA'S ARMS TRADE ... 321

- Chapter 18 -
ARMS DEALER OMAN, JANŠA AND THE MYSTERY OF
FUNDS IN AUSTRALIA ... 337

- Chapter 19 -
DID SREBRENICA DIE FOR THE FREEDOM OF CROATIA? ... 351

- Chapter 20 -
DUELLING SWORD VERSUS MANURE PITCHFORK ... 365

EPILOGUE ... 377

REVIEWS ... 381

ACKNOWLEDGEMENTS ... 391

BIBLIOGRAPHY ... 393

INDEX OF PERSONS ... 397

# KEY CHARACTERS

**Slovenia**

Milan Kučan, President of the Presidency

Matjaž Kmecl, member of the Presidency
Ivan Oman, member of the Presidency
Dušan Plut, member of the Presidency
Ciril Zlobec, member of the Presidency

Janez Drnovšek, member of the Yugoslav Presidency (later Prime Minister and President of Slovenia)

France Bučar, President of Parliament

Lojze Peterle, Prime Minister

Igor Bavčar, Interior Minister
Jelko Kacin, Information Minister (later Defence Minister)
Dimitrij Rupel, Foreign Minister
Janez Janša, Defence Minister (later Prime Minister)

Janez Slapar, Colonel, Commander of the Headquarters of the Territorial Defence (TD)
Miha Brejc, Director of the Security and Information Service (VIS)
Anton Krkovič, Commander of the Moris brigade
Andrej Lovšin, Director of the Security organ of the Defence Ministry (VOMO)
Ludvik Zvonar, Advisor to the Government
Ivan Draušbaher, Head of the Orbis Company

**Federal Yugoslavia and Serbia**

Ante Marković, Yugoslav Prime Minister
Borislav Jović, Yugoslav President
Stipe Mesić, Croatian due to succeed Jović as Yugoslav President
General Veljko Kadijević, Yugoslav Defence Minister
General Blagoje Adžić, Yugoslav Chief of General Staff
Slobodan Milošević, President of Serbia
General Andrija Rašeta (negotiated withdrawal from Slovenia)

**Croatia and Bosnia-Herzegovina**

Franjo Tuđman, President of Croatia
General Martin Špegelj, Croatian Defence Minister
Dobroslav Paraga, president of the Croatian Party of Rights (HSP)
Alija Izetbegović, President of Bosnia-Herzegovina (BiH)
Hasan Čengić, member of BiH Government, provided weapons for BiH

**International arms traders**

Konstantin Dafermos, Greece/Austria
Jerzy Dembowski, Poland
Nikolaj Oman, Slovenia/Australia
Hans-Wolfgang Riedl, Austria
Dmitry Streshinsky, Ukraine
Josip Vukina, Croatia
Walter Wolf, Slovenia/Austria/Canada
Vladimir Zagorec, Croatia
Nikša Župa, Croatia/Slovenia

# FOREWORD

The heroes are not always those we think of. Presumed heroes of Slovenia's independence may very well be the main war profiteers. Matej Šurc's story of Slovenia's independence rightly – and sadly – illustrates this. In the early 1990s, many high-profile actors took advantage of the confusion, secrecy and needs brought about by the conflicts in the Balkans. The beginnings of Slovenian independence were characterised by massive and fraudulent arms trafficking. This very well documented book sheds light on the illegal selling and reselling of weapons that involved powerful men, notably the former Defence Minister Janez Janša.

Several prominent actors in Slovenia indeed used the war to make illegal and huge profits, thanks to the embargo on weapons, to the very high prices of arms, to the desperate needs of neighbouring countries, and to any "good opportunity". Because most of the documents regarding the arms trade disappeared, it is still unknown today where all the money from these sales went. Not to taxpayers – that for sure. It is a shame that the emergence of the new independent state also marked the advent of corruption and abuses of authority in high-level positions.

Beyond the disappearance of most of the sensitive documents, the quest for the truth also suffered from cronyism in the judiciary and intimidation attempts. Several prosecutions were closed without any real justification, nor satisfying epilogue. Even more alarming are the intimidation attempts on those people who know too much or try to learn more about the illegal trafficking. When a Slovenian soldier, Marjan Strehar, expressed more and more curiosity about the money earned from the arms trade, he died in unexplained circumstances in 2012. The date of this murder shows that, even today, this affair remains a secret and sensitive one. An epilogue in a court of law has not happened yet. Some of the prominent suspects of fraud even got back to power in recent years. It is the sign of ultimate impunity.

In this story, the heroes are not those we first think of. They are not those who self-proclaim themselves patriots. The actual heroes are those who devote their entire life to the quest for the truth about these dark episodes of Slovenian history. Those who resist intimidation attempts and death threats. Those who call for justice to be done. Those who relentlessly fight fraudulent activities. Those who constantly refuse impunity. The fight against corruption is all about women and men – not so much about institutions and organisations. It is the story of lawyers, economists, journalists, police officers, officials, activists and any citizens. In sum, it is the story of "ordinary heroes", brave enough to shed light on opaque affairs.

I have had the chance to meet one of these Slovenian heroes during my career, Mr Drago Kos, a former Criminal Police officer who soon became an anti-corruption champion. He is one of the key sources whom the author relied upon to write his book, since he was notably in charge of investigating the illegal weapons trade in Slovenia. His portrait says a lot about the many difficulties encountered when trying to combat impunity.

Fighting corruption requires a lot of courage and pugnacity. As a former anti-corruption prosecuting judge myself, I have seen how difficult it is to carry on a case until the end and to overcome the many obstacles to an investigation. I know perfectly well how much determination is required in order not to give in to intimidation attempts or even death threats. Combating impunity sometimes obliges one to personal sacrifices; a family life is not easily compatible with such efforts. While he was in charge of investigating the illegal arms trade involving Defence Minister Janez Janša, Drago Kos was obliged to live with two bodyguards and to sleep in police buildings, far from his close relatives. If one may bear the presence of bodyguards, it is almost impossible to accept to put the lives of close relatives in danger. Drago Kos' only regret is "to have sacrificed two families"[1] - referring to his two divorces. Against his will, he knows the price of a full and permanent commitment towards more justice and transparency.

Combating corruption often implies looking into the affairs of powerful leaders: authority does not intimidate "ordinary heroes". Several times, because his independence and perseverance were not to the taste of everyone, Drago Kos had to face threats to be removed from his position - "six or seven times in 15 years", he said.[2] His full commitment and his tireless determination to combat corruption led him to become, in 2004, the first elected President of the Commission for the Prevention of Corruption in the Republic of Slovenia. Unfortunately, all his efforts sometimes were destroyed by other personalities under influence. In the present book, Drago Kos notably expresses his frustration that many criminal complaints never reached the courts because of the attitude of the prosecutor in charge: "An investigative magistrate said that with this kind of actions by prosecutors, the suspects don't need defence attorneys."[3]

Fighting corruption also requires looking beyond national borders, as frauds and illicit flows of money are more and more transnational. Drago Kos is also perfectly aware of that, because the illegal arms trafficking that he investigated involved foreign banks, but also because he chaired two international anti-corruption networks, the Group of States against Corruption (GRECO) – the anti-corruption network of the Council of Europe – as well as the OECD Working Group on Bribery. Criminals have taken advantage of the opacity of our financial system and of all tax avoidance possibilities in order to build more and more sophisticated corruption techniques, with an increasing number of intermediaries involved. There are war profiteers; there are also system profiteers. As a result, uncovering the entirety of these complex operations and briberies has become more and more difficult.

Transnational and systemic solutions are urgently needed to fight corruption because everything is connected. You cannot stop corruption without addressing illicit financial flows. In the meantime, you cannot put an end to illicit financial flows without changing the current opaque and secretive global financial system. The illegal pillaging of natural resources, poaching, fishing, arms trafficking … All these lucrative activities take advantage of the opacity of the financial system that we have built. It is now our responsibility to correct our mistakes by bringing back transparency and accountability. How can you effectively dissuade criminals from making use of the opacity of the financial system while we continue to use the system for our tax evasion and tax

---

2) (Ordinary Heroes), Les Arènes, Paris, 2009, p.136.

avoidance activities? It is nonsense. The fight against corruption and the fight against financial opacity must go hand in hand. Tax havens, secrecy jurisdictions, fake corporations, anonymous trust accounts and all these system drifts must end.

Corruption has already caused the loss of considerable amounts of money – not only because of the illegal arms trafficking, and not only in Slovenia. In the whole European Union, the cost of corruption is worth between 179 billion euro and 990 billion euro every year. Public procurement corruption alone costs the EU 5 billion euro per year.[4] Not to mention the cost of corruption and of the illegal pillaging of resources in Africa. Such situations shall not be tolerated anymore. It is more than urgent to put an end to illicit financial flows that are the roots of corruption.

If the fight against corruption is a story of brave men and women, there is no doubt that the solution cannot come from them alone. This battle must conquer the minds of the whole society if it wants to be successful in the end. Changing people's mindsets is probably the greatest challenge ahead. It is only if the whole society relentlessly rejects intolerable practices that stopping corruption may succeed. In order to do that, the truth must be known by all. This book is a key contribution towards this goal. The readers will not be able to tell that they did not know. Everyone can be an "ordinary hero".

*Eva Joly, lawyer and member of the European Parliament*

---

[4] From a study by RAND Europe, "The Cost of Non-Europe in the area of Organised Crime and Corruption", Brussels, European Parliamentary Research Service, March 2016, Annex II, p.115, available here: http://www.europarl.europa.eu/RegData/etudes/STUD/2016/579319/EPRS_

# INTRODUCTION

Just over a quarter of a century ago, Slovenia broke away from Yugoslavia and became an independent state. In a 10-day war in the summer of 1990 – the first war in Europe since 1945 – two million Slovenians defied the redoubtable Yugoslav People's Army and won. David "slew" Goliath and in doing so precipitated the collapse of Yugoslavia.

The newly-independent Slovenians, organised in grass-roots militia, earned a reputation as plucky freedom-fighters rising up to cast off the dead hand of an oppressive communist federal state. Revolution was in the air all across Eastern Europe. Mass demonstrations by subjugated peoples brought the downfall of the Berlin Wall and the collapse of communist regimes in Czechoslovakia and Poland. In the Soviet Union, Mikhail Gorbachev was opening up society to free speech and liberal reforms.

When they voted massively for independence in a referendum in December 1990, Slovenians were thus following the spirit of the times. And in successfully resisting the attempt by the Yugoslav Federal state to crush their autonomy by force in 1991, they demonstrated courage, superior tactics and great negotiating skills.

But the true story of Slovenia's independence is more complex. It is a tale also of incompetence, inglorious compromises, but above all of corruption involving shady arms deals enriching some of those who considered themselves "heroes" of their country's liberation. Unresolved scandals of this massive illicit arms trading corrode the new Slovenian state at its core. The corruption which took root then poisons society in a country which seemed to have everything going for it at its birth.

Slovenia was a "model pupil" among East European states joining the European Union in 2004, eagerly adopting free institutions, a market economy, open borders and the common euro currency. However the inability of its politicians and courts to bring high-level criminals to book has undermined Slovenians' enthusiasm for these forward-looking acts of democracy and solidarity. Cynicism and resentment predominate instead.

Instrumental in this sorry state of affairs is a political party leader who served as Prime Minister before being convicted of serious crimes, then freed on appeal. He remains at large today, and his nickname is *Princ teme* (Prince of Darkness).

*- Chapter 1 -*

# COMRADES AND GENTLEMEN: BETWEEN THE RED STAR AND AN INDEPENDENT SLOVENIA

Slovenians have a rich folk culture, and one of their favourite authors is Fran Milčinski, who wrote an amusing series of nonsense stories entitled *Butalci* (Dim-Witted). One starts like this: "Three hours walk from Shrove Sunday, there lies a village they like to call a town ..."[5] That is how Slovenians felt about their centuries of yearning to be an independent nation. You walk and walk, and dream and walk, but you never get there. Until the closing years of the 1980s, the idea of Slovenians being anything more than a scattered people held together only by a common language and culture seemed as fanciful as the *Butalci*.

Living between the Alps, the Adriatic and the Pannonian Basin, Slovenians could hold off cultural colonisation by larger nations. But despite keeping their own language and culture, they were never able to establish an independent state, controlling their own military. First they were ruled by Austria, then after World War I they were integrated into newly-formed Yugoslavia.

When their lands were occupied by nazi Germany, fascist Italy and Hungary in World War II, Slovenians saw an opportunity to fight for true independence. They rallied in sizeable numbers to resist the foreign occupiers, mounting the first well-planned people's resistance anywhere in Europe, and ending the war controlling their own Partisan army. However, other Partisan leaders had no interest in letting old Yugoslavia split up, and the Slovenian "People's Liberation Army" was forced to incorporate into a Yugoslav People's Army (YPA), led by Marshal Josip Broz-Tito, President of the federal Yugoslav state.

---

5) Fran Milčinski, *Butalci*, Karantanija, Ljubljana, 2001.
6) Yugoslavia became the Socialist Federal Republic of Yugoslavia (SFRY) in 1963, before which it

Slovenian lands were governed by the Communist Party, later renamed the League of Communists, and property was owned by the state.

After breaking with Stalin in 1948, Tito led his country on its own "path towards socialism", establishing a new system of "workers' self-management" to give people more say than under the previous command economy. Significantly, it also gave greater autonomy to Slovenians and the other nationalities making up Yugoslavia.[7]

In response to the 1968 Soviet invasion of Czechoslovakia, which crushed the liberal Prague Spring, the Yugoslav leadership introduced a concept of "armed people" into its defence doctrine and set up Territorial Defence (TD) units. The perceived enemy was no longer the West, but the Soviet Union.

This suited the independent-minded Slovenians, who had become the most developed and economically successful part of the Yugoslav federation. In their part of the country they introduced a complicated theory of "Overall People's Resistance and Social Self-Defence", which they opaquely referred to as "part of the unified armed forces and the widest form of mass organisation of working people and citizens for armed struggle". The bureaucratic language concealed the Slovenians' real purpose, which was to use the force to further their independence.

In the following decades, they quietly and tenaciously transformed the theory into reality. Top positions in the TD were given to trusted Slovenian officials. Companies, municipalities and political organisations even purchased weapons for the Slovenian Territorial Defence directly from Yugoslav arms manufacturers based in Serbia and Bosnia-Herzegovina.

As such, the Slovenian Territorial Defence, even though formally part of the Yugoslav armed forces, gradually morphed into a Slovenian Army. The generals of the Yugoslav People's Army complained that this development would destroy the Yugoslav federation. But nobody, even after the death of the President in 1980, was willing to challenge "Comrade Tito's" will. As the famous Belgrade lawyer Toma Fila put

---

7) Besides Slovenians, Yugoslavia was populated by Serbs, Croats, Montenegrins, Macedonians and Muslims (in 1971 the Muslims received the status of a nation and were one of the constitutive nations in Bosnia-Herzegovina). The equal status of these nations was portrayed by the six torches in the official national emblem of Yugoslavia. The status of ethnic groups was given to Albanians,

it, "the Yugoslav People's Army formed the Territorial Defence as a joke, but the Slovenians formed a serious Slovenian army out of it".[8]

By that time, Slovenian dreams about their own state no longer felt like an endless walk from "Shrove Sunday". Indeed, the realisation of their aspirations was not far away at all. Slovenians only needed to wait for the right moment to break free of the Yugoslav yoke.

Yugoslavia was transforming and breaking apart at the same time. Nationalism was escalating, and the relationship between the federal government and the constituent republics was becoming more and more strained. The last Constitution, adopted in 1974, described Yugoslavia as a "federal community of voluntarily united nations and their socialist republics and autonomous regions".[9] As a result, Slovenia and other republics became de facto economically and politically more independent, and when Marshal Tito died in May 1980, his authority in preventing and resolving internal conflicts died with him.

By the end of the 1980s, cracks were starting to develop in that old stalwart of socialism, the Soviet Union, under the impact of the *glasnost* and *perestroika* reforms of Gorbachev. With the fall of the Berlin Wall in 1989, the Cold War came to an end and Yugoslavia's strategic role as an intermediary between East and West diminished.

Just as world interest turned elsewhere however, the country's inner frictions brought it to the edge of the abyss, all the more since Serbia's ambitions of hegemony began to threaten the equality between nations in the Yugoslav federation. A schism opening up between the developed north of the country and the undeveloped south became insupportable. In Slovenia, more and more people became convinced that the "hard-working" people from the north were supporting "lazy bums" in the south by channeling their money into a "Federal Fund for the Undeveloped Parts of Yugoslavia". After Tito's death, the Yugoslav Federation started to drown in debt and the Federal institutions, led by the eight-member Yugoslav Presidency, were no longer able to cope with the strains.

---

8) Author's conversation with Toma Fila, Belgrade, April 21, 1995. Fila was a top criminal lawyer, who also defended politicians, among them, former Serbian President Slobodan Milošević and Tito's widow Jovanka Broz.
9) Article 1 of the 1974 Yugoslav Constitution: http://www.pisrs.si/Pis.web/pregledPredpisa?

Due to increasing friction among the different Yugoslav nationalities, as well the desire of Slovenia, and now Croatia too, to become independent, the logic of the "armed people" boomeranged on Belgrade, and the country was threatened by the civil war. Defenders of Yugoslav political unity dogmatically insisted on unified Yugoslav armed forces under a single command, and demanded that the Slovenians hand over weapons of the Slovenian Territorial Defence to the Yugoslav People's Army (YPA). But the genie had been let out of the bottle, and in the first months of 1990, everything got worse.

\* \* \*

At noon on May 18, 1990, councillors of the municipal assembly of the small Slovenian town of Šentjur pri Celju were stunned to see a group of soldiers taking weapons from the municipal building and loading them on to military trucks. The locals had never seen anything like this. The weapons belonged to Slovenia and to the municipality of Šentjur pri Celju, because they had paid for them. But now, in front of their eyes, the Yugoslav People's Army was taking weapons away from them. In their eyes, the perpetrators were no longer a people's army. The Army was forcing on them an unacceptable and obsolete policy of centralism, and was rejecting all moves towards democratisation, on the grounds that it would bring about the disintegration of the Yugoslav federation.

The oldest councillor of the municipal assembly, who half a century earlier personally experienced the horrors of the World War II, stepped in front of a sergeant of the Yugoslav People's Army wearing three golden stripes on his epaulets and said: "You know sergeant that these are our weapons?"

The Yugoslav sergeant yelled at the soldiers to hurry up loading the semi-automatic and automatic rifles, M-48s, machine guns and sniper rifles. He sternly looked at the old delegate and said: "I understand that the weapons are ours."[10]

"No, you don't understand, these weapons are ours, they belong to the Slovenian Territorial Defence. Slo-ve-ni-an! Do you understand?" exclaimed the old councillor.

"Of course, I understand. Even though I don't speak Slovene very well because I have only been here four years, I get what you're saying," said the sergeant.[11]

"Don't screw with me, the language is not important. It's the weapons. Why are you taking them?" continued the old man.

"We received an order from the commander of the Territorial Defence to secure the weapons and move them," the sergeant tried to explain.[12]

"Which commander? Yours?" asked the old man.

"Yes, ours and yours, General Hôčevar, born a Slovenian," answered the sergeant, mispronouncing the name.[13]

"It's pronounced Hočévar, but that's not important. If he really did order that then, he's no longer ours," said the old councillor uneasily.

It is true that the commander of the Slovenian Territorial Defence, Lt-General Ivan Hočevar, was a Slovenian, but at this time he was obeying commands of the Yugoslav authorities in Belgrade. The outcome of this dichotomy was an irreconcilable difference of opinion, not only in Šentjur pri Celju but throughout Slovenia.

The old councillor grumbled that Hočevar should go to hell and then turned around and joined his colleagues, who had retreated from the midday sun into the municipal building. For two hours they contemplated what to do, and at 2 pm they received an order from the Slovenian authorities in Ljubljana that they were not allowed to hand their weapons over to the Yugoslav People's Army. But it was too late. The Yugoslavs had already taken away 400 to 500 infantry weapons.

"God dammit, two hours too late!" exclaimed one of the municipal councillors. The president of the municipality, who had started a new term of office that very day, shrugged his shoulders. He wasn't the only one. Many other municipal councillors throughout Slovenia were helplessly observing how the Yugoslav People's Army, in front of their eyes, was disarming the Slovenian Territorial Defence. Dreams of an independent Slovenia were turning into a nightmare.

The municipal commander of the Territorial Defence in Šentjur pri Celju had received an order from the Territorial Defence Republic Headquarters to hand over weapons two days earlier, and had

---

11) Ibid.

informed the municipality's president. But they were still puzzling what this really meant when the Yugoslav People's Army trucks rumbled in. Events were moving faster than their thoughts.

As it happened, only a few hundred old weapons remained in the warehouse, mostly pistols and a few thousand rounds of ammunition. But Šentjur pri Celju, in Slovenia's Štajerska region, was not the only place to be raided. Another 46 of the country's 62 municipalities lost their weapons to the Yugoslav People's Army on that days.

\* \* \*

Events unfolded differently in Jesenice, a town in northwestern Slovenia primarily known for its steel works. On May 15, 1990, the commander of the municipal Territorial Defence received an order from Ljubljana. In a top-secret dispatch badly translated into Slovene, point 1 stated: "Transfer for safekeeping into YPA facilities of all weapons, ammunition and ordnance which are currently held outside the YPA's facilities."[14]

This was the order issued by the aforementioned Lt-General Ivan Hočevar, signed in his name by the chief of staff of the Territorial Defence Republic Headquarters, Major-General Drago Ožbolt. General Hočevar never explained why he delegated signature of this fateful command to his subordinate, which left a big mark on Slovenia's history and on him as well.

At 10 am next day, the head of the Jesenice municipal government received this order. His faced turned pale. He did not know what to do, and nervously paced around his office for 15 minutes before yelling to a secretary in the neighbouring office: "Get me Dolanc!"

Stane Dolanc, who at the time had been retired for a year, was still among the best-informed people in Yugoslavia. During his 45-year political career, he had held nearly all the most visible positions in the Yugoslav government and the Communist Party. He was feared by Yugoslav dissidents, liberals and emigrants, who were hostile to the communist government.

---

14) Order of the Territorial Defence of the Republic of Slovenia to remove weapons from the

Fellow politicians considered him to be a denationalised Slovenian, an indispensable member of Tito's "court," and a sworn enemy of Tito's wife Jovanka. In the decade following Tito's death, Stane Dolanc was probably the single most important Yugoslav politician, and was the *éminence grise* of the Yugoslav Federal Secret Service (UDBA). In Serbia, on the other hand, he was suspected of supporting Slovenian "separatism".

Until his death in 1999, Dolanc lived like a recluse in his "beehive", located in a small village named Gozd Martuljek under the magnificent Špik mountain group, one of the most beautiful places in the Julian Alps, and just along the road from Jesenice. Because of his expertise in mushrooms, some jokers nicknamed him "The Great Mushroom Picker".

The municipal official from Jesenice did the right thing. At quarter past 10 in the morning, the phone rang in Dolanc's vacation home in Gozd Martuljek. A few minutes later, at 10:30 am, a call came into the office of Slovenian President Milan Kučan,[15] and the old political fox Dolanc was the first to inform him of Hočevar's order to disarm the Slovenian TD.

These were watershed moments for Yugoslavia. President Kučan had been in office only six days (he assumed his post on May 10). In elections on April 8, 1990, he had run with the support of reformed communists, defeating Jože Pučnik, leader of the centre-right union Demos and a former political prisoner who had spent many years in exile in Germany.[16]

The first non-communist government in the history of Slovenia, led by Christian Democrat Lojze Peterle, was sworn in only a week after Kučan became President.[17] On the same day in April

---

15) From a report by the TD commander of the Jesenice Municipality, Bojan Šuligoj, to the Republic's Secretariat for People's Defence, no. 31/2-91, Jesenice August 12, 1991.
16) In the second round of elections, on April 22, 1990, Milan Kučan received 58.6% of votes and Jože Pučnik 41.4%. Four additional members of Presidency were elected: Matjaž Kmecl, Ivan Oman, Dušan Plut and Ciril Zlobec.
17) Along with presidential election on April 8, 1990, Slovenes also elected members of the three chambers of the Slovenian Assembly. Even though the old laws were still in place, these were the first multi-party elections in Slovenia after the World War II. The elections were won by Demos, the right-of-centre coalition with 55% of votes. The most votes within the coalition went to Christian Democrats, so their leader Lojze Peterle, received a mandate to form a new government. Second were reformed Communists (United List) and third were Socialist Youth – Liberals, later

as in Slovenia, elections also took place in Croatia, where Franjo Tuđman, a former YPA Staff General and historian, and leader of the right-wing nationalist Croatian Democratic Union (HDZ), won an overwhelming majority. Meanwhile, the post of Yugoslav Federal President with a one-year mandate was transferred from the Slovenian Janez Drnovšek to the Serbian Borislav Jović.

The military and political leadership in Belgrade deliberately chose this time of flux to disarm those who they believed were destroying Yugoslavia. They considered the new Slovenian and Croatian leaders as "separatists" who had won key elections and were now moving into positions of power in Ljubljana and Zagreb. Neither republic had yet formally declared independence, but the Security and Counter-intelligence Service of the YPA, also known as the KOS,[18] reckoned this was the smart moment to seize weapons from the Territorial Defences in both republics.

But 13 months later, when Slovenia finally did declare independence, Belgrade's intelligence chiefs got their evaluation completely wrong. When Yugoslav army tanks appeared on Slovenian roads on June 26, 1991, most of the generals in Belgrade believed that the Slovenians would greet them with flowers and flags with a red star …

Exactly the opposite was true. As the 10-day war of 1991 unfolded, Yugoslav intelligence officers were receiving the bulk of their information from the Slovenian Agency for State Security (SDV), which they thought was loyal but was deceiving them by forwarding highly selective information.

Back on May 15, 1990, Slovenian President Kučan had every confidence in the tip-off he received from Stane Dolanc, his old colleague and Communist Party comrade. The two had stayed in touch but tried to keep this secret so that people would not equate Kučan, the presidential candidate of "reformers", with old communist forces. When Kučan received news from authorities of the town of Slovenj Gradec that they too had received an order from the Territorial Defence Republic Headquarters to hand over their weapons, he knew a catastrophe was in the offing.

---

18) KOS (Counter-Intelligence Service of the YPA), was formed in 1946 and was officially in existence

But Kučan was in a quandary and had to be very careful. Yugoslavia was a powder keg, with a fuse slowly burning away. Any reckless move could trigger intervention by the Yugoslav People's Army under the pretence that the separatists in Ljubljana and Zagreb were violating the Constitution. The President decided to double-check the order, which said Territorial Defence weapons should be transferred to YPA facilities for "safe keeping". First he called General Hočevar, commander of the Slovenian TD, and then Federal President Jović in Belgrade.

"Their explanations were comforting but, as it soon turned out, untrue," Kučan drily recalled several years later.[19]

General Hočevar tried to trick Kučan by telling him that the whole thing was only a "technical" operation of exchanging old weapons and that these measures were necessary to safeguard the weapons of the TD. However, in Slovenia and Croatia, there had been only a few thefts of weapons. In all these cases, the culprits were either passionate hunters or collectors, with no political motives.

General Hočevar gained valuable time with his deceit, and the YPA was able to collect weapons from the Slovenian municipalities unobstructed. All the while Slovenian officials were busy checking the situation on the ground and contemplating what to do. Kučan later paid dearly for his caution, gullibility and procrastination: his political enemies accused him of indecisiveness and worse.

The most vocal among them was Ivan Janez Janša, who the following day, May 17, assumed his duties in the Defence Ministry as a member of Peterle's government.[20] He first attracted attention to himself with his very critical views of the Yugoslav People's Army (YPA) in the early 1980s. He was then an active and zealous communist, but due to conflicts with his superiors, he was expelled from the League of Communists in 1983. As a youth functionary, he also sowed discord in the Organisation of Socialist Youth.

In his new post, Janša began receiving alarming reports from members of the Territorial Defence in the field informing him that the Yugoslav People's Army was taking their weapons away. So that same day, the

---

19) "I'm still convinced that the weapons affair is the mother of all affairs," interview with Milan Kučan, *Delo*, Ljubljana, April 23, 2011.
20) The official name of the Defence Ministry at the time was the Republic's Secretariat for People's

newly minted Defence Secretary issued an order to municipalities not to hand over to the YPA any weapons in their possession. In particular, they should keep arms belonging to the Slovenian leadership,[21] because these weapons were owned by Slovenia and were bought with Slovenian money. General Hočevar later described these as "grey" weapons.

The Slovenian Presidency was still struggling to find information to throw light on the confusion. Two days after General Hočevar's order, the Presidency still did not have a single official document to confirm it. To surprised and disturbed civilians it looked like sabre-rattling by the Yugoslav military. Word spread around the country that "Belgrade is disarming us and the federal authorities are stealing our weapons". Regional and municipal officials in charge of people's defence in towns, villages, and regions, were baffled and scared. They did not know whether or not they should take action, and were reluctant to act against the officers of the YPA, who were unconditionally executing orders from their superiors. So in those hot days in May, the personnel of the Slovene Territorial Defence often succumbed to inflexible rules of the military hierarchy and acted according to the saying: "File a complaint only after you execute an order."

As with the Army sergeant and the old municipal councillor in Šentjur pri Celju, people in other towns all over Slovenia were now confronting one another. For decades, the Yugoslav People's Army and the local populations had lived and worked together. Their children attended the same schools, they rooted for the same soccer teams, and they drank together in pubs.

"Will we have to shoot at each other now?" they asked themselves in fear.

The Yugoslav People's Army took Slovenian weapons away in two stages over two days. Most municipalities which obeyed the order to hand over their weapons on the first day did so because they did not know what was happening. But some municipal officials refused from the start to obey the order, with the result that 16 municipalities kept Territorial Defence weapons under their control.[22] Among them was

---

21) Some weapons were held by municipalities, companies and socio-political organisations (DPO). According to the Yugoslav Constitution, DPOs consisted of Communists and satellite organisations, Socialist Union of the Working People (SZDL), Union of the Socialist Youth of Slovenia (ZSMS),

Jesenice, which thanks to, amongst other people, the Great Mushroom Picker himself, comrade Dolanc, kept several hundred rifles, dozens of pistols and even four mortars and 150,000 rounds of ammunition.

A report from the YPA's Zagreb Command to the military leadership in Belgrade stated that the military could not seize weapons and ammunition from 13 Slovenian municipalities or from Territorial Defence battalions in the towns of Kranj and Kočevska Reka. According to this report, Slovenes kept almost 16,000 light infantry weapons and 126 metric tons of ammunition.[23]

Among those who also did not hand over their weapons was the 30th Development group (30th Rsk), which represented the peacetime core of the Territorial Defence's 27th Protection Brigade in Kočevska Reka. This key military unit had been set up in a remote area to protect the Slovenian political leadership in case of war, at a time when Yugoslavia felt threatened by the Soviet Union.

By the 1980s, this unit had confirmed the beliefs of those who thought that of all the Yugoslav republics, Slovenia took Territorial Defence the most seriously. In 1982, the YPA's inspection team (GINA),[24] held military exercises to check the preparedness of the 30th Rsk. GINA was surprised to witness the unit's professionalism and battle-readiness and named it the best Territorial Defence unit in all Yugoslavia. Ironically, the YPA helped to train it, but as its battle-readiness increased, its relationship with the YPA worsened.

The leaders of the 30th Rsk showed courage and resourcefulness when confronted with General Hočevar's order to hand over their weapons. Former chief of brigade staff Lt-Colonel Jože Polovič remembers the tense moments: "We moved a lot of our weapons from the military warehouse Borovec near Kočevska Reka, where they could be seized by the YPA, to a safer location at Gotenica.[25] We started to move weapons by trucks on the evening of May 18 and continued to do so through

---

articles and discussions, Union of Police Veterans' Societies Sever, Ljubljana, April, 2011, page 107.
23) From a report from the 5th Army Region Command to the General Staff of the Armed Forces of the SFRY about executing orders to disarm TDs in Slovenia and Croatia, Zagreb June 5, 1990.
24) GINA stands for *Generalna inspekcija narodne armije* (General Inspection of the People's Army).
25) Gotenica is in the previously "closed area" of Kočevska Reka stretching over 200 square km. In 1948, after the dispute with Soviets, it became a temporary refuge for the highest Slovenian Communist Party officials. They built a system of underground tunnels, with a capability to host 100 people. In

the night. We had six military trucks available, and, as I find especially amusing, we were helped by Yugoslav Army soldiers from the garrison in the town of Ribnica. In the meantime, the local folks were taking care of soldiers' superiors in the local pub."[26]

Anton Krkovič, who comes from the little village of Kaptol in the middle of the Kočevje forest, would soon be appointed commander of the 30th Rsk, which was transformed into a Brigade and in the independence war acquired a controversial reputation under the nickname Moris.[27] Back then, as a local people's defence official, he caught sight of the order to transfer weapons from the Territorial Defence to the Yugoslav People's Army warehouses, but the issuer of the order had been obliterated, and Krkovič realised its importance too late. In front of his eyes the YPA took away all of the TD's weapons in Kočevje, which had been bought with Slovenian money. Kočevje was one of five municipalities that could not even keep its own weapons, and Krkovič understandably felt depressed.

He later explained: "In the Kočevje municipality courtyard I saw YPA trucks, uniformed people, and municipality officials. I assumed that this was an ordinary operation in which the Federal military was helping members of Kočevje's Territorial Defence. I asked the commander of the Kočevje municipality TD what was going on and he answered tellingly that it's better for the weapons to go to the military facility than to be taken by Janša."[28]

The episode was embarrassing for Krkovič when he was sporting a Brigadier's rank on his military uniform. Despite his efforts to avoid blame for the weapons' seizure, Krkovič was later mocked by his colleagues, who remarked that "he's a Brigadier after the battle".

Twenty years later, Ludvik Zvonar, Advisor to the Slovenian Government, told him: "Unfortunately, your municipality handed over its weapons. You were in charge of the municipal office for people's defence, which means you were part of the leadership.

---

26) Author's conversation with Jože Polovič, Ribnica, May 15, 2011.
27) At the end of 1990, the 27th Protection brigade changed its name to the 30th Development group (30 Rsk). Even though this military formation already got the nickname "Moris" during the 1991 10-day war in Slovenia, Defence Minister Janez Janša issued a decree renaming it the "Special" Brigade Moris on October 13, 1992.
28) From Albin Miklič (Rebels With a

You were the second most responsible person for making sure that weapons were not handed over to the YPA ..."²⁹

* * *

On Friday, May 18, President Kučan, Defence Secretary Janša and other members of the Presidency sat at a table with General Hočevar. The commander of the Slovenian Territorial Defence, who sided with Federal Yugoslavia, later disclosed that President Kučan wanted him to sign a statement saying that he personally opposed the order that Slovenian Territorial Defence hand over its weapons. "I simply could not do it. If I signed, I would have gotten 20 years in jail, or even worse. I told Kučan and members of Presidency that they could not solve their problems with Belgrade on my shoulders. They simply wanted to sacrifice me without any conscience. I was a general in the wrong place at the wrong time."³⁰

General Hočevar was a Slovenian born in the picturesque town of Kostanjevica na Krki. He had a brilliant career as a military pilot, but as Yugoslavia began to break up, he found himself caught between a rock and a hard place. Should he follow the Yugoslav Constitution, or should he favour nascent Slovenian sovereignty? The decision was indeed difficult, but in the end, Hočevar stayed faithful to the red star of Federal Yugoslavia and his general's epaulets, and because of that Slovenians branded him a Janissary – which means renegade.³¹

In the following months, other Yugoslav People's Army officers throughout Yugoslavia were confronted with similarly difficult decisions. For decades, they had served one country in one army. Then, almost overnight, they had to make decisions whether to stay with the Yugoslav People's Army, which afforded them a more or less decent life, or to switch to the "opposing" military forces, for which no one could guarantee success in achieving independence. And how would

---

29) Ludvik Zvonar, "Response to the letter of Brigadier Anton Krkovič", *Vojnozgodovinski zbornik*, no. 41, Logatec 2010.
30) "At the wrong time in the wrong place", interview with Ivan Hočevar, *Mladina*, Ljubljana, January 26, 1993.
31) A Janissary was a member of the elite unit of the Ottoman Turkish military during the middle ages. Most Janissaries were non-Muslim children kidnapped in the Balkans and turned into Turks.

they be accepted by the new state, if it ever came to exist, since they had worn the epaulets of the enemy? At the same time, they were worried that crazy politicians and generals would push the Yugoslav People's Army into, God forbid, a war pitting brother against brother.

These were questions that people in the military were asking themselves in 1990. Most were ambitious for success and promotions, while their wives longed for safe homes. Their children were torn apart by the awful predicaments into which the families were forced.

\* \* \*

On the afternoon of May 18, 1990, the Slovenian Presidency issued an order to stop moving weapons into Yugoslav People's Army warehouses. The Presidency sent an encrypted message to the municipal headquarters of the Territorial Defence. Decoding this message took much precious time, but with this risky and daring move the Slovenian political leadership showed that it did not intend to obey orders from the Yugoslav Federal leadership.

To press this home, on May 21, 1990, Milan Kučan and Presidency member Dušan Plut traveled to Belgrade to meet the Federal President, Borislav Jović.[32] They were joined at the meeting by the Federal Secretary of People's Defence, Veljko Kadijević, and by the Slovenian member of the joint Federal Presidency, Janez Drnovšek, who until that time knew nothing about the order to seize the weapons.

Kučan and Plut firmly held to their viewpoint that weapons of the Slovenian Territorial Defence were Slovenia's property, and that the Yugoslav People's Army should give them back. They demanded the replacement of the commander of Slovenian TD, General Hočevar, who issued the order to seize the weapons.

It was now a full week since the order to hand over weapons, and the top Slovenian leadership still had not yet seen the actual document giving the order. Kučan and Plut did not get to see it in Belgrade either, but they were got a clear idea of the intransigence they faced.

---

32) Dr. Dušan Plut is a geographer and ecologist and Professor at the Philosophic Faculty of the

In an evening meeting on that same day, May 21, Kučan and Plut informed other members of the collective Slovenian Presidency that Borislav Jović saw no reason why the Yugoslav Presidency should discuss such a key issue as the seizure of weapons from the Slovenian Territorial Defence. Kučan told his colleagues: "Kadijević confirmed to members of the Slovenian delegation that the Yugoslav Presidency did not make the decision to move the Slovenian TD's weapons to warehouses of YPA, because safeguarding of the weapons was a mandatory responsibility of the armed forces."[33]

It turned out later that Jović and Kadijević were not telling the truth on this matter either. In his diary, Jović wrote: "We will not allow the misuse of weapons in Slovenia and Croatia to result in possible clashes or a violent secession … We practically disarmed them. Formally, it was done by the Chief of General Staff of the YPA, General Blagoje Adžić, but essentially it was done on our order."[34]

Kadijević was no more sincere in his claim to be worried that weapons would end up in the hands of the "destroyers" of Yugoslavia. A few years later, he admitted in a book that the real reason was that "we had to disable the TD in those parts of the country where it could represent a core for forming military forces of the secessionist republics. With that purpose, we disarmed all units of the TD before the outbreak of military conflicts in Yugoslavia. With the help of some TD officers, we tried to ensure that the TD would not come under the control of secessionist political leaderships. We were successful only partly, more so in other republics than in Slovenia".

Significantly he revealed the underlying pro-Serbian motive behind the action, adding: "It's understandable that along with the YPA we used members of the TD and their weapons in those regions of Croatia that were populated by Serbs."[35]

Thus Kadijević confirmed that rebel Serbs in Croatia and Bosnia-Herzegovina were supported not only by the YPA but also by the Territorial Defence in both republics. Even before any military actions, the TD in those areas was supplying weapons to local Serbs and helping them with military organisation.

---

33) Stenographic record of the meeting of the Presidency of the Republic of Slovenia, May 21, 1990, page 5.
34) Borislav Jović, *Poslednji dani SFRJ* (The last days of SFRY), Kompanija Politika, Belgrade 1995, page 144.
35) Veljko Kadijević,                         (My Point of View About the Breakup:

\* \* \*

Besides economic recession and discord between the constituent nations, Yugoslavia in the late 1980s and early 1990s also suffered from a crisis in leadership, one of the consequences of a loose and unsettled Federal Constitution. Distrust and personal animosities grew among politicians, who publicly threatened each other, called each other names and lied to each others' faces. They infected the Yugoslav nations with their intolerance and exclusiveness, pouring oil on the fire flickering among the Yugoslav nationalities.

The incompetence of those Yugoslav politicians can be traced back to Tito's era, when Tito systematically prosecuted all who thought with their own heads and who he suspected of moving away from his cherished principle of "brotherhood and unity". As a result, after Tito's death, key positions in the Federal government, Communist Party and the military were held by colourless and incompetent – but obedient – bureaucrats.

One of them was Borislav Jović, who upon becoming Yugoslav Federal President, hardened his view against the "problematic republics", Slovenia and Croatia. Jović was widely seen as a person with limited abilities,[36] and more or less a puppet of the most influential Serbian politician, Slobodan Milosević.

As for Veljko Kadijević, he was born between the two World Wars in a poor region in the middle of Dalmatia's hinterland between Croatia's Biokovo Mountains and the border of Bosnia-Herzegovina. At the age of only 18, he joined the Partisan movement but did not make a big impression in fighting. He even deserted once, so those who remembered him from the national liberation struggle were surprised by his rapid promotions following the war.[37]

When confronted by Kučan and Plut on May 21, these mediocre Yugoslav personalities resorted to the flimsiest of excuses for

---

36) Viktor Meier, *Zakaj je razpadla Jugoslavija* (Why Yugoslavia Disintegrated), Znanstveno in publicistično središče, Ljubljana 1996, page 216.
37) *"Veljko Kadijević je 1943 dezertirao iz partizana"* (Veljko Kadijević deserted from the Partisans in 1943), a conversation with Kadijević's fellow combatant, Ivan Andrijić, *Slobodna Dalmacija*, Split, April 3, 2007. Croatia later accused Kadijević of war crimes against civilians. He escaped to Russia,

the action against the Territorial Defence. General Kadijević implausibly asserted that the order was prompted by a riot before a soccer match between Red Star Belgrade and Dinamo Zagreb at the Maksimir stadium in Zagreb eight days earlier.

Before the beginning of the match, fans of the Belgrade team burst on to the pitch and went on the rampage. They called themselves *Delije* and were led by the infamous Serbian Željko Ražnatović, nicknamed Arkan, a professional killer and war criminal. Little more than a year after the disturbance at Maksimir, Arkan was to take command of the Tigers, a paramilitary unit recruited largely among the *Delije* football fans, and lead them on a killing spree of the non-Serbian population of Croatia.

On that May Sunday, fans of the Dinamo team known as the "Bad Blue Boys" also broke the barriers that separated spectators from players and burst on to the pitch to engage in an hour-long brawl with Belgrade fans. Dozens of policemen and fans were injured and the match was cancelled. Arkan was arrested and charged with violence, but the Croatian authorities released him.

As the arguments went to and fro at the meeting in Belgrade on May 21, 1990, they uncovered the essence of the dispute between Slovenia and Yugoslavia: the Federal state represented by Generals Kadijević and Hočevar considered the Federal Army and the Territorial Defence to be unified armed forces, whereas the Slovenians, arguing that they were following Marshal Tito's will, were gradually and systematically transforming their Territorial Defence into an army of the Republic of Slovenia.

\* \* \*

The same order that was issued in Slovenia was also used to disarm the Territorial Defence of Croatia. As in Slovenia, the command of the Croatian TD was completely on the side of the Yugoslav People's Army. Through the same deception that removal of Croatian TD weapons was just to safeguard them, almost all weapons and ammunition in TD warehouses, municipal buildings and companies were taken or locked away. But the disarming of

Croatia, unlike the disarming of Slovenia, took place without any resistance and went almost unnoticed.

Allowing the easy confiscation of the weapons was one of a string of strategic mistakes made by the new Croatian President, Franjo Tuđman, who was apparently over-awed by the prestige of the "mighty" Yugoslav People's Army and afraid of its wrath. Croatian military analyst Fran Višnar also stresses the important role of the Security Service of the YPA (KOS): "Members of KOS were in only two nights able to transfer to police stations in Croatia the whole quantity of Territorial Defence weapons. In the spring of 1990, at least half of the police officers in Croatia were Serbian nationals. Tuđman's electoral triumph really disturbed the Serbs living there, so they enthusiastically and without hesitation collaborated with the YPA and its security service. They carefully masked the weapons and secretly transported them to Bosnia-Herzegovina."[38]

Not all Croatian politicians agreed to "write off" the weapons that had belonged to the Croatian Territorial Defence. Among them was the future Croatian President, Stipe Mesić, who vehemently opposed Tuđman's despondent acquiescence. "If Croatian municipalities in Bosnia-Herzegovina were successful in preventing the handing over of their weapons, why wasn't this possible in completely Croatian regions like Zagorje and Dalmatia?" he asked in November 1994.[39]

But such protests were in vain. In May 1990 Croatia lost almost 200,000 infantry weapons, many of them mortars.[40]

The Croatian Territorial Defence had been gutted, and the repercussions were deadly: Serbian forces caused the most casualties among Muslim civilians in Bosnia-Herzegovina using mortars taken from Croatian TD warehouses. "When in the spring of 1992 the war also broke out in Bosnia-Herzegovina, these weapons were seized by military forces of the breakaway Republika srpska, led by Ratko Mladić," says analyst Fran Višnar.

---

38) Author's conversation with Fran Višnar, Zagreb, May 4, 2008.
39) Viktor Meier, *Zakaj je razpadla Jugoslavija* (Why Yugoslavia Disintegrated), Znanstveno in publicistično središče, Ljubljana 1996, page 214.

Thus, the infamous Serbian General Mladić could use weapons that had been seized from Croatia to start his murderous campaign in Bosnia-Herzegovina, a campaign that would bring him to The Hague's International Criminal Tribunal for former Yugoslavia 19 years later.[41] He was helped in this by the Army's seizure of the weapons of Territorial Defence units in Muslim areas of Bosnia-Herzegovina already in 1990.

\* \* \*

Let's return to Slovenia. After the abortive talks by Kučan and Plut in Belgrade, the Slovenian leadership decided to inform the Slovenian public about the true intention of the order in which the Yugoslav People's Army had tried to seize the weapons of the Slovenian Territorial Defence. Reports that the YPA was disarming the Slovenian people appeared in the public media.[42]

The evening meeting of the five-member Slovenian Presidency headed by Kučan on May 21 was joined by the Secretary of People's Defence Janez Janša, who briefed members about the confusion in the field caused by General Hočevar's order issued to the regional headquarters of the Territorial Defence. The Presidency set itself to repair the damage caused by the order, secure warehouses and then try to have weapons returned to members of the Slovenian Territorial Defence. They invited General Hočevar to attend, but were by no means sure that he would respond, nor did they entertain hopes he would be helpful. President Kučan remarked that talking to the General was like "kicking the wall with a bare foot".[43]

Hočevar turned up and addressed members of the Presidency as "gentlemen and comrades," which was a sign of the times. Under socialism, people addressed each other as a comrade, but in increasingly independent Slovenia, this word was being replaced by the "bourgeois" gentleman. Socialism was giving way to a new order.

---

41) International Criminal Tribune for the former Yugoslavia – ICTY, was formed with United Nations resolution no. 857, and was established on May 25, 1993. The court, headquartered in The Hague, Netherlands, is authorized to put on trial only individuals, not countries or organisations. The highest sentence it can adjudicate is life in prison.
42) Article written by Territorial Defence reserve officer Janko S. Stušek in the Slovenian newspaper

Despite the superficial courtesy, the General was offended and reacted emotionally. He let everybody know very clearly that the Federal Constitution had to be respected, and that "as long as Yugoslavia exists, he will behave like an officer of the Socialist Yugoslav armed forces".[44]

He castigated the participants for possessing 16,000 "grey weapons".[45] These were weapons and ammunition that Slovenian administrative organisations, including the League of Communists and the Union of Socialist Youth, had purchased for their needs. General Hočevar said Yugoslav law allowed only the military, Territorial Defence and police to possess weapons. He accused the Slovenian Secretary for Defence, Janez Janša, of criminal responsibility for protecting these "grey weapons". Janša had already informed members of Presidency that with very few exceptions,[46] the headquarters of the municipal TD had not handed over their weapons to the Federal military.

As the meeting reached deadlock, Ciril Zlobec, a great Slovene poet and now a member of the collective Presidency, complained in exasperation: "The more we talk about Territorial Defence, the more it becomes evident how complicated the situation really is. We cannot even talk about the TD, let alone be able to claim that we know how it's functioning."

"But it's not functioning," added President Kučan.[47]

Zlobec told Hočevar that the political problem only arose because he did not inform the Slovenian leadership about his order to remove the weapons. Had he done that, there would only be a technical question as to how to store and safeguard weapons. There was no reply, and Hočevar left the meeting.

When he was gone, the first to speak was Ivan Oman: "We should insist on a smart commander for the Territorial Defence."

"Typical apparatchik, without an ear for politics," added Matjaž Kmecl.

"I feel a little bit sorry for him," said Kučan. "You put the man in a position that's over his head and then you get this …"

---

44) Ibid., page 36.
45) Slovenian authorities later estimated that the quantity of "grey weapons" (owned by municipalities, companies and socio-political organisations) was significantly lower, only about 10,000 to 12,000 pieces along with accompanying ammunition. Source: 20th Anniversary of the Manoeuvre Structure of National Protection: a collection of articles and discussions, Union of Police Veterans' Societies Sever, Ljubljana, April, 2011, page 59.
46) Weapons that were held by municipalities and socio-political organisations were handed over to the

Responded Zlobec, "I feel sorry for him too. I tried to make him show more intelligence so that he would understand the situation in Slovenia ... If only he would say: – I understand you, but the military logic is such – ..."[48]

Even before this meeting, General Hočevar had realised that he would be replaced. When in the hallway he had asked President Kučan whether this summons means that ...? Kučan merely nodded slightly.

The Federal Yugoslav President refused the Slovenian demand to dismiss General Hočevar that May 1990. But it was only a short reprieve. He was finally removed at the end of September after the Slovenian National Assembly passed a new constitutional law on defence.

\* \* \*

Zlobec's irritated reference to the complex structure of the Territorial Defence went to the core of the dispute between Ljubljana and Belgrade. Each side interpreted in its own way the bureaucratic, dogmatic language and the loopholes in the complex legislations governing the Territorial Defence. The Yugoslav Constitution adopted in 1974 only made matters worse: at the time of adoption it was the longest in the world. It complicated rather than clarified.

In Belgrade, General Kadijević exclaimed that the concept of Territorial Defence was a great "fraud", but things had moved too far. The attempt of the Yugoslav People's Army to take away the Slovenian TD's weapons had only limited success, and relations between the Yugoslav capital and the rebellious republic thereafter took a decisive turn for the worse. It became clear that there was no way back. Nobody could stop Slovenia on its path towards independence.

Perhaps because it had been so easy to rein in the Slovenian military at the end of World War II, Serbian and Yugoslav federal officials underestimated Slovenia's resolve. They appreciated Slovenians for their industrious work habits and capabilities, but thought they were a nation caring for accordions, skiing and a glass of wine more than fighting. The Slovenians however had another quality: they knew how to outfox their opponents.

* * *

To prevail against the dogmatic, reactionary forces heading Federal Yugoslavia, it was important for the Slovenian political leadership to stay unified. In the meeting with the Slovenian delegation in Belgrade, Kadijević lost no time in trying to drive a wedge between them. He complained at their appointment of Janez Janša, who had recently been sentenced by a Military Court in Ljubljana, as Defence Secretary.[49]

President Kučan emphatically sided with Janša. He insisted to Kadijević that the naming of the Defence Secretary was exclusively a Slovenian prerogative and that Janša was the only person who provided reliable information what was going on in the field after the order to seize the Slovenian TD's weapons.

Janša afterwards wrote that by ordering a halt to the handing over of weapons to the Yugoslav Army, the Slovenian Presidency "established itself as the de-facto commander of Slovenian Armed Forces".[50]

So far so good, but the Slovenian politicians were counting their chickens before they had hatched. Their "unity" during the crisis over removal of the Territorial Defence weapons was only tactical and temporary. Slovenia was in for a hot summer, with the threat of war in the air, but in the background there was a hard fight for power among the Slovenian politicians. Right-wingers were ideologically opposed to the communists, who had given up their political monopoly and were repositioning themselves as democrats. Thus, it was not long before the history of these last few tense weeks was being rewritten.

In the fall of 1993, Janša wrote in a report that "in May 1990, the Slovenian Territorial Defence was left without 80 per cent of its weapons and ammunition and without a large portion of its military equipment and vehicles".[51] Careful analysis shows this was an exaggeration, but Janša was set on distorting the events in order to settle scores with his political opponents.

---

49) In spring 1988, Janša and three other defendants were arrested on suspicion that they had divulged military secrets. The so called JBTZ affair (named after the first letters of the accused's last names: Janša, Borštner, Tasić and Zavrl) broke out. The trial against the four took place in a Military Court and touched off the "Slovenian Spring", a national revolt which turned into demands for fundamental political reforms and independence for Slovenia.
50) Janez Janša, *Premiki* (Manoeuvres), Mladinska knjiga, Ljubljana, 1992, page 50.

According to realistic estimates, the Slovenian side before May 1990 had approximately 75,000 infantry weapons, plus just over 20,000 old rifles, machine guns and pistols, mostly originating from World War II. The Yugoslav Army (YPA) did not touch these weapons because they were obsolete, unusable and lacked ammunition. These weapons were left in various municipal and industrial facilities.

Veterans' associations estimate that after May 15, 1990, the Slovenian Territorial Defence was able to keep approximately 12,500 usable weapons, four million rounds of ammunition, large quantities of ordnance, and even a few tanks and a smaller number or anti-armour and anti-aircraft guns.[52] Most of their light anti-armour and anti-aircraft weapons were taken away from them by the YPA. About 20 per cent of Slovenian weapons were housed by companies and about a tenth was kept in the TD regional headquarters. The rest was with the police.[53]

At least two-thirds of the Territorial Defence's weapons were kept in facilities under the control of the Yugoslav People's Army, so when the TD weapons were "seized" there was very little actual moving. Mostly the locks on YPA warehouses were simply changed. That left plenty of scope for Slovenian patriots to outwit their opponents. The municipalities in the Gorenjska region were able to keep most of their weapons by playing tricks, as was related later by the former commander of the Territorial Defence in Gorenjska, Janez Slapar: "With officers in the regional headquarters and commanders of municipal headquarters, we agreed on a bluff manoeuvre. Even though we signed the order to hand over our weapons, in reality we acted in our own way. We tried to gain time. We informed our superiors and then waited for the decisions of the Slovenian leadership. Everybody knows for himself what he did and did not do."[54]

According to Slapar, the Yugoslav People's Army placed large-calibre ammunition, anti-armour weapons, anti-aircraft weapons and artillery which they had seized in its warehouses. However in the 13 months from May 1990 to the beginning of the military conflict in

---

52) *20. obletnica Manevrske strukture narodne zaščite – zbornik prispevkov in razprav*, (20th Anniversary of the Manoeuvre Structure of National Protection - a collection of articles and discussions), Union of police Veterans' Societies Sever, Ljubljana, April, 2011, page 116.
53) Several authors, *Prikrita modra mreža* (Hidden Blue Net), Inštitut za novejšo zgodovino, Ljubljana,

June 1991, daring Slovenians were able to take back a lot of them. Most were taken legally on the pretence that they were needed for Territorial Defence training, but were then never returned to the Yugoslav People's Army. Slovenians moved them to secret locations, in this way obtaining several thousand infantry weapons, including anti-armour and anti-aircraft weapons, just before the outbreak of war.

Slapar and other war veterans are thus convinced that "Slovenia kept enough weapons to defend its sovereignty."[55] The dispute over the handover of the weapons however has never let up. In April 2011, Janša told the Slovenian National Assembly:[56] "The disarming of TD was one of the most shameful events in the history of Slovenia."[57]

Twenty years after Slovenian independence, Janša was joined in his campaign of distortion by the former Interior Minister, Igor Bavčar, and by the former commander of the Moris Brigade, Anton Krkovič. They accused Kučan and his aides – most of them former members of the communist elite – of knowing about the disarmament of TD in advance and doing nothing in response, thus making themselves guilty of high treason.

Milan Kučan, who by that time had ended his term as Slovenian President, retorted that Janša was picking on the Slovenian Presidency in order to excuse his own clumsy and inappropriate actions at the time.[58] More to the point, the constant manipulation of the story concerning the seizure of the arms was a convenient excuse for the illegal arms trafficking which Janša would be deeply engaged in later on, and which will be described in later chapters.

Bavčar, later a businessman involved in questionable affairs, showed where he stood on the question in his own statement in 2011: "The disarmament of the TD was the beginning of an infamous story. It was an outrageous action, and later armament was necessary only because of this disarmament. For me, the only affair is disarmament of the TD."[59] Janša and Krkovič took the same line. They were "three brothers in arms" and knew why they had to talk like this.

---

55) "Slovenia kept enough weapons to defend sovereignty", *Delo*, Ljubljana, May 20, 2011.
56) The National Assembly is a 90-member parliament.
57) Transcription of the recording of the special session of the National Assembly, Ljubljana, April 12, 2011.
58) Author's conversation with Milan Kučan, Ljubljana, July 8, 2014.

## - Chapter 2 -
# AS STORM CLOUDS GATHER, SLOVENIA RE-ARMS

It would be hard to find a more remote place than Lašče nad Dvori. This little village in the heart of the Slovenian region of Suha Krajina is situated on a hill above the picturesque Krka River. At the time of independence, Slovenian arms merchants purposely chose the only inn in this forgotten village for their dealings.

On September 10, 1990 the drivers of two trucks with Croatian licence plates and logos of the Croatian freight transport company Čazmatrans had a hard time finding their way to this obscure place. Only in the evening did they draw up in front of its inn. During the next few years, this location would become a favourite meeting place for the traders in arms.

When day turned to night, two shadowy figures quietly changed licence plates. A few minutes later, the same two trucks, accompanied by a passenger car, headed towards the Slovenian town of Kočevje. They made a first stop a few kilometres to the north, where the policemen on guard recorded the arrival of three vehicles into the "closed area" of Kočevska Reka.

The vehicles stopped next to wooden houses underneath a spruce tree. A few trustworthy and carefully-vetted men worked hard for several hours, loading ordnance, ammunition and weapons. The heaviest and the most precious were two anti-aircraft guns with a calibre of 20 millimetres, worth more than 200,000 German marks. Documents show that the two trucks also loaded 1,000 grenades for anti-aircraft guns, four machine guns with 14,000 rounds of ammunition, 200 anti-tank mines and 41,000 rounds of various ammunitions.[60]

---

60) "Weapons, ammunition and other assets, given to individual regions - municipalities", date and place unknown. At the bottom of document there is only this following annotation: "After the transfer, lists of material were signed by warehouse employee and the commander of 30th Rsk, Mr. Krkovič." The interlocutor, who wanted to stay anonymous said, that on September 10, 1990, the following items

The two loaded trucks then drove off towards the town Bregana on the Slovenian-Croatian border and were followed by a Mitsubishi passenger car, occupied by four armed members of the Slovenian Special Police Unit, dressed in civilian clothes and tasked with following the weapons all the way to the Croatian border.[61] It is interesting that even now, these men refuse to talk about this mission. The passenger car stopped, but the trucks drove on. On the other side of the border, they were taken over by Croatians in uniforms sporting red-and-white chequered badges.

So on September 10, 1990, during preparation for Slovenian independence and under increased pressure by the Yugoslav People's Army, Croatians received anti-aircraft weapons from Slovenians, even though the Slovenian Territorial Defence would desperately need them in the subsequent armed conflict. When war did break out, attacks by the Yugoslav Air Force represented the greatest danger. By secretly shipping the weapons south to their Croatian neighbours, ministers professing to be virtuous "defenders" of their homeland were actually weakening Slovenia's defence capability.

Neither the Prime Minister nor Slovenia's Presidency gave permission to sell weapons to Croatia. The weapons were being sold by the Secretary of Interior, Igor Bavčar, and the Defence Secretary, Janez Janša, both of whom were negotiating on their own with the Croatian Interior Minister, Josip Boljkovac, and the Defence Minister, Martin Špegelj.

A document about arms shipments around Slovenia shows that another personality involved in this transfer of weapons to Croatia was Anton Krkovič, who some time later together with Bavčar and Janša accused the Slovenian Presidency of allowing disarmament of the Territorial Defence.

According to the agreement between the Slovenian and Croatian ministers, Krkovič sold weapons to Croatia kept in the "closed area" of Kočevska Reka. These weapons belonged to the special unit of the Interior Ministry set up earlier to protect Tito's closest communist comrade, the Slovenian Edvard Kardelj – Krištof, and other leaders of the Slovenian Communist Party and government.

---

of ordnance, 376 hand grenades, 40 tromblon mines, 40 Zolja anti-tank rocket launchers and two

At the beginning of summer 1990, these weapons and ammunition were kept by a Special Police Unit of the Territorial Defence under the command of Vinko Beznik, and were later taken to a safer location in the facilities of the Command of 30th Rsk of the Territorial Defence in the hamlet of Primoži.[62] In September the same year, new defence officials transferred weapons to several Slovenian regions and municipalities to fill the gap caused when the Federal army seized two-thirds of weapons that belonged to Slovenian Territorial Defence.[63] But as mentioned above, some of these also found their way to Croatia.

\* \* \*

Ivan Hočevar, who in the meantime effectively became a general without an army, persisted in his post as Commander of the Slovenian Territorial Defence. While the Yugoslav Presidency did not want to fire him, the Slovenian Presidency could not. So General Hočevar spent most of his time sitting around in his house doing nothing.

Meanwhile the Slovenian leadership was looking for new ways to free themselves from the Yugoslav yoke. They were carefully developing their own defence structure, while taking care not to poke a finger into the eye of the Federal authorities.

Their thoughts turned to the so-called National Protection, a Slovenian peculiarity that grew out of the anti-fascist struggle during World War II. The National Protection operated as an urban guerrilla force against German and Italian occupying forces and domestic collaborators. After the liberation of Yugoslavia, it was put in charge of protecting local communities, public property and companies. With time, it started to lose its significance, but it remained active.

In early summer 1990, the National Protection took on broader responsibilities and its scope of activities increased. It started to transform into a so-called Manoeuvre Structure of National Protection (MSNZ) composed of police and Territorial Defence.

---

62) Primoži is a little village between the towns of Gotenica and Kočevska Reka, about one km away from the main road connecting them.
63) Notranjska region received 120 shells, Koroška 300 (September 15, 1990), while Western Štajerska

It would be hard to allege that the new structure operated outside the law, but to cover their backs, the Slovenian leadership obtained a written opinion by Serbian General and politician Petar Gračanin, who certified that "the Slovenian National Protection is not contrary to the Constitution".[64]

In the meantime, the Slovenian Parliament passed a basic legislative act for Slovenian independence, which provided a constitutional and legal basis to reorganise the Territorial Defence so that it was completely Slovenian. The contentious General Hočevar was finally fired and moved to Belgrade.

On September 28, 1990, the Slovenian Presidency accepted Janša's suggestion to name reserve infantry Major Janez Slapar as acting Chief of the Republic's Staff of the Territorial Defence. About four months before his nomination as Territorial Defence commander of the Gorenjska region, Slapar successfully defied the Yugoslav People's Army by refusing to hand over weapons. However, this time, Slapar was facing an even greater challenge without any of the necessary material and human assets.

"Can you imagine if today, when no one is threatening with a military intervention, a chief of staff of the Slovenian military would for one week work completely alone, without officers, secretaries, driver, telephone or a vehicle? And the week after that, he'd start working with only one assistant, who was named the previous evening?"[65] That is how Slapar remembers his first days as the head of Slovenian Territorial Defence.

To make things more difficult, Slapar's reporting lines within the Slovenian government were confused. For battle-readiness and command of the Territorial Defence, Slapar was responsible to the collective Presidency of Republic of Slovenia, which under a new constitution became the Supreme Commander of the Slovenian Armed Forces. But already before Slapar's appointment, Defence Secretary Janez Janša[66] had been given responsibility for "certain matters regarding

---

64) Petar Gračanin's letter, Belgrade, May 16, 1985. Source: Archives of the Republic of Slovenia.
65) Major-General Janez Slapar, *"Vodenje in poveljevanje v TO med osamosvojitveno vojno"* (Leading and Commanding in TD During the War of Independence), *Vojnozgodovinski zbornik,* no. 1, Logatec,

commanding of the Territorial Defence".[67] To underline his authority, Janša took to appearing in public dressed in a military uniform.

Slapar and Janša soon fell out. Frictions between them appeared over the naming of the new regional commanders of the Territorial Defence. Slapar wanted to nominate those already in the TD structure of command, on the grounds that "people in key positions of the TD must also have military education and a commander's authority, not just *esprit de corps*". The trouble was that Janša had to approve commanders before their names were sent to the Slovenian Presidency. For the two biggest Slovenian cities, Ljubljana and Maribor, they compromised: Miha Butara to command Ljubljana's TD and Vladimir Milošević Maribor's.[68]

It was not the best way of filling key posts. When a few months later Slovenia found itself under the attack of the Yugoslav People's Army, it turned out that Butara was a bad choice because the Ljubljana region was militarily ill-prepared.

In the beginning of October 1990, soon after Slapar took over as the Chief of the Republic's Staff of Territorial Defence, the force's previous command structure was disbanded. There were two other key figures in the new defence structure. One was Vinko Beznik, head of the Special Police Unit. The other was Anton Krkovič, still smarting from having allowed the Yugoslav People's Army to take away weapons under his jurisdiction, who was named head of National Protection reporting to Slovenia's Secretary of Interior.[69]

Apart from these two, Janša was successful in bringing his own people to other key positions of the Territorial Defence. In moving some aside, he argued that he was cleansing the emerging Slovenian

---

67) Decree no. 830-01-16/90 (*Official Journal*, Ljubljana, October 5, 1990) gave Janez Janša following authorities concerning the TD: 1. Directing and organising the TD as a military formation and formation of a permanent military force; 2. Implementation of the defence-development plan of the TD, determining the direction of annual plans for the TD's equipment and training; 3. Responsibility for replenishing units and headquarters and other personnel issues in the TD; 4. Directing and coordinating supply of materials to the TD; 5. Directing and organising training of the TD units and their staff personnel; 6. Organising security matters in the TD with the assistance of the Security Organ of the Republic's Secretariat for People's Defence; 7. Monitoring and evaluating battle-readiness of the TD. For all this, the Republic's Secretariat for People's Defence issues guidance and monitors its implementation. For all of these responsibilities, the Secretary of Defence reports directly to the Presidency of Slovenia.
68) Author's conversation with Janez Slapar, Radovljica, April 24, 2013.
69) Decree naming the Chief of Staff on Slovenia's TD, no. 0001-1-Z-2/13-90, RSNZ, Ljubljana,

military of influences of the old regime. Of the newcomers, some of whom were awkward personalities or lacked military qualifications, he demanded absolute loyalty to himself.

The Yugoslav People's Army was busy too. On the evening of October 4, 1990, a military police battalion, commanded by a Major Rajko Meh[70] from a barracks on the outskirts of the capital, seized the Republic of Slovenia Territorial Defence Headquarters building in the middle of Ljubljana. Even though they found the building empty and without electricity, they stayed there for six months.

The seizure of the building by armed Yugoslav soldiers led by a major who was himself Slovenian raised a public outcry and turned the Slovenian public decisively against the Federal Yugoslav government.

\* \* \*

In the lead-up to the 1991 war with the Yugoslav People's Army, it was not just Janša's Defence Ministry orchestrating the dispatch of weapons from Slovenia to Croatia. The Interior Ministry headed by Secretary Igor Bavčar had a hand too. Momčilo Slavnič, who at the time headed the Ministry's department for equipment and armament, admitted to criminal investigators that in 1990 they sold to the Croatians 170 automatic weapons, machine guns and 80 hand-held RB-57 anti-tank rocket launchers.[71]

Slavnič said that a warehouse employee from the department drove the weapons by truck to the "closed area" of Kočevska Reka, where he was not allowed to be present during unloading. He was allowed to approach his truck only when it was reloaded with other weapons to be taken away to Gotenica.

Never before or after were politicians of Slovenia and Croatia so closely connected as they were in the last months before the breakup of Yugoslavia. But this "close friendship" was based only on similar political ideology and material interests. These collaborators played

---

70) YPA Major Rajko Meh was since May 1990 already a secret agent of Slovenia, code-named Pagat. He was associated with the commander of the Special Police Unit, Vinko Beznik. Meh escaped from the YPA on July 1, 1991.

71) Official record of the criminal investigator of a conversation with Momčilo Slavnič, UKS MNS,

leading roles in the Slovenian Democratic Union (SDZ)[72] and the Croatian Democratic Union (HDZ), two right-wing parties which won elections in spring 1990 and came to power in their respective countries.

In April and May that year, leaders of both parties met twice in the Slovenian castle Otočec near the Croatian border. Then on August 23, 1990, Janša and Bavčar met with their Croatian counterparts, Defence Minister Martin Špegelj and Interior Minister Josip Boljkovac, in the village of Podstene in the heavily-forested region Kočevski Rog. They exchanged information and coordinated defence against possible attack by the Yugoslav People's Army." The consequence of these talks was substantial aid by Slovenia to Croatia in 1990 in the form of arms and military equipment,"[73] admitted Janša's associate, Anton Krkovič.

Nineteen years later, Boljkovac was still talking gratefully of the helpfulness of Slovenian minister Bavčar: "In Čabar,[74] we established a joint headquarters, which was also visited by the President of Croatian government, Stipe Mesić. The Slovenians gave us everything that we needed. They promised us everything: the logistics, buildings and even Slovenian territory if we had to retreat during the fighting."[75]

At the time, Boljkovac was not to know that his Slovenian counterparts were acting on their own. However in a television interview in May 2009, Boljkovac asserted that he informed Slovenian President Kučan's aides who in Slovenia was "receiving millions from the sales of arms".[76] He also made another loaded allegation: that a high-level Slovenian politician who in May 2009 was a member of the opposition in Slovenian Parliament worked for the Yugoslav Counter-intelligence Service of the Yugoslav People's Army (KOS).

Boljkovac, who died in 2014 at the age of 94, did not mention any name in the interview, but there seems little doubt that he was talking about one and the same person, Janez Janša.

---

72) In December 1989, the Slovenian Democratic Union (SDZ) joined a coalition named Demos. Leading members of SDZ later became important politicians: Janez Janša, France Bučar and Dimitrij Rupel. In the 1990 election Demos won and formed a new government.
73) "Conversation with Brigadier Anton Krkovič", *Ampak*, Ljubljana, August-September, 2006.
74) Čabar is a small Croatian village in Croatian region Gorski Kotar, bordering Slovenia.
75) Discussion with former Croatian Interior Minister Josip Boljkovac, on the Croatian TV show Sunday at Two,

* * *

In summer 1990, Croatia and Serbia were on the verge of major conflict, thanks largely to the baleful influence of one man: Slobodan Milošević. Before styling himself as the Serbian *Vožd* – or Leader – Slobodan Milošević was a banker and a Communist Party apparatchik. In the middle of 1980s, he skillfully took advantage of the anger of Serbian academics about alleged injustices against Serbs in Tito's Yugoslavia and made Serbian nationalism a pillar of his aggressive politics.

With one sentence, *Niko ne sme da vas bije* – no one shall dare to beat you – uttered in a fiery speech to Serbs in neighbouring Kosovo in April 1987, he fed their fears that Kosovo's Albanian majority was threatening a land Serbs considered "the cradle of the Serbian nation". In 1989, Serbia re-asserted its direct rule over Kosovo, which had had autonomy within Yugoslavia, and Milošević made another inflammatory speech on a Kosovo battlefield where Serbs had been defeated 600 years before and came under Turkish rule.[77]

Milošević swept aside Serbian reformers who disagreed with his strong-arm policies, and fomented an "anti-bureaucratic revolution" against ineffectual communist leaderships. He abandoned his Communist past and replaced it with nationalism. To begin with, he preached a stronger Federal Yugoslavia, but in the face of pro-independence tendencies in Slovenia and Croatia, he turned to a "contingency" plan to unite all Serbs in one country – Greater Serbia. Slovenia was not affected because it had no Serbian minority, but the ambition was a major threat to Croatia and Bosnia-Herzegovina, where there were large numbers of Serbs.

In Croatia, President Franjo Tuđman did nothing to enamour himself to the 600,000 Serbs living there. In a new constitution, he demoted their status from "nation" to "minority" and double-crossed a moderate local Serbian leader who sought an accommodation. From then on, Serbs in Croatia would talk to Tuđman's regime only

---

77) On Saint Vitus Day, June 28, 1389, armies of the Kingdom of Serbia and of the Ottoman Empire fought a huge battle on Kosovo polje. Both sides suffered tremendous losses and the commanders, Turkish Sultan Murat and Serbian Prince Lazar, were killed. After the Kosovo battle, Serbia was

through the barrels of their guns. The most influential Croatian Serbs aligned themselves with Milošević's extremism, and three of them later ended up on trial at the International Criminal Tribunal for the Former Yugoslavia.

In the middle of August 1990, Serbs organised a referendum on autonomy in the region of Knin, where most of them lived. The Croatian authorities refused to recognise it, and the Serbs reacted by blocking roads with boulders and tree trunks – a rebellion they called their "Tree Trunk Revolution".

\* \* \*

Clear of the animosities building up in Croatia, in autumn 1990 Slovenia took a decisive step towards statehood: it decided to hold a referendum on its sovereignty and independence. First proposed by a small Socialist Party, it was eventually backed by the mainstream Demos movement. When the referendum took place on December 23, almost 90 per cent of Slovenians positively answered the question: Should Slovenia become a sovereign and independent country?[78]

People in other Yugoslav republics, and regions of several other European countries, were soon asking similar questions themselves. It seemed that Slovenia with its assertiveness and unity was setting an example to be followed. But at home, Slovenians began to feel nervous at the small size of their country's territory and population. How would a nation of only two million people squeeze itself into the European family? They tried to give themselves courage with a propaganda poster: A country with four million working hands should not fear an independent future.

The Slovenian political leadership set itself the task of implementing the referendum decision within six months. Yugoslavia was peaceful, but it was only the calm before the storm. Events were moving fast elsewhere too. In that same month of December 1990, Milošević and his Serbian Socialist Party of former communists easily won a Serbian

---

78) According to data of the Slovenian election commission, 1,361,728 people, or 93 per cent of

election. Emboldened by his victory, Milošević raided the Yugoslav monetary system and borrowed from the National Bank of Yugoslavia Yugoslav dinars worth the equivalent of about 1.4 billion U.S. dollars.[79] This money was never returned to the Yugoslav treasury, but many of Milošević's cronies suddenly became noticeably wealthy.

In the first days of January, Slovenia and Croatia reacted to the Serbian seizure by announcing they would no longer recognise any obligations towards the Federal government, not even their debts. Authorities in Ljubljana ceased sending their foreign currency reserves to Belgrade.

The Serbian theft of Federal funds and the response of Slovenia and Croatia represented a fatal blow to the government of the Yugoslav Prime Minister, Ante Marković. Ironically Marković, a Croat from Bosnia-Herzegovina, had done rather well since his appointment as Prime Minister in March 1989. He lowered rampant inflation, introduced economic reforms, and the standard of living in Yugoslavia began to rise. However, Milošević's destructive nationalism and Slovenia's move towards independence undermined his achievements. Marković was a gifted politician, but as some astute observers remarked at the time, he was suitable to lead any government except Yugoslavia's.

\* \* \*

With the decree of September 1990 putting Janša in charge of "certain matters regarding command of Territorial Defence",[80] the collective Slovenian Presidency made a step in the wrong direction, but they could scarcely have foreseen how he would use his new authority.

In 2006, a Parliamentary Commission in charge of supervising intelligence and security services (KNOVS) concluded that "with the incorporation of the Territorial Defence into the Republic's Secretariat for People's Defence, Janša gained greater authority regarding training of Slovenian soldiers, especially for anti-sabotage actions. He also increased the collection of money from municipal funds for purchasing weapons and military equipment".[81]

---

79) The amount was 18 billion 243 million Yugoslav dinars.
80) Decree no. 830-01-16/90 (*Official Journal of the Republic of Slovenia*), no. 35 Ljubljana, October 5, 1990.
81) Report of the KNOVS for 2006, no. 020-02/93-36/26, National Assembly of the Republic of

The Slovenian Territorial Defence needed to reinforce its arsenal, but under increasing pressures by the Yugoslav People's Army it was unable to purchase arms elsewhere in Yugoslavia. So its headquarters turned to arms suppliers abroad. In late 1990 and early 1991 it acquired 1,000 automatic weapons and ammunition, as well as Armbrust hand-held rocket launchers.[82] At the same time, the police bought smaller quantities of weapons for their own use.

During the war which followed in summer 1991, it turned out that Armbrusts had only propaganda and psychological value because they were too weak to penetrate the armour of Yugoslav People's Army tanks. Slovenian TD soldiers were also not sufficiently trained in operating them. According to a former military intelligence officer, many unreported mistakes were made in firing Armbrusts, some of them with deadly consequences.

SAR-80 automatic rifles arrived in Slovenia at the beginning of December 1990 and 14 days later, during a military celebration in Kočevska Reka, soldiers were sporting these rifles in front of photographers. Prime Minister Lojze Peterle enthusiastically exclaimed that for the first time, it really smelled like a Slovenian military. But it also started to smell like corruption, war profiteering and fraud.

Most experts subsequently found the SAR-80 automatic rifles were unreliable weapons of low quality. Ludvik Zvonar, a Slovenian advisor in charge of obtaining weapons abroad, scorned them as "plastic clubs". Apparently, they had never been used in the sort of conflict which Slovenia faced in summer 1991. Made in Singapore, they were a bad copy of American M-16s. Asian countries soon started to abandon them in preference to real American M-16s or Russian Kalashnikovs.

A representative of the SAR-80 manufacturer, Singapore state company Chartered Industries of Singapore (CIS), refused to talk about selling these weapons to Slovenia. He said the company sold the licence for making the gun to an Israeli Company, Israel Military Industries, but representatives of this company would not talk either.[83] Slovenia

---

82) According to the KNOVS report for 2006, the following weapons were purchased for the TD from November 1990 to March 1991: 600 Armbrust hand-held anti-tank rocket launchers, 1,000 Kalashnikov automatic rifles, 900 SAR-80 automatic rifles of 5.56 mm calibre, and 900,000 rounds of ammunition of 5.56 mm calibre.

probably procured this first, "non-Yugoslav" shipment of weapons for its Territorial Defence through unofficial, contraband channels.

Croatian military analyst Fran Višnar, referring to Croatian military and political sources, claims that "Janez Janša, by agreement with Croatian Defence Minister Martin Špegelj, sold to Croatia several hundred Singapore SAR-80 automatic rifles and Ultimax-100 light machine guns[84] at prices that were three to four times higher than what he paid to the Singapore supplier".[85]

It seems clear that in Slovenia and Croatia SAR-80s were used only for propaganda and psychological purposes. Members of the Croatian Guard carried them while guarding government buildings, the Presidential Palace and other strategic points in Zagreb. With these new weapons, they let the Yugoslav People's Army know that Croatia was also arming itself.

But more sinister in this affair is the way they were procured, paid for and smuggled into Slovenia. In the middle of December disturbing news appeared in the Slovenian media. Media outlets reported that these weapons were unloaded from a ship in the Italian port city Genoa and then brought to Slovenia with help from the Union of Slovenian Sports Societies in Italy, headquartered in the Italian city Trst (Trieste). The whole operation was coordinated by an employee of the shipping and forwarding company Intereuropa from the Slovenian port of Koper. However, instrumental in bringing the weapons into Slovenia was a man named Sandi Grubelič, who in autumn 1990 was implicated in a scandal that reverberated throughout Yugoslavia.[86]

Sandi Grubelič was owner and chief executive of a company named Trend in the small town of Grosuplje not far from Ljubljana. The company attracted 7,500 customers who paid upfront fees in western currencies to lease cars made in Western Europe. However, only about 1,500 actually received cars. The company returned money to 800 other customers, but the rest got nothing, and losses reached 30 million German marks. At that time, one of the Trend employees was

---

84) The Ultimax light machine gun was also a product of Singapore's company CIS, but unlike the SAR-80 it was a very precise and modern weapon.
85) Author's conversation with Fran Višnar, Zagreb, May 4, 2008.

Rajko Janša, brother of Defence Secretary Janez Janša. He also lived in Grosuplje, where Grubelič was once a municipal official. In November 1995, Grubelič was sentenced to six and a half years in prison for fraud, but he was released after a year and a half, and a Ljubljana court dismissed a lawsuit filed by creditors against him in 2014.[87]

A few weeks after the disclosure of Grubelič's involvement in the import of SAR-80 automatic rifles to Slovenia, a member of the Slovenian Presidency, Dušan Plut, questioned Defence Secretary Janša about the financing of purchases of weapons for the Slovenian Territorial Defence.

Janša answered that purchases were made according to budget plans. Some money was left from the previous year because weapons ordered were never delivered, and some became available after Slovenia's decision to no longer contribute funds to the Federal government.

Janša stated that funds were also collected from municipalities and municipal councils for people's defence, as well as from Slovenian trading companies. He complained of police harassment of two of the entities involved in the import of the weapons – "Union of Slovenian Sports Associations in Italy" and the Slovenian freight forwarding company Intereuropa.[88]

Ciril Zlobec, a member of the Slovenian Presidency, asked Janša if the National Assembly has already discussed "the dirty dealings regarding weapons and the problem with Grubelič" and whether this issue had already been clarified.

Janša: "What does it mean – clarified?"

Zlobec: "I mean dirty dealings regarding weapons and Grubelič and whether somebody defrauded people and used this money to purchase weapons."

Janša: "Military intelligence service has all these names. But if the names become public somebody will say that we are covering something up."

Zlobec did not give up. He proposed that the Justice Minister accelerate legal proceedings against Grubelič and that a Parliamentary commission should look into the matter. At which point, he was

---

[87] For more information go to: http://www.siol.net/novice/crna_kronika/2014/11/Grubelič.aspx (May 14, 2015).

interrupted by the President of the National Assembly, France Bučar, who protested: "You cannot ask the government to reimburse you if somebody defrauds you."

Zlobec: "I didn't mean that, but somebody who's suspected of a large fraud usually appears in court, but Grubelič did not. So I have the impression that somebody is protecting him. The interest of the Slovenian Presidency and other political structures should be that this kind of suspicion is dealt by prosecutors and courts."[89]

President Milan Kučan agreed that the Presidency should make a public statement regarding the Grubelič affair, but nothing further came of it. It had negative consequences for Slovenia later, but at the time nobody except Zlobec paid much attention to it.

Behind Zlobec's warnings lay discomfiture that people who would later style themselves as "heroes" of independence would turn out to be not unfamiliar with fraud. Meanwhile by January 1991 the deadline given by the Yugoslav People's Army (YPA) for disarming and disbanding all paramilitary units had passed. The Federal army threatened with extraordinary measures, but Slovenia decided it would no longer send recruits to the YPA.[90]

\* \* \*

Slovenia had to sort out its relationship with Croatia, and Defence Secretary Janša and Interior Secretary Bavčar took it upon themselves to do this. January, 20, 1991, they met with their Croatian counterparts, Defence Minister Špegelj and Interior Minister Boljkovac. They decided on an eight-point system of cooperation in case the Yugoslav People's Army attacked the two countries.[91]

---

89) Ibid.
90) On March 8, 1991, the Slovenian Parliament passed a law abolishing mandatory military service in the YPA for citizens of Slovenia. From August 1990 onwards, summons to report for military duty were sent only to those recruits who were to serve their military duty in the 5th Army Region, which included Slovenia and parts of Croatia.
91) In case of an attack by the YPA, Slovenia and Croatia will: 1. Declare total independence of both republics and recall their representatives in the Federal government; 2. Stop financing the Federal government; 3. Inform the United Nations Security Council and request an intervention by a peace force; 4. Make an appeal to their respective citizens to leave the YPA; 5. Seize all Federal property on their respective territories; 6. Stop sending supplies to command facilities, units and institutions

In his book, Špegelj wrote: "Based on this agreement among the four ministers and an accord between the two presidents of the republic, our standpoint was that an attack on one republic meant an attack on both."[92]

Janša and Bavčar signed this document without the approval of the rest of the Slovenian political leadership. If implemented, this military pact would turn Slovenia and Croatia into some sort of "Siamese twins". But two days after the agreement was signed, the Slovenian Presidency decided that it would not accept it or make it public. When the Yugoslav People's Army (YPA) later embarked on armed intervention in Slovenia, Tuđman's Croatia did not help Slovenia militarily. On the contrary, in the naval port of Pula, Croatian authorities handed over Slovenian deserters to the YPA.[93]

In March 1991 armed conflict started in Croatia, first in the town of Pakrac in the Western Slavonia, then in the Croatian National Park of Plitvice, where a Croatian and a Serbian police officer were shot dead. Involved were the Serbian warlord Arkan and Goran Hadžić, who later headed of a Serbian para-state in Croatia. Croatian authorities arrested them but they were quickly released.

Croatian politicians remained blissfully complacent, except for Defence Minister Martin Špegelj, who had been a general in the Yugoslav People's Army. In autumn 1990, he organised the first secret procurement of large quantities of weapons from Eastern European countries, in particular from Hungary until Hungarian Prime Minister Josef Antall banned it.[94] Špegelj later declared that from the beginning of October 1990 to the middle of January 1991, he procured more than 30,000 rifles and machine guns and large quantities of anti-aircraft weapons, ammunition and ordnance.[95]

---

order and sovereignty - use of these measures and means is agreed upon and coordinated, together with the mutual assistance; 8. Judicial authorities in both republics will start legal proceedings against those YPA members who participate in violent actions. Source: "Attack on Savudrija?", *Mladina*, Ljubljana, June 11, 2010.

92) Martin Špegelj, *Sjećanja vojnika* (Soldier's Memories), Znanje, Zagreb, 2010, page 219.

93) Tape recording of the meeting of the Slovenian Presidency, July 1, 1991. Božo Repe, *Viri o demokratizaciji in osamosvojitvi Slovenije, III del: Osamosvojitev in mednarodno priznanje*, (Sources about Democratisation and Independence of Slovenia, Part III, Independence and International Recognition), Arhivsko društvo Slovenije, Ljubljana, 2004, page 131.

94) From the book of Viktor Meier: *Zakaj je razpadla Jugoslavija* (Why Yugoslavia Disintegrated), Znanstveno in publicistično središče, Ljubljana, 1996, page 220.

The Yugoslav military secret service KOS detected several truck shipments crossing the Hungarian-Croatian border, as well as ships unloading light weapons in the Kvarner area of the northern Adriatic, but it did not react. Instead it set a trap for Špegelj, secretly filming him in conversations with aides which were edited to portray him as a weapons smuggler, a mortal enemy of Serbs and a destroyer of Yugoslavia. With this, the Croatian leadership was compromised.

About six weeks later, in the beginning of March 1991, he was told by President Tuđman that three other officials would replace him in weapons procurement.[96]

Špegelj had immediately to cut contacts with suppliers in Eastern Europe, who were selling off weapons no longer needed by the disintegrating Warsaw Pact, the Soviet bloc's military alliance. The result was that from March until July 1991, when it received first deliveries from Slovenia, Croatia did not import any significant quantities of weapons. Later Špegelj remembered how he was tantalising close to large deals. Sellers regretted his exclusion, since they could no longer sell off the old weapons profitably rather than send them for scrap.

This halt in arms procurement at a crucial time was a consequence of Tuđman's contradictory and confusing politics. The Croatian President was at the time sceptical about radical moves against the Yugoslav People's Army (YPA). He overestimated its actual power and mistakenly trusted his former colleague from the General Staff of the YPA, Federal Defence Minister Kadijević, who assured him that YPA would not attack Croatia.

Tuđman also did not care too much for the military alliance with Slovenia. He found a more useful partner in Milošević, whom he met in Karađorđevo in Serbia, where the two discussed dividing up Bosnia-Herzegovina. At their next meeting in the Croatian town of Tikveš, Milošević aggressively demanded that Bosnia-Herzegovina be divided between Serbia and Croatia as soon as possible, but the meeting did not produce any tangible results. In the following

---

pieces, 80 anti-aircraft machine guns, 40 Strela 2M rocket launchers with 400 rockets, and even 20 American Stingers with 200 rockets. They imported 70 metric tons of ordnance and procured 30 tons of explosive, several tons of different calibre ammunition, and 12 armoured vehicles. Source: Martin

months, Croatian and Serbian representatives met at least 48 more times to discuss the same topic.⁹⁷

On March 9, 1991, massive anti-government protests broke out in Belgrade. Angry opposition rose against Slobodan Milošević's rule and his propaganda machine. Demonstrations were led by Vuk Drašković, a controversial Serbian writer and leader of the opposition Serbian Renewal Movement. Drašković too wanted a Greater Serbia, but when armed conflict broke out in Yugoslavia he transformed himself into a determined opponent of war and ethnic cleansing.

Police violently crushed the March 9 demonstrations, and in the evening, Yugoslav People's Army tanks appeared on Belgrade's streets. The Yugoslav Federal President, Borislav Jović, later claimed that he won agreement from Macedonia's and Bosnia-Herzegovina's members of the collective Yugoslav Presidency that they would side with the Serbian and Montenegro representatives in approving this army intervention. So Milošević had proof of the loyalty of his generals, and the generals felt they had approval to hold Yugoslavia together by force.⁹⁸ Federal Defence Minister Veljko Kadijević was determined not just to defend Yugoslavia but to preserve socialism across Eastern Europe by force. He considered Soviet leader Mikhail Gorbachev as a traitor who had allowed Eastern Europe to be taken over by NATO.

The Belgrade demonstrations gave a pretext for Kadijević and his generals to demand that the Federal Presidency declare a state of emergency. On March 12, 1991, the military put members of the collective Presidency on military buses and took them to an underground Army bunker in the prestigious Belgrade district of Dedinje.⁹⁹ The Slovenian member of the Presidency was absent, but this is how Croatian member Stipe Mesić remembers those events: "We were cold in that bunker because the military turned off the central heating. Even though they 'generously' offered us military coats, they were, in fact, hoping that one of us would break down and vote for what military wanted."¹⁰⁰

---

97) This information was given by historian Božo Repe in the article: "Attack on Savudrija?", *Mladina* Ljubljana, June 11, 2010.
98) Macedonia at that time was represented in the Yugoslav Presidency by Vasil Tupurkovski and Bosnia-Herzegovina by Bogić Bogićević.
99) Dedinje is considered as the most elite part of Belgrade. President Milošević lived there, as did

Kadijević's ploy failed because only four of the eight Presidency members voted for the state of emergency.[101] The most pressure was put on the member from Bosnia-Herzegovina, but he did not give in. The most he would do was quip: "It's true that I'm a Serb, but not by profession."

The following day, without telling anybody, Kadijević flew to Moscow to visit Soviet Defence Minister, Marshal Dmitry Yazov.

"He had been preparing for this trip to Moscow in advance as a contingency plan, in case the Yugoslav Presidency did not approve the state of emergency," wrote his advisor for international relations, Colonel Dragan Vukšić. Kadijević told Yazov that conditions in Yugoslavia were catastrophic, but the Soviet Marshal responded that things in the Soviet Union were even worse, because six Soviet republics no longer recognised Moscow's authority and that "only a blind man cannot see that the Warsaw Pact and the Socialist community no longer exist".[102] Gorbachev did not want to meet with Kadijević, so the general returned to Belgrade disappointed. His last attempt to bring the "brotherly" Soviet Union on to his side had failed.

According to Colonel Vukšić, Kadijević did however indirectly influence the final stand of hardliners in the Soviet Union. During an earlier visit by Yazov to Belgrade in 1989, Kadijević suggested to him to stage a *coup d'état* in order to "defend Communism in the Soviet Union".

On August 20, 1991, Marshal Yazov and a group of drunken supporters did indeed try to seize control of the Soviet Union, but the coup failed and the conspirators were arrested. Soon afterwards, Gorbachev himself was removed and the Soviet empire came irrevocably to an end.

Kadijević had still not given up however. As soon as he returned from Moscow, on March 14, 1991, he insisted on another extraordinary session of the collective Yugoslav Presidency, this time joined also by the Slovenian member, Janez Drnovšek. General Kadijević demanded they vote for a declaration of a state of emergency and introduction

---

101) This time, the state of emergency was supported only by the "Serbian bloc", comprised of the members from Serbia, Montenegro and the "autonomous" regions of Vojvodina and Kosovo, even though Serbia had abolished their autonomy already in 1989.
102) Dragan Vukšić, (YPA and the breakup of Yugoslavia), Tekomgraf,

of a military combat-readiness; police should transfer its authority to the military and all laws contrary to the Yugoslav Constitution should be abolished.[103]

However, these measures were only supported by three members of Presidency loyal to Milošević. This time, the Kosovo member of the Presidency, Riza Sapunxhiu, also voted against the measures – and was sacked immediately because of it.[104]

After this failure Kadijević did not dare to try further strong-arm tactics. Had he succumbed to the temptation, the world would have considered it a military coup and the Yugoslav People's Army would have lost its legitimacy already before the civil war inside Yugoslavia got underway.

However, it soon lost it forever with its aggression against Slovenia.

\* \* \*

In the meantime, Slovenia was secretly accumulating weapons, but not without internal dissent. In February 1991, despite the deteriorating situation in Yugoslavia and the menace of an armed conflict, two opposition parties plus several social movements prepared a so-called Declaration for Peace. It demanded a halt to the further arming of Slovenia and Yugoslavia. Except Ivan Oman, the Declaration was signed by all members of the collective Slovenian Presidency, including its President, Milan Kučan, as well as by the member of the Yugoslav Presidency, Janez Drnovšek.

The governing Demos Coalition condemned the Declaration, and the most vocal in his criticism was Defence Secretary Janez Janša. With the dispute over the Declaration, the relationship between the Slovenian liberal left and conservatives worsened dramatically. Sharp polemical exchanges occurred during the passage of new legislation to build up Slovenia's defence and arm the Territorial Defence. In the end, the legislation passed in parliament with a small majority of votes.

---

103) Viktor Meier: *Zakaj je razpadla Jugoslavija* (Why Yugoslavia Disintegrated), Znanstveno in publicistično središče, Ljubljana, 1996, page 235.
104) Riza Sapunxhiu was removed by the Serbian Assembly. The decision was supposed to be made by the Kosovo Assembly but it was already disbanded. As his replacement, Serbs nominated Sejdo

And so the re-arming continued. A Parliamentary Commission in charge of supervising intelligence and security services (KNOVS) later wrote in a report that "between May of 1990 and June of 1991, the Slovenian Territorial Defence acquired 170 MGV-176 machine guns".[105] The weapons were paid for with funds from the regional headquarters of the Territorial Defence.

This machine gun was developed by a Slovenian company named Puškarna Kranj and was produced by Orbis, a subsidiary of Gorenje, a large Slovenian manufacturer of household appliances. Some military experts claim that the weapon with its ammunition drum holding special rounds was of a bad quality, because it often got stuck and was more dangerous to the user than the enemy. The joke went around that the acronym MGV stood for *mali gospodinjski vrtalnik* (small household drill).

Serbian sources recall that when Yugoslavia was still united, businessmen from Gorenje brought this machine gun to the headquarters of the Yugoslav People's Army in Belgrade to show it off. A colonel who was holding the gun "accidently" dropped it on the marble floor and the MGV-176 broke into pieces. The embarassed Gorenje businessmen had to pick up the bits from the floor.[106]

This machine gun should not have been allowed for military use because of its insufficient calibre, unreliable functioning and especially because it needed uncoated bullets banned for military purposes by international convention. However, the Orbis salesmen were not bothered by that, and they also sold MGV-176 to Croatia.

Military intelligence officer Zvonko Murko remembers that "in the middle of May, with the knowledge of director of VOMO[107] I drove several cars with Croatian licence plates to Puškarna Kranj in Slovenia. There we took possession of about 20 MGV-176 automatic guns which we took to the Croatian town of Vinica".[108] He noticed

---

105) Report about the work of KNOVS for year 2006, No. 020-02/93-36/26, National Assembly RS, June 28, 2007. MGV-176 or *Maschinegewehr* (machine gun) Velenje.
106) From the newspaper section *Atlas organiziranog kriminala na Balkanu* (Atlas of the Organised Criminal in the Balkans), editor of the section Miloš Vasić, Belgrade, 2005.
107) VOMO is the acronym for *Varnosti organ Ministrstva za obrambo* (Security organ of the Defence Ministry). In 1994 VOMO changed its name to *Obveščevalno-varnostna Služba* (OVS), Intelligence and

that three trucks with licence plates from Osijek in Croatia drove away from Puškarna Kranj fully loaded with these machine guns and ammunition.

This was probably not the only shipment of Slovenian machine guns to Croatia. Some of the first Croatian volunteers were seen guarding strategically important facilities armed with MGV-176 machine guns. After the outbreak of war, the Croatian army replaced these Slovenian weapons with other standard issue weapons. However in the following years, the director of Orbis, Ivan Draušbaher, closely cooperated with Janša and his advisor Ludvik Zvonar, especially when they were selling off weapons seized from the Yugoslav People's Army warehouses.

On December 22, 1994, Slovenian criminal investigators requested from Croatian Interpol documents regarding Orbis' operations in Croatia, because Draušbaher was suspected of keeping for himself all the cash made from selling weapons to the Croatians. All the Croatian criminal investigators could come up with was that on August 30, 1991, the Zagreb's company Industrogradnja paid Orbis 3 million Yugoslav dinars for weapons that Orbis priced extravagantly in German marks. After 10 days, Orbis delivered 10 MGV-176 machine guns, priced at the very high amount of 3,500 German marks per gun.

Together with this first shipment, Orbis also delivered five laser scopes for infantry weapons at 2,200 marks per piece. In the second shipment 14 days later, the price for these laser scopes increased to 5,500 marks.

On October 24, 1991, Orbis delivered the last recorded shipment containing 30 AK-47 Kalashnikov automatic rifles to the Croatian company Industrogradnja. Orbis set a price of 1,300 marks per gun,[109] even though on average, the cost of a Kalashnikov was only around 200 marks. Even in times of highest demand, its price never exceeded 500 marks.

Slovenian and Croatian criminal investigators later agreed to exchange all the documents regarding suspicious weapons dealings, but the exchange never happened. The head of the Croatian

---

109) Answer from the Criminal Police of the Croatian Ministry of Interior to the Slovenian request

criminal investigators, Marijan Benko, admitted to his Slovenian counterpart, Mitja Klavora, that the exchange was blocked by Croatian President Tudman.[110]

\* \* \*

On June 17, 1991, shortly before the outbreak of military conflict, Slovenia imported Racal military radio communication equipment from Britain. After hard negotiations that Slovenian officials say even involved Buckingham Palace, and thanks to extensive efforts by a mysterious Paul Wickey, the British government approved the deal.[111]

But the radio equipment had to travel by land because the Yugoslav military threatened to shoot down the plane that was supposed to deliver the equipment to Slovenia. Slovenia's Territorial Defence received the radio equipment too late to learn how to operate it, so during the armed conflict it was mostly useless.

These stories, especially the ones connected with the MGV-176 and SAR-80, throw a poor light on the people tasked with ensuring that Slovenia was ready for armed conflict after its declaration of independence. Numerous military analysts are convinced that weapons imported into Slovenia before June 20, 1991 did not significantly increase the capability of the Territorial Defence and Slovenian police. Among the defenders the new weapons only increased confusion and distrust.

Importantly, the first deliveries of weapons from abroad established smuggling channels enabling high government officials to make money. The question what would have happened had Slovenian government tried to stop the "smugglers" from Grosuplje divides Slovenian people 25 years after. Some people say that there would be no Slovenia, while others insist that Slovenia would be in a better shape than it is today.

---

110) Author's conversation with Mitja Klavora, Borovnica, November 9, 2009.
111) Report of the Parliamentary Commission in charge of supervising intelligence and security services (KNOVS) for 2006, no. 020-02/93-36/26, National Assembly of the Republic of Slovenia,

- *Chapter 3* -

# SLOVENIA DECLARES INDEPENDENCE: YUGOSLAV ARMY INVADES

Thursday June 20, 1991 was an important day for Slovenia. However, nobody except a few carefully selected people knew that. Early that morning the freight ship Herman C. Boye anchored in the Slovenian port of Koper. The ship and its seven-member crew under the command of a Danish captain, Martin Olsen, was loaded with 16 heavy containers weighing altogether 193 tons, according to its bill of lading.[112] This mysterious cargo was closely connected to the fate of a small nation of two million people, which was on the threshold of a historic move to establish its own independent state.

The Herman C. Boye left the Bulgarian Black Sea port of Burgas on June 14. According to the bill of lading, it was carrying "special equipment". Officially, this equipment was meant for the security forces of the United Arab Emirates, or at least that is what was written in the documents of the company that coordinated the shipment. However, the shipment never made it to Abu Dhabi. The ship's containers were unloaded in the Port of Koper, which at the time was still part of Yugoslavia, and then loaded on to eight articulated trucks belonging to the Slovenian companies Viator and Intereuropa.

The convoy of trucks, accompanied by members of the Special Police Unit and the Moris brigade, drove off towards the "closed area" of Kočevska Reka. Only when the shipment got there was it considered safe, because prior to that there was a danger that it would be intercepted by the Security Service of the Yugoslav People's Army, which was on the lookout for contraband shipments of weapons. When the convoy arrived in Kočevska Reka, the people who ordered the cargo could breathe a collective sigh of relief.

---

112) Information from the ship Herman C. Boye bill of lading, Burgas, June 14, 1991.

The containers were loaded with automatic weapons, advanced anti-tank weapons and anti-aircraft weapons.[113] Slovenia bought all these weapons for a bit more than 7.8 million German marks, or almost 4.4 million U.S. dollars. At the time that was a sizeable outlay, but because of various blockades set up by the Yugoslav authorities, the Slovenian weapons buyers did not have many options to find more favourable deals. They were also in a hurry, because the time for implementation of the plebiscite decision was soon approaching, and Belgrade was threatening to use force if Slovenia "seceded unilaterally".

This first "official" shipment of weapons to Slovenia[114] was ordered and paid for on June 1, 1991, using funds from the budget that the Slovenian Parliament approved in March 1991.[115]

Austria and Bulgaria were also connected with these weapons. In a report, Defence Secretary Janša specifically thanked Manfred Hofer, the unofficial representative of the Austrian Government, and a representative of the Bulgarian Government, whom he did not name. Janša admitted that the weapons were bought very late, for which he later blamed his superiors, "because in the time of highest tension, several Slovenian politicians, including the President of the Slovenian Presidency, Milan Kučan, and three other members of the Supreme Command, signed a Declaration for Peace that went up against the formation of the Slovenian army".[116]

The weapons that arrived from Bulgaria were immediately distributed to the Territorial Defence, which strengthened their confidence. There was a belief that from then on, it would be easier to confront the Yugoslav People's Army.

---

113) The shipment contained 5,000 AK-47 Kalashnikov automatic rifles, 3.2 million rounds of ammunition, 30 RPG-7 rocket propelled grenade launchers together with 180 anti-tank rockets, 12 Fagot anti-tank launchers with 82 rockets, 10 Strela 2M anti-aircraft launchers with 60 rockets, 1,000 Spanish-made Llama handguns and 1000 Argentinian-made Bersa handguns. Source: Adviser to the Government Ludvik Zvonar in his report on weapons for the Territorial Defence and Police imported from June 1991 to December 1992.
114) Adviser to the Government Ludvik Zvonar, in his report dated September 21, 1994, mentions five official shipments.
115) Report of the Parliamentary Commission in charge of supervising intelligence and security services (KNOVS) for 2004, no. 020-02/93-36/26, 213-05/90-1/47, National Assembly of the RS, March 18, 2005.
116) Janša's weapons report for the period between May 15, 1990 and September 1, 1993, no.

Together with the official shipment, there was also a "commercial" shipment designated for the Orbis company. The shipment included 1,000 Spanish Llama handguns and 1,000 handguns manufactured by the Argentinian Bersa Company. Janša later wrote that these handguns were "samples, meant for testing". In reality, they were meant to be resold.

\* \* \*

Slovenia settled the 7.8 million German marks cost of purchase in advance and in two instalments. The first instalment was paid by the Slovenian company Iskra Commerce and the second by the Ministry of Defence. The Austrian company Stalleker & Co. GmbH from Vienna was chosen as a middleman and had to pay for transportation costs, containers, taxes and other expenses. However, the Slovenian government never reimbursed the Austrian company for those costs.

Four years later, the debt including the interest had increased to more than 800,000 German marks, and the Austrian company tried whatever it could to collect it. It repeatedly requested clarification from Minister Janša and from his advisor Zvonar, but to no avail. The company also tried diplomatic channels but without success. Slovenia has since become an independent and internationally recognised state, but even today the Defence Ministry denies any knowledge of this business.[117]

In March 1994, Janša was removed from his ministerial position because of interference by the military in the civil sphere, and was replaced by Jelko Kacin. A year later, Stalleker sent the Slovenian Defence Ministry all its available documentation on the delivery of weapons. During an investigation in April 1995, the military Intelligence and Security Service (OVS) established that the archives of the Defence Ministry contained no contracts or any of the other documents that Stalleker claimed it had sent. "This

---

117) On March 29, 2011, the Strategic Communication Service of the Defence Ministry informed this book's author, who requested information on business dealings between the Defence Ministry and

means that documents were deliberately hidden or destroyed," the investigators wrote in their report.[118]

In the autumn of 1995 Stalleker[119] and his lawyers came to Ljubljana, but again they heard the same answers that none of the Slovenian ministries had any contracts or any other documents regarding the deal. Stalleker took a great personal risk for Slovenian independence, advancing his own capital, but instead of appreciation the Slovenian side trampled over the "gentleman's agreement" and refused to settle the debt.[120]

With this cheating of Stalleker, the Slovenian arms merchants put Slovenia in a bad light, at least in Austrian business circles. But that shipment with its unsettled debt was nothing compared to the additional "commercial" containers loaded with arms that were taken away from Koper on June 20, 1991, after the arrival of the ship Herman C. Boye.

The bill of lading shows that the ship brought 16 containers, but that day the trucks drove away from Koper towards the "closed area" of Kočevska Reka with 57 containers – that is 41 more than what was written in the ship's documents. The figure of 57 was stated by the commander of the Special Police Unit, Vinko Beznik, in a report to the Criminal Police. His police force members escorted this and other truck convoys transporting weapons, and they definitely did not make this figure up.[121]

In July 1994, Criminal Police officers briefed members of the Parliamentary investigative commission about the discrepancy, but none of the members did anything about it. It looked as if those "extra" 41 weapons containers did not interest anybody in Slovenia.

In September 1994, under interrogation during an internal investigation at the Defence Ministry, the Assistant Commander

---

118) The report of the ISS Director Marjan Miklavčič to Defence Minister Jelko Kacin about dealings with the company Stalleker GmbH (findings), no. 881-600/9043, Ljubljana, April 25, 1995.
119) The company owner's last name was Stalleker. Source: cooperation of the Minister of Defence with company Stalleker. The report of the OVS Director Marjan Miklavčič to Defence Minister Jelko Kacin, no. 881-600/230/8184, Ljubljana, January 27, 1995.
120) Minutes from the meeting between representatives of the Stalleker Company and the Defence Ministry, October 27, 1995.
121) A Report of the Special Police Unit Commander, Vinko Beznik to the Criminal Police

of the Republic of Slovenia's Territorial Defence, Col. Peter Zupan, who was in charge of the rear area, estimated that the first shipment might have contained as many as 76 containers. He said he was aware that a small portion – the 1,000 handguns mentioned before – went to Orbis and some of the weapons were redirected to Slovenian military.

But he added: "Some of the weapons we gave to Croatia with the help of middleman Josip Vukina. Beside him, one of the buyers was also Ivan Bećir. I don't know how these weapons were paid for, but Minister Janša and Director of the Security organ of the Defence Ministry (VOMO) Lovšin should know more about it."[122]

The statements of Colonel Zupan and Beznik corroborate each other, leading to the conclusion that a large portion of the first weapons shipment "for Slovenia" that arrived in Koper on June 20, 1991 found its way into Croatian hands. It seems that the money from the sale of these weapons simply "evaporated".

\* \* \*

The first shipment of weapons to Slovenia came from a Bulgarian state-owned company Kintex, which was involved in large-scale dealings of arms and drugs. At the end of the 1980s, through the Italian Bank BNL it received two unsecured loans amounting in total to 41 million U.S. dollars to buy computer and electronic equipment for the armament industry of Saddam Hussein's Iraq.[123]

By 1990, Kintex had become one of the largest companies in Bulgaria in terms of revenue earned in foreign exchange. Three years later, with the help of western intermediaries, it started to supply arms to the African country Liberia, which was caught up in a bloody civil war and was paying for weapons with money from the sale of so-called "blood diamonds".[124]

---

122) Official minutes by the VOMO's commission interrogation of Peter Zupan, MO Ljubljana, September 23, 1994.
123) BNL (Banca Nazionale del Lavoro) headquartered in Rome. In 2006, it was taken over by a French Banking Group BNP Paribas, which in 2014 was the sixth largest bank in Italy. In 1980s, BNL

Ludvik Zvonar does not know, or does not want to know anything about the questionable "reputation" of the first suppliers of weapons to Slovenia: "We badly needed weapons and we didn't care whether they would be delivered to us by the Pope or the devil himself."[125]

In the months following the first unloading of containers from the ship Herman C. Boye, at least four more ships brought weapons and military equipment from Bulgaria to Koper. Much controversy surrounded these shipments. One contained a vast number of unneeded and unusable gas masks. Another consignment was intercepted by the Slovenian police on its way to Croatia, and a third vessel brought a huge over-supply of weapons which had to be stored over a lengthy period at Ljubljana's Brnik Airport.

\* \* \*

From the last days of January to mid-February 1991, the Slovenian leadership including President Milan Kučan held a number of meetings with leaders of other Yugoslav republics. There was talk of transforming Yugoslavia into a confederation, but the Slovenian leadership finally realised this did not make sense anymore. So the Slovenians accelerated their efforts to set up their own statehood and pushed for a peaceful break-up of the Socialist Federal Republic of Yugoslavia.[126] From spring into summer, presidents of all six Yugoslav republics met each other several times. Their meetings resembled a touring political circus that had no chance of finding a solution acceptable for all.[127]

While Tuđman and Kučan were making plans for independence, Serbian President Milošević and his Montenegrin counterpart and ally, Momir Bulatović, insisted on keeping a unified Yugoslavia on the grounds that foreign countries recognise only the Yugoslav government.

---

125) Author's conversation with Ludvik Zvonar, Radovljica, December 29, 2010.
126) The term "break-up of the SFRY" was later adopted by the Badinter Arbitration Commission (its president was a French lawyer, Robert Badinter). The commission operated within the framework of the Peace Conference on Yugoslavia. It came to the conclusion that Yugoslavia broke apart and it no longer existed and that none of the successor states had a right to call itself the exclusive successor to Yugoslavia.
127) The six presidents first met on March 27 in the Croatian city Split, then on April 4 in the Yugoslav capital Belgrade; April 11 in Slovenian's Brdo pri Kranju; April 18 in Ohrid, Macedonia;

This attitude of foreign countries was confirmed during a meeting of the republics' presidents in Belgrade at the beginning of April. On the same day as that meeting, a so-called troika of foreign ministers from the European Community visited the Yugoslav capital. However they met only with President of the Yugoslav government Ante Marković, even though it was already clear that Marković had been pushed to the margins of political decision-making. They ignored the presidents of the six Yugoslav republics.[128]

The biggest commitment to a resolution of the Yugoslav crisis was shown by Bosnia-Herzegovina's President Alija Izetbegović and Macedonia's President Kiro Gligorov. At the last meeting of presidents of the six republics on June 6 in Sarajevo, Izetbegović and Gligorov presented a compromise proposal to create a hybrid political system between a federation and a confederation. But they were turned down by the other republics' presidents.

\* \* \*

In the meantime, a shocking incident took place in the village of Borovo selo near the Croatian town Vukovar. On May 2, 1991, raiders from Serbia ambushed a bus with Croatian police officers and killed 12 of them. After the massacre, the Yugoslav People's Army arrived at the scene under the pretext that it had to protect local Serbs. Indeed, only a few days before the massacre, extremists from the Croatian political party HDZ led by a future Minister of Defence, Gojko Šušak, had been provoking the local Serbian population.

Four days after the bloody incident in Borovo selo, a massive demonstration broke out in the Croatian port city of Split. Croatian nationalists killed a 19 year-old member of the Yugoslav People's Army military police of Macedonian nationality, who had been trying to calm down the situation.

Because of quarreling among the different nationalities, the Yugoslav Federal Presidency was incapable of dealing with the ever worsening crisis. Its paralysis became total on May 15, when it was the turn of

---

128) Viktor Meier, (Why Yugoslavia Disintegrated), Znanstveno in

Croatia's Stipe Mesić to take over the revolving post of President from Serbia's Borislav Jović. Four members of the collective Presidency who supported Serbia categorically refused to let Mesić take up the post.

Four days later 78 per cent of Croatians voted for independence in a referendum for independence, but this only exacerbated ethnic divisions, since Serbs from the Croatian region of Krajina boycotted the vote.

\* \* \*

In the middle of May 1991, the Slovenian government started to send Slovenian recruits to the Slovenian Territorial Defence instead of to the Yugoslav People's Army (YPA). But the Yugoslav People's Army was not idle either. On May 23, after the Slovenian Territorial Defence captured two of its soldiers, the YPA encircled one of two TD learning centres situated in the village of Pekre near Maribor, where the first Slovenian recruits were being trained.

The Territorial Defence soon released the two YPA soldiers, but the YPA was not satisfied. During negotiations to end the encirclement in the night of May 23 to 24, soldiers of the YPA handcuffed Colonel Vladimir Miloševič, commander of the Eastern Štajerska region's Territorial Defence Headquarters and took him to a YPA barracks. (The colonel had a Montenegrin father and Slovenian mother and was not related to Serbian President Milošević).

The Slovenian Security and Information Service (VIS) had already learned in advance that the Yugoslav People's Army was going to kidnap a high ranking officer of the Territorial Defence. However, nobody did anything about it – and this later turned out to be a wise decision.

Colonel Miloševič described what happened: "A few hours before the negotiations, the head of VIS Operations Region Maribor, Silvo Komar, informed me that they had intercepted information that the commander of the Maribor YPA Corps, Major General Mićo Delić, received an order for my arrest during the negotiations. I called Secretary Janša, who already knew about the possible kidnapping and he answered something to the effect that – You have soldiers and have to protect yourself –."[129]

After that, a few units of the Territorial Defence moved around the Maribor region, but did not intervene, and nor did Slovenian police officers guarding the Maribor municipal building where the negotiations were taking place.

What was the reason for this inactivity? Twenty-five years after the events it is easier to understand. Had the Territorial Defence or police tried to prevent Milošević's kidnapping violently, it would have likely led to a bloody confrontation. Who could then have stopped it and what would the consequences have been? Moreover, who in Yugoslavia or abroad would have believed the Slovenian authorities that in the middle of negotiations they resorted to armed action merely to prevent a purported kidnapping? Not for the only time, Slovenian passivity prevented a tense situation getting out of hand. Silvo Komar, the head of the Maribor's Security and Information Service, said he told only two other people besides Colonel Milošević about the kidnap plan. "We didn't tell others about the plan, so we would not cause a panic. This was a very risky move, but we were successful."[130]

Colonel Milošević was then still in the thick of it however. Arriving shortly after 1 am at the meeting place, he left his gun outside and sat down around a table with Yugoslav People's Army General Mićo Delić, who like him had studied at the Military Academy in Belgrade. A few minutes later, a Special YPA Unit from Zagreb barged into the room, arrested Milošević, handcuffed him and took him off to a nearby army barracks.[131]

He was kept in handcuffs overnight, and the next day General Delić refrained from coming to see him despite their Military Academy friendship. Milošević concluded from that that the highest command of the Yugoslav People's Army (YPA) had ordered his kidnapping. He was subjected to hours of degrading interrogations.

Then news of the kidnapping reached the president of the YPA military court in Slovenia, Col. Alojz Ferlinc, who was shocked to hear about it over the radio. He saw no reason to prosecute Milošević and decided to arrange his release as soon as possible. However Ferlinc had to be very careful because he was a Slovenian and it was

---

130) Author's conversation with Silvo Komar, Ljubljana, June 23, 2015.

likely that the supreme command of the YPA did not completely trust him: "Any suspicion by the supreme command in Belgrade could have triggered an appeal against my decision, and the YPA would have named military judges to decide differently. So I named an investigating judge and a prosecutor. According to the laws of the time, the investigating judge had to interrogate the suspect within 24 hours and then extend his detention or release him."[132]

So that same morning of May 24, the President of the Military Court, the investigating judge and the prosecutor all went to the Vojvode Mišića barracks in Maribor, where Milošević was held. The investigating judge was a Serb, Mirko Stojanović, a loyal and trusted member of the YPA and in Ferlinc's view an honest man.[133]

Stojanović too did not find any evidence of a criminal offence. After only 15 minutes of interrogation, the investigating judge signed a decision that the Slovenian Territorial Defence was cleared of any suspicion. But Ferlinc had a hunch that things were not over yet, and decided to stay in the Maribor barrack until he saw Milošević exiting.

Now the real drama began. Colonel Ferlinc went to inform General Delić and others about the acquittal, but he quickly noticed they were uneasy about it. "Delić called the 5$^{th}$ Army Region Command in Zagreb. It could be discerned from the phone conversation that Zagreb's military commander, General Konrad Kolšek, rudely ended the conversation saying that he had nothing to do with Milošević's detention. All this could mean only one thing: the kidnapping of the Slovenian Territorial Defence officer was ordered by the supreme command in Belgrade."[134]

When the next phone call came through, this time from Belgrade, General Delić turned pale. Ferlinc assumed from picking up some ear-piercing yelling that the person on the other side of the telephone was the hardline General Blagoje Adžić, Chief of the YPA's General Staff. The commander in Maribor was ordered to detain Colonel Milošević until a military helicopter could pick him up and take him to Zagreb or maybe Belgrade, in other words, somewhere outside Slovenia.

---

132) Author's conversation with Alojz Ferlinc, Ljubljana, August 10, 2015.
133) Mirko Stojanović is the father of the soccer expert Slaviša Stojanović, former coach of the

Moreover Colonel Ferlinc was himself to board the helicopter, together with the investigating judge Stojanović and the prosecutor.

There was a deafening silence as the Army commanders in Maribor decided what to do: follow through with the acquittal or obey a senseless order from one of the highest military authorities. In the meantime, Maribor city authorities turned off electric power to the Vojvoda Mišića barracks, and enraged inhabitants of Maribor gathered outside, their numbers increasing by the hour. They were shouting that the YPA was an army of occupation.

Tension rose and Colonel Ferlinc realised time was not on his side. He threatened General Delić that he was going to tell the angry crowd that the YPA was detaining Colonel Miloševič without justification. Then the phone rang again and Delić was even more frightened than when he talked to Belgrade. After he rang off he said with a trembling voice that Slovenian Interior Secretary Igor Bavčar had just threatened him that the Slovenian defence was going to shoot down the Army helicopter if it tried to fly out of Slovenia, regardless of who would be on board.[135]

Bavčar was heading the Republic's Coordination Group tasked with coordinating military preparations, but as shown in the next chapter, it operated as a "state within a state".

Bavčar's deputy in this group was the Defence Secretary Janša. Also in this group was Lt-Colonel Danijel Kuzma, head of the Regional Operations Command of the Territorial Defence.

According to Miloševič, Kuzma gave a verbal order to shoot down the helicopter to a Territorial Defence officer in Maribor, Major Vladimir Pravdič, who passed it on to another commander, Lt-Colonel Alojz Štajner.[136] A humming noise could be heard as a Mi-8 helicopter approached from the direction of Croatia in the southeast and prepared to land. When Lt-Colonel Štajner spotted the helicopter from the roof of the local Territorial Defence headquarters, he ordered his subordinates located on a lower floor, to shoot down the helicopter.

Pravdič remembers these moments: "Beside me, Štajner's shouting was also heard by Bojan Petan, a member of our staff and Anton Železnik,

---

135) Ibid.

Commander of the Municipal TD Command Maribor - Rotovž. However, we did not forward his order to the soldiers who were manning the anti-aircraft guns at the TD training centre in Pekre where the helicopter was preparing to land. Instead of ordering fire, I decided to count to 10 while the helicopter flew away. Based on my experience from military exercises, I figured that the helicopter was then too far away and that our anti-aircraft guns would not be able to hit it."[137]

His colleague Branko Petan said: "We the three of us mutually agreed not to follow Štajner's order. If we were called upon to explain our decision we would have said that we acted wisely and that we prevented a potential tragedy."[138]

The helicopter then turned and moved towards Maribor. This time, it flew over the Drava River, even further from the TD's anti-aircraft guns. Because of dense trees, it could not land close to the Vojvode Mišića's barracks, where Colonel Milošević was detained. It landed not far away, at the Maribor's Police Academy.

The threat to shoot the helicopter down worked. In the afternoon of May 24, Colonel Milošević was released from the Maribor barracks and greeted by a large crowd of city inhabitants.

However, what would have happened had the Army tried to take Milošević out of Slovenia, as demanded by the YPA Headquarters in Belgrade? Consequences could have been terrible had the Slovenians shot down the army helicopter. This would have rubbed salt into an already open wound and given angry Yugoslav People's Army generals a pretext to launch a full-scale military attack on Slovenia. Bavčar's Coordination Group was keen to show off the military muscles of the Slovenian Territorial Defence, but in the event the hawks in both Belgrade and Ljubljana were thwarted.

In this war psychosis, many illogical orders were given in strange circumstances, but in this case reason prevailed, mostly at the lower levels of command on both sides. On the Slovenian side, they refrained from shooting. The Yugoslav People's Army's officers based in Slovenia also helped defuse the crises. The president of the military court,

---

137) Author's conversation with Vladimir Pravdič, Ljubljana, June 16, 2015.

Colonel Ferlinc, engineered the acquittal of Milošević, while Maribor Army Commander General Delić held back when caught between the order from Belgrade to fly the kidnapped officer out of Slovenia and Minister Bavčar's threat to shoot down the helicopter.

However, not everything had a happy ending. On the same day, May 24, at around 7 pm, a group of demonstrators jumped on an armoured personnel carrier which was leaving the Army barracks. One of them, Maribor resident Josef Šimčik, fell awkwardly while trying to climb on the vehicle, landed under its wheels, and died when it drove over him.

Šimčik, a Croatian by birth, was the first recorded victim of the Slovenian struggle for independence. By coincidence, the driver of the military vehicle that killed him was also a Croatian.[139] Therefore, the driver and the victim were compatriots, except that the former was wearing the Yugoslav People's Army uniform and the latter died in the struggle for Slovenian independence.

After the Šimčik incident, the Yugoslav People's Army found itself under even greater pressure from the Slovenian public. Moreover its cunning plan to force the Slovenian Territorial Defence to close down the Pekre training centre in return for Milošević's release misfired completely. As Silvo Komar, the head of the Maribor's Security and Information Service, put it: "We thwarted the YPA's plans to derail the independence process. At the same time, we got practically all of the Slovenian public on our side, which gave us strength to use all means to protect our path towards independence."[140]

\* \* \*

Shortly afterwards, Ludvik Zvonar stopped by the office of Eastern Štajerska Territorial Defence Command in Maribor to see commander Milošević.

After discussing future plans, Zvonar pulled a Colt gun with a silencer from his bag, as if to say, "you see, what's awaiting those who would try to stand in our way"? At least, that is how the Colonel understood the gesture. Zvonar also accidently dropped a paper which

---

139) The information was confirmed by Maribor's

Milošević found the following day. "The piece of paper, which was in fact an invoice, contained a list of weapons with unusually high prices. Everything looked very suspicious. I photocopied it and gave the original to Silvo Komar, who ... handed it to his superior, Miha Brejc, Director of the Security and Information Service (VIS), who gave it to Defence Minister Janša. The Minister yelled at me, – why are you messing with these invoices? –" Milošević remembers 24 years later.[141]

The invoice contained a list of 17 different types of infantry weapons and ordnance. Among them were 200 hand-held Zolja anti-tank launchers, 30 mortars with 500 mortar mines, more than 7,000 mines, almost 6,000 hand grenades, 1.2 million rounds of ammunition, 700 older M-48 rifles, and 20 machine guns.

The invoice that Zvonar lost in Milošević's office also shows that the price charged for a 120 mm mortar mine was an astronomical 970 German marks and for an 82 mm mine 558 marks.[142] By comparison, six months later they were selling the same calibre mines to Croatians three to five times cheaper.[143]

"I handed the paper to Komar as evidence that in the spring of 1991, the Secretariat of Defence[144] was already selling to Croatia weapons at predatory prices. Maybe because of the photocopy of this invoice, the police carried out a house search at my home. They didn't find it because I hid it in my slipper. That's why this piece of paper is so crumpled," explained Milošević laughingly.[145]

Ludvik Zvonar explained that this invoice was not for an actual sale of weapons and the prices were only used as a starting point in negotiations between governments of Slovenia and Croatia on helping Croatia.[146] However, this was not true. Zvonar lost his invoice in May, but the Slovenian and Croatian governments only began their negotiations after the outbreak of armed conflict in Slovenia a month later.

---

141) Author's conversation with Vladimir Milošević, Murska Sobota, May 20, 2014.
142) The invoice also shows other overpriced items: the price for a Zolja hand held rocket propelled launcher was 2,000 DEM, M57 hand held mortar was 3,500 DEM and a 7.9 mm machine gun was 4,200 DEM. The grand total for all weapons was 7,579,708 DEM.
143) Documents about sale of weapons from the former YPA warehouse in the town of Zgornja Ložnica show that in July, 1991, a Croatian Major Ivan Bećir paid 186 DEM for a 120 mm shell (which is 5.2 times cheaper) and 173 DEM for a 82 mm shell (3.2 times cheaper).
144) The Secretariat of Defence was renamed Defence Ministry only on June 27, 1991.

Experts confirmed that the weapons listed on the invoice came from the "closed area" of Kočevska Reka or Gotenica. It is known that in June of 1990, members of Beznik's police unit took control of these weapons. After that, they received an order to distribute them to the Headquarters of Territorial Defence in Slovenia. However before May 26, 1991, a large portion of weapons found its way to Croatia through secret channels.

The proceeds of more than seven million German marks shown in Zvonar's invoice don't appear in any balance sheets. At the time, Yugoslavia was on its deathbed and Slovenia was feverishly preparing for independence and possible aggression by the Yugoslav People's Army. This invoice is evidence that in those uncertain times Defence Minister Janša and his aides were disarming Slovenia.

Members of the Slovenian Presidency, as well as the Prime Minister at the time, Lojze Peterle, did not know anything about all this – at least that is what they claim now.

When asked if Janša, perhaps on the urging of a foreign country, was sending weapons to Croatia on his own, a former aide who wants to stay anonymous immediately answered: "You can be sure of that".

"I suppose he was pressured by European countries?"

"Mostly by only one."

"Which one?"

"Which one do you think was the most supportive of Slovenians and Croatians?" By which, one assumes, he meant Germany.

\* \* \*

After the unsuccessful kidnapping of Colonel Milošević and the turbulent reaction of Maribor's residents, the military leadership in Belgrade was reinforced in its belief that the Slovenian government would not give in. However, Serbian leader Slobodan Milošević was not really interested in Slovenia, where there were no indigenous Serbs. He was concerned how he could seize as much of Croatia as possible on the pretext of protecting the Serbian population, and how he could dismember Bosnia-Herzegovina.

In the meantime, the visit of U.S. Secretary of State, James Baker to Belgrade on June 21, 1991 gave Yugoslav Federal Prime Minister Ante Marković new hope that Yugoslavia could be preserved and continue with reforms. He understood the support of Washington as a silent consent for "disciplining" Slovenia – like collecting customs duties with the help of tanks, as he envisioned it.

The United States was following the events in the Balkans with only one eye and without any particular interest. The U.S. was too busy with other crises points and President George H. W. Bush was concerned how the disintegration of Yugoslavia could influence the Soviet Union, which was also breaking apart at the seams.

The European Community, preoccupied with its own problems, took little notice of the deteriorating situation in Yugoslavia and plumped for the status quo: it decided on June 23, 1991 not to recognise either Slovenia or Croatia as independent states.

Three days before the declaration of the Slovenian and Croatian independence, the leaderships of both republics met in Zagreb. The Slovenian delegation asked the Croatians about their preparations for independence and was stunned that the Croatians just looked away and avoided answering. After a while, they admitted with some uneasiness that they were not prepared at all. They were going to declare their independence, but that would be all …

The Slovenian President Kučan tried to lighten the painful atmosphere with a playful quip: "Will you, at least, put some sugar in the fuel reservoirs of the tanks going to Slovenia?"

"At this moment, I don't feel like joking," sternly answered Croatian President Tuđman.[147]

\* \* \*

Slovenia thus found herself alone on the threshold of a major conflict with a much stronger opponent. What was to be done? For 46 years, there had not been any armed conflict in Slovenia. However the people remembered how they had been caught at the

---

147) From the notes of Miha Ribarič, member of the Slovenian delegation in the meeting of both

intersection of great powers fighting each other in two brutal world wars which spread into Slovenian lands. Slovenian soldiers who were drawn into the conflicts did not lack bravery: they fought in World War I, and in World War II.

Then, in the early summer of 1991, it looked as if they would have to fight again to expel yet another occupier who dared to try to subjugate Slovenian lands. The prospect was harrowing after so many years of intervening peace. Everything looked senseless and ridiculous, and could not be compared to anything – or could it? Maybe it could ... to the Slovenian writer Fran Milčinski's humorous stories about the *Butalci* which started this book. He wrote how 500 years ago, *the Turk, that heretical trash, was frightening people around the country and stealing their livestock and youngsters.*

So the *Butalci* decided that something had to be done:

> *The men decided they would not surrender without a fight. On the road going to their world-famous Butale town, they would put a signpost saying: "Road prohibited to Turks!"*
> *And then another voice piped up: "Let's not be frightened rabbits! Let's write – Road strictly prohibited! –"*
> *And then they all enthusiastically supported the mayor's ruling that there should be no mercy towards the Turks, and that the sign should say: "Road most strictly prohibited!"*[148]

Strengthened thus in their fortitude, on June 25, 1991 the Slovenian Assembly passed a law implementing the Basic constitutional document regarding the sovereignty and independence of Slovenia. With this action, Slovenia declared independence.

On the same day, the Federal government in Belgrade decided that the Yugoslav People's Army (YPA) and the Federal police should take control over Slovenia's borders with Italy, Austria and Hungary.[149] With this move, the Yugoslav Prime Minister Marković effectively gave the green light for military intervention in Slovenia (though he denied he actually gave the order).

---

148) Fran Milčinski, *Butalci*, Karantanija, Ljubljana, 2001.

There was no official declaration of war by the Yugoslavs, and Slovenia for its part declared neither war nor a state of emergency. "We did not do it because the Yugoslav government also did not declare war against us. Even before the YPA's intervention, we warned Belgrade that we will defend our decision reached at the referendum on independence, with weapons if necessary, if the military shoots at us," said former Slovenian President Milan Kučan.[150]

Slovenia fought the battle not only by military means but also by propaganda, and it was "an uphill battle", as the former Slovenian Foreign Minister, Dimitrij Rupel, remembers.[151]

At first, the Yugoslav authorities planned to "secure" the Slovenian border with only 1,990 troops and about 150 tanks and armoured vehicles. The main thrust of the attack on Slovenia was carried out by the 5th Army Region and 13th Corps of the YPA. The 5th Military Region, headquartered in Zagreb, encompassed Slovenia and a part of Croatia. It was the worst armed military region in Yugoslavia and was under the command of a Slovenian General Konrad Kolšek. The 13th (Reka) Corps of the YPA was also under the command of a Slovenian General Marjan Čad.

Fate really played an ironic game with the small Slovenian nation, because the Deputy Defence Minister of Federal Yugoslavia, Admiral Stane Brovet, was also a Slovenian, as was Živko Pregl, Deputy Federal Prime Minister in Belgrade. Many of General Kolšek's Army colleagues suspected him of secretly collaborating with the Slovenian Territorial Defence, and he was replaced after only three days of fighting. In Slovenia, his action on behalf of the Yugoslav People's Army was not held against him – his only "sin" being that he was perhaps thinking more like a Yugoslav than a Slovenian. Back in the spring of 1990, Kolšek warned the Slovenian leadership at a meeting that Slovenia outside Yugoslavia "doesn't have any chance for national existence and it would lose its national identity in a few decades".[152] Events however would prove him wrong.

---

150) Author's conversation with Milan Kučan, Ljubljana, July 8, 2014.
151) Author's conversation with Dimitrij Rupel, Ljubljana, June 24, 1991.
152) Discussion of the Commander of the 5th Army Region, General Konrad Kolšek with President

Even though the whole officer corps objected to the independence of Slovenia, the generals were not united in how to prevent it from happening. General Adžić wanted to attack Slovenia with a strong military force and subjugate its leadership, but in the end the softer option of seizing Slovenian border crossings put forward by General Kadijević prevailed. He believed a show of force would convince the Slovenian leadership to cancel independence, and would also scare Croatia.

The Yugoslav People's Army units participating in the intervention in Slovenia were given only limited authority to act. They employed only a few per cent of their technical capabilities, they could not use artillery or mortars, and military aircrafts were not deployed until later in the conflict, with the explicit order to fire only in self-defence. Helicopters were not allowed to open fire. They were used only for transportation of Federal customs officers and military officers, and to transport soldiers wounded in the fighting.

In launching the intervention, Kadijević however apparently had not been authorised by Federal Prime Minister Marković. As a witness at the Hague Tribunal 12 years later Marković described the day when Slovenia was attacked: "I called a meeting of the government and accused Kadijević of using my name without my approval. I told him that I would have opposed the military intervention in Slovenia, even though I could not have prevented it. Kadijević answered that this was exactly why he did not inform me and left the meeting. The YPA's action took place with the approval of the President of Yugoslav Presidency Jović, who told me that it was his duty to secure borders of Yugoslavia."[153]

General Marjan Čad was ordered to seize all border crossings with Italy in the region of Goriška and Primorska by 3 pm on June 27. However he did this already by 7 pm the previous day, with only 350 soldiers and about 12 tanks. Even today, people try to figure out whether this commander of Slovenian origin jumped the gun in order to confuse the Slovenian defence or his superiors in the Yugoslav General Staff. In either case, he showed an incredible skill

---

153) Ante Marković as a witness at the Hague Tribunal (ICTY) during the trial against Slobodan

and cunningness and proved that he was well versed in the *blitzkrieg* strategy of the German Army between 1939 and 1941. General Čad fulfilled his task admirably, almost without complications, and most importantly without incurring any casualties.

The Defence Secretary at the time, Janez Janša, later alleged that the main culprit for General Čad's easy advance to the western border was a member of the Slovenian Presidency, Ciril Zlobec. Janša accused him of treacherously disclosing the date of independence to a foreign diplomat beforehand.[154] However Janša ignored that he and other politicians had also been publicly talking about this date. Zlobec denied the accusation and objected that Janša must have had access to a phone-tap of his conversation with the foreign diplomat.[155] This was one among many attempts by Janša to humiliate and shame Ciril Zlobec, an upstanding man, a former Partisan, a poet and a novelist, who frequently challenged the lies and arbitrariness of his tormentor.

General Čad's surprise military action showed that the Slovenian defence was incapable of stopping the incursion of the Reka Army Corps. "Members of the Territorial Defence were not adequately prepared, they were insufficiently equipped and without communication abilities," remembers Dušan Moljk, former chief of police in the Slovenian port of Koper.[156] He added that the police performed a bit better, but they received inadequate orders from the Slovenian leadership.

Slovenian politicians at the time were busy with preparations for a historic moment, the ceremony to declare the independence of the Republic of Slovenia on June 26, in Ljubljana on Republic Square in front of the Slovenian Parliament. Even today, the people of the Primorska region bordering Italy are sore that the pre-occupied leadership in Ljubljana "forgot" that a sizeable part of Slovenian territory was occupied the day before.

---

154) War for Slovenia, collection of documents, Janez Janša's Editorial, Nova obzorja, Ljubljana 2014, pages 499 and 501.
155) From the book of Ciril Zlobec, *Lepo je biti Slovenec, ni pa lahko* (It's nice to be Slovene but it's not easy), Mihelač, Ljubljana, 1992, pages 141 and 142.
156) "A day which was perhaps the worst in the war for independence",

On that June 26, 1991, a damp and a fateful day, the people of the Slovenia had mixed feelings: euphoria and faith in the new-independent Slovenia, mixed with unease and fear about what the morrow would bring. That same evening, a Yugoslav Air Force plane flew low over the capital to scare the population. But that did not deter tens of thousands of people from flooding the centre of Ljubljana to welcome the birth of the Slovenian state.

"Slovenia doesn't threaten anybody so there's no need for military aircraft above our cities or tanks on our streets. Only a person who lacks arguments and a sober judgment resorts to force," said the main speaker at the evening celebration, President Milan Kučan. "Tonight, dreams are allowed. Tomorrow is another day."[157] His ominous words have gone down in history.

Sure enough, in the early morning hours of the following day, the aggression of the Yugoslav People's Army against Slovenia began. At around 2:30 am, tanks from the YPA's military garrison at Vrhnika[158] moved towards the main Slovenian Airport of Brnik near Ljubljana. Writing in broken Slovene, General Kolšek on the morning of June 27 sent Slovenian Prime Minister Lojze Peterle a stern warning: "Any resistance will be crushed." Despite this intimidation, the YPA incongruously stated that it counted on the cooperation of all the institutions and the citizens of Slovenia.[159]

That was not to be. The generals had miscalculated. Wherever they showed up, the Army's soldiers and tanks encountered rage and open hostility from almost all the Slovenia's citizens. However, there was no military resistance on that day, not even a passive one.

Where were all these young men? Where were the officers who only a few hours earlier were toasting the new state with glasses full of wine? Where was the future Brigadier Krkovič, who only a few hours earlier saluted the President in front of a guard of honour, with sword in hand? Was he perhaps looking for his sword?

---

157) The whole speech is available at: http//www2.gov.si/up-rs/2002-2007/bp-mk.nsf/dokumenti/26.06.1991-90-92 (April 5, 2015).
158) Vrhnika is a small industrial town about 20 km southwest from Ljubljana. The 1st armoured battalion of the YPA, which was the largest tank unit in Slovenia, was stationed in the military garrison of that town.
159) Written order from the Commander of the 5th Army Region of the YPA, General Konrad

And where was the Slovenian military, that is to say, the Territorial Defence? For several months they had been claiming they had been collecting weapons and moving them to secret locations, even selling "surplus weapons" to Croatia. But then, on the first day of the attack, not a single round was fired, not a single automatic weapon rattled. What went wrong?

Even today participants in those events are loath to admit that they were astonishingly ill-prepared for armed conflict with the YPA. In many units, they had not even un-wrapped their weapons to get them ready for action. Mobilisation did not take place as it should, and the communications system did not function. For this ill-preparedness, nobody was ever called to account, not at the time of the attack and not afterward.

Army tank units from Croatia were moving into Slovenia from three directions: from Zagreb towards Novo mesto, from Varaždin towards Ormož and from Karlovec towards Metlika. In Croatia, nobody stood up against the column of tanks except for a few civilians with occasional barricades.

"Slovenians unrealistically accelerated things and we will not get involved," said Croatian President Tuđman at a meeting of the Croatian leadership. Defence Minister Špegelj reminded Tuđman about the agreement on common defence, friendship and cooperation with Slovenia and asked him what to say to the Slovenians.

"Nothing," replied Tuđman coldly.[160]

Janša later confirmed that Špegelj ignored that advice, and was informing Slovenian defence forces in detail about military movements towards Slovenia and about military aircraft taking off from the Croatian airfields.[161]

So how did they solve the problem with the Turks in the famous old story of *Butale*? There was a conundrum, wrote Milčinski playfully – *because in Butale, there is daylight only during the day, and during the night it is dark, they were justified in fearing that the Turk would come during the night and would not see the fearless prohibition.* So they decided to put a guard with a halberd and a spear who would stop the Turk. And if he did not listen, the guard would call the mayor and other men. Here is how the story ends:

> *So it happened that the guard felt his guts stirring, and he stepped aside to relieve himself. He drove the halberd into the ground and squatted. However, a blackberry bramble was growing there and when he wanted to get up, the bramble caught in his pants. In mortal fear, he bellowed: "Ouch, ouch! The Turk is holding me. Help, help!"*
> *When the mayor and the men heard the dreadful yelling, they jumped up and rushed together to discuss what to do. They decided that right now there was no time for discussion or conclusions. They just ran up the hill. On the hill they waited until morning. And when day broke, and there was neither smell nor sound of the Turk, they decided that the writing on the sign was good, the writing should stay, but the blackberry bramble should be cut down and burnt.*[162]

The warriors for independent Slovenia can take offence at this parallel if they want. But Fran Milčinski deserves a bow and a humble acknowledgment, for he figured out the how ingeniously Slovenian territory could be defended. It apparently worked with the Turks, and it worked also against Yugoslav General Kadijević.

- Chapter 4 -

# SLOVENIAN IS KILLING BROTHER SLOVENIAN

June 27, 1991, two days after independence, was a humid, stuffy and tense day from the early morning on. Alarming news about the approaching Yugoslav People's Army caused alarm among the people. They quickly pulled on to the roads trucks, buses, construction machinery and everything else which could stop the tanks.

However, if they were serious, the defenders should also have placed steel roadblocks and hedgehog barriers on the roads, and put explosives underneath. But they did not. Tanks leaving the barracks at Vrhnika squashed the fragile cars like empty beer cans. The Yugoslav People's Army distributed propaganda leaflets urging Slovenian residents to cooperate. However, the text was written in such bad Slovene language that many people just laughed despite the seriousness of the situation:

> *"Residents of Slovenia,*
> *Stay at home and in your workplaces. Don't allow yourselves to be abused against your vital interests.*
> *We invite you to peace and cooperation!*
> *Units of the Federal Secretariat of Internal Affairs and the YPA will fulfill their duties consistently and energetically. Any resistance will be crushed."*[163]

The Gods of fortune in those days were favourably inclined towards Slovenia. Based on General Čad's experience, General Kolšek assumed that his other mechanised armoured units would "stroll" to the border crossings without a problem.

Things however did not unfold according to his plans. As the situation rapidly worsened, the Slovenian defence finally started to wake up. Surprise was quickly followed by a sobering up. The Republic's Coordination Group set up by the Slovenian Presidency in March on Janša's suggestion started to function. It was led by the Interior Secretary Igor Bavčar with Janez Janša as his deputy. During the war, it was only supposed to coordinate activities of the Territorial Defence, police and civil defence and to cooperate with other coordination sub-groups around the country.

The Supreme Commander was the Presidency of the Republic of Slovenia. However as soon as the first fighting broke out there was a disconnect between the Presidency and the Republic's Coordination Group. One might even ask whether a relationship between the two ever existed … Who was the one person who really commanded Slovenians in uniform? The question remains unanswered to this day.

Bavčar gave the impression of a determined and even-tempered man. He was successful in covering his own fear and uncertainty. Janša had the look of a deeply worried, and at times even confused and lost man. Was it because he had wrongly believed reports that the Yugoslav People's Army would not leave its barracks? On purpose or not, he neglected warnings from Slovenian informers that the YPA was preparing to move.

The Army's movements in Slovenia followed the script of a not completely refined and even less tested military exercise called *Bedem 91*,[164] which had been approved by General Kolšek. It foresaw military operations based on various military scenarios, but was full of shortcomings that made it unworkable in a real situation.[165] Although the Army was attacking Slovenia, the plan was designed as a defence against a foreign enemy. Specifically, it envisioned Italian forces backed by Anglo-Americans charging into Slovenia through the "Postojna Gate" on a line that follows today's motorway between Koper on the Adriatic coast and Ljubljana.

A third phase of the battle plan foresaw a complete mobilisation of Partisan units, comprised of military reservists from the 5$^{th}$ Army Region, to fight alongside the Army troops. But these reservists

---

164) *Bedem* is a Serbian word for a trench.
165) The author has in his possession the Operational plan for the military exercise

were already integrated into the Slovenian Territorial Defence and were now looking at the Yugoslav People's Army soldiers down the barrels of their guns.

By the evening of June 27, the Army had occupied or blocked about 30 border crossings and reported that its mission was completed. However it was not over. That day Slovenian forces clashed with Army troops in the village of Trzin[166] near Ljubljana. Part of the Army's armoured brigade comprised of three armoured personnel carriers and a communications vehicle, heading towards Brnik Airport from Vrhnika, got stuck on a little bridge crossing the creek just before entering Trzin. The convoy was surrounded by the Territorial Defence soldiers, who were joined in the afternoon by Special Police Units under the command of Vinko Beznik. There followed a battle of time and nerves ...

This is how Beznik remembers those events today: "The psychological pressure was tremendous. It's very hard to fire the first shot because when you do, you can expect that your opponent will shoot back 10 times more strongly. However, the commander has to keep control over his units even in an unpredictable situation like this one."

"Why didn't you obey the order of the top leadership to attack?"

"Had we attacked immediately after receiving the order, we would have risked an air attack. Air Force fighter jets could have caused a massacre, also among the civilians. In the afternoon, Bavčar called me and demanded blood. Literally! He yelled at me, called me names ..."

"Called you names? Bavčar?"

"Yes, back then. Usually – luckily for Slovenia – he acted more reasonably then Janša. However, he used such bad words that I told the phone operator who established the phone connection to immediately disconnect it.

"You only attacked towards the evening, at 6:37 pm to be exact. Why then?"

"Troops from encircled armoured vehicles from Vrhnika were about to receive reinforcement from the Yugoslav People's Army anti-sabotage squad that flew in with two Mi-8 helicopters. At that moment, we couldn't delay the attack anymore and had to prevent the Army breaking out of the blockade."¹⁶⁷

When the two Army helicopter landed, heavy gunfire broke out for an hour. Slovenian units destroyed several military vehicles, and the Army troops were surrounded and forced to surrender before nightfall. One member of Slovenian Territorial Defence was seriously wounded and died on the way to the hospital, and on the Yugoslav side four soldiers were killed. Several people were wounded on both sides, among them two civilians.

Meanwhile 26 Yugoslav anti-sabotage squad members retreated to the village Depala vas, where they terrorised the local residents during the night, but in the morning they surrendered without a fight to Beznik's police officers.

At the time of these armed clashes in Trzin occurred an event which threatened to bring the disaster of all-out war much closer. In the centre of Ljubljana, Slovenian Territorial Defence soldiers shot down an unarmed Gazelle SA 341 military helicopter of the Yugoslav People's Army, number 664, even though it did not endanger anybody. The helicopter was piloted by an Army Captain of Slovenian origin, Toni Mrlak.

Unbeknown to the Territorial Defence soldiers, Mrlak was secretly collaborating with the Slovenian side and was about to switch camps. For some time, important members of the Slovenian government had known that Mrlak was preparing to move two or maybe three Gazelle helicopters to the Slovenian side. However, nobody ordered members of the Slovenian armed forces not to shoot at the three unarmed Army Gazelles, numbers 660, 664 and 718.

Toni Mrlak was killed on June 27 at 7:19 pm, together with a flight technician from Macedonia, Bojanče Sibinovski, when the two were delivering bread to YPA soldiers in Vrhnika. That TD soldiers had killed one of their own compatriots under such circumstances caused a wave of misgivings on the Slovenian side.

The death of the Slovenian pilot in a Yugoslav helicopter reminded many Slovenians of a verse written by the famous Slovenian poet, France Prešeren:

*For six months, the bloody river has wetted the ground,*
*Slovene is killing his brother Slovene –*
*How awful is man's blindness!*[168]

Prešeren's epic poem *Baptism at Savica* describes the forcible Christianisation in the eighth century of people in lands later to be Slovenian. The poet speaks of "a butcher's slaughter" of pagans – a holy war to enforce submission, with no ethics, morals or piety. Twelve centuries later, in the decisive moments of the Slovenian independence, a few deluded and eccentric people were set on similarly enforcing their ideology. Some independence warriors were possessed by fanaticism. They dogmatically took advantage of the moment of Slovenian independence to challenge the enemy, to show their muscles and to start a war. In this frantic fever, they ordered the shooting down of a helicopter piloted by their fellow countryman, friend and co-worker. He wore the enemy's insignia, but this unarmed helicopter normally used for training was likely to come into Slovenian possession very soon.[169]

On that June 27, Toni Mrlak spent most of the day flying around Major-General Marjan Vidmar, Deputy Commander of the YPA Ljubljana Corps. Three times they took off in the helicopter from the Marshal Tito barracks in Ljubljana and flew over the Slovenian regions of Notranjska, Gorenjska and Primorska to watch tank movements, seizure of border crossings by the YPA and blockades of several Army barracks.

General Vidmar, also a Slovenian by birth, later confirmed that the Gazelle pilot did not have his radio turned on and, therefore could not have heard the public appeals by the Slovenian Presidency to soldiers of the YPA to desert their units and refrain from participating in the aggression against Slovenia.[170] Mrlak probably did not listen

---

168) France Prešeren, *Poezije* (Poems), Prešernova družba, *Krst pri Savici* (Baptism at Savica), Ljubljana, 2008, page 204. Prešeren (1800-1849) is considered the greatest Slovenian poet.
169) Pilot Jože Kalan and aviation technician Bogomir Šuštar, Slovenes, working for the Yugoslav Army, proved that this was possible when on June 28, only a day after shooting down of Toni Mrlak's helicopter, they escaped to the Slovenian side with Gazelle helicopter no. 660 from the Maribor Cadet school.
170) (A Planned Operation), interview with Marjan Vidmar,

to the radio out of caution, because his flight technician on board, Sibinovski, was pro-Yugoslav.

That morning on June 27, at Mrlak's behest, a radio technician of the Army helicopter squadron covertly set up a secret frequency in all three Gazelle helicopters so that crews could communicate amongst each other when they flew over to the Slovenian side.[171] Toni Mrlak had meanwhile secretly met with representatives of the Slovenian government and received their approval to prepare for defection.[172]

Towards the evening, he dropped off General Vidmar and flew towards Brnik Airport to pick up a colleague, who was subsequently ordered to bring bread to the troops in Vrhnika. However Toni insisted that he would fly there himself because he was afraid that his less experienced colleague could be shot down by the Slovenian TD.[173] He believed that his compatriots would not shoot at him, because the authorities knew of his impending defection.

At around 7 pm, Mrlak and Sibinovski departed to pick up bread in the middle of Ljubljana. After that, they headed towards Vrhnika barracks using the normal and safest helicopter route along the railway track.

Two members of the 30th Development group (30th Rsk) of the Territorial Defence launched two Strela 2M anti-air missiles from a terrace on the top of the TR3 skyscraper in Republic Square in the middle of the city. The first rocket missed and landed without exploding next to the Gallery of Modern Art and the Serbian Orthodox Church, but the second one hit the tail of the helicopter from a distance of 1,300 metres. The Gazelle crashed 200 metres away from where it was hit – on the Rožna dolina road close to the railway.

---

early retirement which was granted only on July 7, the last day of armed conflict.
171) Radio technician Josip Volf, a Croatian by birth, set up a secret frequency 122.05 MHz in Gazelle helicopters no. 660, 664 and 718. Besides Mrlak, the secret frequency was also known to the helicopter crew Kalan-Šuštar which escaped to the Slovenian side on June 28.
172) The information was confirmed by Jože Kalan, a Major in the Slovenian Armed Forces, who until his retirement was commander of the Slovenian Armed Forces helicopter unit.
173) That pilot was a Macedonian national, Ljubomir Sandevski, who personally was against the YPA aggression against Slovenia. He later admitted that he himself contemplated fleeing with his

"This was like shooting a man in his back, deceitfully and dreadfully," said his widow Emilija Mrlak.[174] Moreover, shooting down a helicopter in the centre of the capital potentially violated the 4th Geneva Convention and international law regulating the protection of civilians in armed conflicts.

The anti-aircraft unit which was stationed on the top of the skyscraper TR3 was under the command of Colonel Krkovič, commander of the 30th Rsk of the Territorial Defence, later renamed the Moris Brigade.

"Krkovič issued several orders to the soldier of 30th Rsk who was monitoring the surroundings with a colleague to shoot at the Gazelle," says a member of the brigade.

"They talked on the radio and Krkovič harshly demanded that the soldier on top of the building should destroy the helicopter. I stood next to the commander and could hear the soldier's hesitant words that the helicopter might fall into the residential area and cause casualties. However, the commander wouldn't budge and kept on shouting – Shoot! Shoot –!"[175]

\* \* \*

People in Slovenia have long been served with platitudes how downing of the helicopter was an "error," a "mistake" or an "accident". Military officers looking for excuses spread the false notion that the incident was just an "unfortunate series of events". In order to cover up the truth, some authorities and colleagues of Mrlak even attributed the greatest responsibility for his death to the victim himself – he was accused of being naïve, indecisive and lacking good judgment.

Janez Janša wrote in his book that "the TD shot down the helicopter, which twice flew at low altitude over the building of the Slovenian Presidency".[176] However, nobody has ever come forward to confirm this statement. Eyewitnesses who saw the last flight of Gazelle helicopter

---

174) "He was betrayed by his own country", interview with Emilija Mrlak, *Nedeljski Dnevnik*, June 24, 2015.
175) A conversation with a member of the 30th Development group of Ljubljana TD. The author knows the name of the soldier but does not want to disclose it out of concern for soldier's safety.
176) Janez Janša, *Premiki* (Manoeuvres), Mladinska knjiga, Ljubljana, 1992, page 319. Despite the fact that nobody ever confirmed seeing the helicopter flying at the low altitude over the Slovenian

number 664 over Ljubljana say that it flew at a low altitude over the railway track that cuts across Ljubljana's Tivoli Park and continues towards Vrhnika.[177] That is not the location of the Presidency building.

About 20 years after the downing of the helicopter, writer and actress Draga Potočnjak, who was Mrlak's sister-in-law, wrote a book entitled The Secret Order, in which she concluded that Toni was sacrificed intentionally for propaganda purposes: "They sold him as a bad propaganda article in the armed conflict between the Yugoslav Army and the Slovenian Defence Forces," she wrote.[178]

Janša, as one of the participants in the undertaking, speaking a year after Slovenian independence, said: "This was a psychological turning point, especially in a sense that the TD and police units realised that the enemy was also vulnerable. We didn't have helicopters, tanks or large-calibre artillery pieces. Only after missiles were used and one their aircraft was downed – unfortunately with a Slovenian on board, with whom we had contacts, and it was agreed that he would change sides, but who unfortunately had not yet made his mind – only then was the psychological barrier crossed."[179]

As a "hawk" in the crisis, Janša was not satisfied with neutralising the enemy military force and negotiating with the Yugoslav People's Army to withdraw from captured border crossings. He needed blood, fire and ruin, probably because foreign TV networks in Slovenia were broadcasting pictures of the first real war in the heart of Europe since the fall of nazism. The intention was to wake up the foreign community.

The propaganda battle was one of the most important factors in the Slovenian 10-day war. In this regard, Slovenia's propaganda machine was much more effective than the Yugoslav one, as was testified by the foreign journalists who came to witness the birth of the new Slovenian state. Because Slovenian airports were closed, many foreign journalists were stuck in Ljubljana and were more or less forced to follow the flow of news fed to them by the astute Information Minister, Jelko Kacin, a close friend of Toni Mrlak.[180]

---

177) In order to protect eyewitnesses the author names eyewitnesses under the initials of their first and last names: G.F., B.U., P.Š., A.R., D.N., and J.D.
178) Draga Potočnjak, *Skrito povelje* (The Secret Order), Sanje, Ljubljana, 2013, page 355.
179) Janez Janša in the TV Show *Žarišče* (Focus), *TV Slovenija*, Ljubljana, June 25, 1992.
180) Most of the halls in Cankarjev dom (Cankar's House) in Ljubljana are underground. During the

At the same time, foreign intelligence services were active on the ground. The German BND[181] was trying to find out whether the Yugoslav military really wanted to subjugate Slovenia and prevent its independence. The logic of the Slovenian hawks was that the fiercer the battles were going to be, the sooner would Germany recognise Slovenia's independence. Janša firmly believed in the favourable disposition of the German government, and in pursuit of this goal nothing was sacrosanct. Not even the life of his colleague.

The Slovenian hawks, who afterwards boasted they were heroes of independence, issued secret orders that "activities should become more forceful". The Yugoslav People's Army had to be challenged. They urged attacks on Army barracks in Slovenian towns, regardless of the casualties this lunacy would cause among civilians.

Fortunately, the homeland defenders in the field were capable of sound judgment. At hot spots around the country, they preferred to negotiate, wait and even pull back, instead of using firearms. Regional commanders of the TD, police officers and YPA officers mostly knew each other and several times calmed hotheads on both sides. In the end, Slovenia's war of independence lasted only 10 days and had relatively few casualties. But at the time, many feared the conflict with the Army would develop into a lethal and ruinous war – as indeed happened later in Croatia, Bosnia-Herzegovina and Kosovo.

However these accommodating and calming moves did not suit Defence Secretary Janša. He was unhappy that he heard repeated reports on June 27 about helicopter landings, "but we haven't yet shot down any of them. [...] I called Tone Krkovič – his unit had anti-aircraft machine guns and also several anti-aircraft missiles were positioned on Ljubljana roofs – and asked him what on earth his crews were doing. He promised he would personally inquire about it".[182]

Krkovič understood Janša's hint and acted on it. Twenty years after the tragic incident he bragged: "Our members under my direct control shot down the first helicopter in Ljubljana."[183]

---

181) BND (*Bundesnachrichtendienst*) is the German Federal Intelligence Service.
182) Janez Janša, *Premiki* (Manoeuvres), Mladinska knjiga, Ljubljana, 1992, pages 169-171.

The affair showed how confused Slovenian lines of command were at this crucial moment in the country's destiny. Supreme Command of the Slovenian Armed Forces was, according to the legislation at the time, in the hands of the Slovenian collective Presidency headed by Milan Kučan. The Coordination Group headed by the Interior and Defence Secretaries was only a consultative body, charged also with forwarding the Supreme Commander's orders to regional and local units.

It is true that the Slovenian Presidency, in its extended formation, on the morning of June 27 issued an order to the Slovenian Territorial Defence to protect the facilities and communications it was guarding.[184] However, the order referred to using firearms only for defence and not for any other purpose. Going beyond this limited action, the Coordination Group that same day issued the following order: "Open fire with every weapon on all YPA helicopters, regardless of their position." It was signed by Janez Slapar, Janez Janša and Igor Bavčar.

Milan Kučan, after the publication of the order of the Coordination Group, admitted that he did not know about it and added that relations between the Slovenian Presidency and the Coordination Group were "complicated".[185] All of which leads to the bitter realisation that in this case, the highest collective political body in Slovenia allowed itself to be misled by Slovenian "hawks", as happened also regarding the sale of arms.

Vinko Beznik was probably the only member of the Coordination Group present on a battlefield. He said that cooperation with the Slovenian Presidency was poor or completely non-existent: "Bavčar and Janša led the war. I think that the Presidency was not included in the decision-making at all. Prudence and mostly rational moves by Interior Minister Bavčar were acting as a counterweight to the Minister of Defence, who received his information mostly from Krkovič."[186]

The Coordination Group knew of Mrlak's plan to defect. One member of the Group, Elo Rijavec, was Mrlak's uncle and knew that

---

184) Božo Repe, *Viri o demokratizaciji in osamosvojitvi Slovenije, III. del: Osamosavojitev in mednarodno priznanje* (Sources about Democratisation and Independence of Slovenia, Part III, Independence and International Recognition). Arhivsko društvo Slovenije, Ljubljana, 2004, page 41.
185) Former President Milan Kučan in his statement upon presentation of the book, *Skrito povelje* (The Secret Order), Slovenska akademija znanosti in umetnosti, Ljubljana, June 19, 2013.

his nephew and other helicopter pilots were planning to flee to the Slovenian side. They discussed details of the risky operation – including the secret sites in Slovenia where articulated trucks would be waiting to take the helicopters away to another location.

The next day, June 28, YPA pilot Jože Kalan and his technician, Bogomir Šuštar, did succeed in defecting with their helicopter. The two Slovenians made a dangerous flight from the Maribor Cadet School to a secret site on the Golte Mountain, where they were met by the Slovenian TD members and taken to a safe place together with their helicopter.

Mrlak and Sibinovski were shot down with a Strela 2M anti-aircraft rocket which was most likely part of the first "official" delivery of weapons to Slovenia that came by ship to Koper from the Bulgarian port of Burgas on June 20. The acquisition of these weapons was led by Ludvik Zvonar, a member of the Coordination Group. By another sad irony, Zvonar's deputy was the same Elo Rijavec. Moreover Mrlak's uncle was a subordinate of Krkovič, who gave the order to shoot the helicopter down. Today Rijavec does not want to discuss what he did or did not do for his nephew.

Another influential person who worked closely with Toni Mrlak was Jelko Kacin, who at first was deputy to Janša at the Defence Ministry and since April 1991 Information Minister and a member of the Coordination Group. Mrlak and Kacin shared a common interest in aircraft and helicopters. They flew together many times. In the spring of 1991 they secretly filmed border crossings and the last time they were together was May 6, about a month and a half before the outbreak of the armed conflict. At the time, the Yugoslav Security Service of the YPA was getting dangerously close to Mrlak.

Unfortunately Kacin did not think to warn anybody, in the Coordination Group or the rest of the defence structure, not to shoot at Gazelle helicopters that were taking off from the Brnik Airport. He was reportedly only a few seconds too late to prevent his friend's death: when he heard the explosion he ran to the telephone but the next moment he saw black smoke over Rožna dolina.

Almost 9 years later, in March 2000, Andrej Lovšin, former Director of the Security organ of the Defence Ministry (VOMO), rattled Kacin's conscience. He sent a letter to the Slovenian media accusing Kacin of being squarely responsible for the downing of the helicopter and the death of its pilot: "We developed a system of interpersonal information which was effective. This system informed all the YPA pilots that things were serious and that they should not take any risks. The late pilot Mrlak bypassed this system and was in contact with the Deputy Defence Minister at the time, Jelko Kacin. Even after Kacin became Information Minister, Mrlak stayed in contact with Kacin, who did not inform him that flying over Ljubljana was forbidden and that TD units had received an order to launch missiles."[187]

\* \* \*

Lovšin had a reason for this attack. During the spring of 2000, a special Parliamentary commission investigating illegal arms trading found a trace that led to former Defence Minister Janša and Lovšin. The investigators ascertained that during the war in Croatia, the two were selling arms to Croatia for cash at unusually high prices, conducting business without invoices to cover their tracks.

Kacin was well aware of their activities. After Janša's removal as Defence Minister in spring 1994, Kacin took his place and launched a detailed internal investigation which uncovered colossal illegal arms trading at the Defence Ministry between 1990 and 1993.

Members of Parliamentary investigative commission headed by Liberal Democratic Party (LDS) deputy Rudolf Moge constantly complained that the Defence Ministry was not handing over documents they requested, but the resourceful Kacin found another way to obtain them. He asked his Croatian colleague, Marinko Gašpić-Kljaković, the former defender of the Croatian coastal city of Split, to send him a copy of the document Gašpić-Kljaković received in Ljubljana from Lovšin in 1991, when he purchased a large quantity of weapons.

When Kacin showed the investigative commission a certified copy of the document received from Split, Defence Ministry officials were

---

187) Andrej Lovšin, "A theft of the school's chalk and peeing into the sandbox", Delo, March

forced to find the same document and hand it over to the investigators. It contained a list of weapons prepared by the Deputy Commander of the Slovenian Headquarters of Territorial Defence for the Interior, Colonel Peter Zupan, to which Lovšin had added extremely high prices by hand.[188] The most expensive were anti-aircraft and anti-tank weapons needed most by the Croatian defenders.

In handing over the Defence Minister post to Kacin in March 1994, Janša turned over his own weapons report.[189] Janša included his own price list, and he claims this was the sole basis for sales of weapons to Croatia and Bosnia-Herzegovina by the Defence Ministry. In reality, it bore no more relationship to the truth than the findings of the Administration of U.S. President George W. Bush in 2003 that Iraqi leader Saddam Hussein possessed weapons of mass destruction.

Lovšin sold the Croatian buyer Marinko Gašpić-Kljaković rocket launchers for Osa anti-tank missiles for 3,000 German marks apiece, even though Janša in his report assigned a value of only 900 marks to that weapon. Osa missiles were sold for 1,000 marks each, while Janša listed only 190 marks, and he sold Zolja rocket launchers for 2,000 marks, while the price given by Janša was barely a third of that – 650 marks. The buyer from Split overpaid the most for M57 mortars, which cost him 1,500 marks apiece, compared with only 200 marks on Janša's list. Strela 2M anti-aircraft missiles were, according to the Janša price list, worth 5,000 marks apiece, but Lovšin sold them for five times more – a whopping 25,000 German marks.

These discrepancies became known thanks to information provided by Croatia's Gašpić-Kljaković. In January 1996, in a conversation with a member of the select investigation group of the Slovenian military Intelligence and Security Service (OVS), Gašpić-Kljaković stated that Lovšin inserted prices by himself into a prepared list of weapons, and that as an expert he found them incredibly high. When he requested the list, Lovšin handed him only a copy which carried neither a header nor a signature.

---

188) The list included five M57 mortars, accompanied by 100 shells, 52 Osa anti-armour rockets, together with three launchers, 100 Kalashnikov automatic rifles with 100,000 rounds of ammunition, 100 Zolja anti-armour rocket launchers for one-time use, 50 Strela 2M anti-aircraft rockets and 20 Maljutka rockets. The document is without a date and markings.

The total purchase price for the weapons was 2,503,500 German marks. Gašpić-Kljaković's colleagues withdrew cash in various foreign bank notes at the Privredna banka Zagreb and they brought it on the same day to the Defence Ministry in Ljubljana.[190]

"Upon delivery of the money, I did not receive any receipt," stated Gašpić-Kljaković. Lovšin's men counted the money only the following day. The Croatian took the purchased weapons to Split and handed them over to Admiral Sveto Letica, who at the time was Commander of the Croatian Navy.[191] At the request of his superiors, the admiral prepared a report on the weapons purchase in Slovenia and sent it to Gojko Šušak, Croatian Defence Minister at the time.[192]

In Slovenia, there is no record of a permit to transfer weapons across the border, and the amount of 2,503,500 marks received was not recorded in any of the Slovenia's official balance sheets for that year.

Gašpić-Kljaković approached Lovšin twice more. When he was looking for weapons to defend Dubrovnik on the Dalmatian coast, the prices were the same as for the first purchase. However, when he wanted to buy weapons for Croatia's Navy, the price for the most sought-after weapon, the Strela 2M rocket, increased incredibly from 25,000 to 75,000 German marks apiece. Lovšin subsequently dropped the price to 50,000 marks, but it was still too expensive for Croatians, and the purchase did not take place.[193]

A comparison between the outrageously high prices for the arms that Lovšin sold to defenders of Split and prices in Janša's price list is one of the most important pieces of evidence produced by Moge's investigative commission. In his last report, Moge wrote "that all this casts a very bad light on what was happening at the time at the Ministry of Defence".[194]

---

190) A receipt for withdrawal of foreign currency from *Privredna banka Zagreb, direkcija trezorja*, no. 9980103 Zagreb, November 1991.
191) Official note of VOMO investigator Damjan Režek of a conversation with Navy Captain Marinko Gašpić-Kljaković, Ljubljana, January 10, 1996.
192) *Izvješče o primljenim sredstvima* (A report about received assets), Split Defence Command, no. 218-04-04-91/19 Split, November 26, 1991.
193) Printed list of weapons for the Croatian Navy with related prices, without a date or signature. Lovšin later wrote in lower prices. (Documents concerning Gašpić-Kljaković are kept by the author.)
194) The report on the direction and findings of the investigation and reasons why the commission

After all of this, nothing happened. Members of the Slovenian Parliament did not have enough courage, or political will, to bring the case to a conclusion. The findings of Moge's commission remained in a Parliamentary drawer, and prosecutors closed their eyes, apparently hoping that the statute of limitation would end the whole issue.

Andrej Lovšin still denies involvement in selling weapons for extraordinarily high prices for cash only, and calls on anybody to sue him if they want. Under questioning on oath before a Parliamentary commission, Lovšin said he could not remember anything and made fun of the members of Parliament.

However Jelko Kacin had no qualms about talking. He told the Parliamentary investigators that he talked to Lovšin about the weapons-for-cash trade in August 1994, and Lovšin told him: "All of this stuff relates to Janez Janša. He was taking care of all of the documents so ask him."[195] Which may explain why Lovšin later tried to discredit Kacin over the shooting down of the helicopter.

For nine years, Toni Mrlak's widow Emilija strove in vain for her under-age daughter and son to be granted a special allowance available to people disabled in war or the relatives of soldiers killed in the conflict of 1991. After years of presenting evidence to bureaucrats, lawsuits, appeals, mediations, quarrels and delays, on March 15, 2000, Emilija finally received a letter from the Ministry of Labour, Family and Social Affairs announcing that her daughter and son were granted the allowance.

"But this only happened after I appeared on TV and urged politicians to stop playing dumb and throwing mud on Toni's name," Emilija Mrlak says today.[196] In the meantime, a statue has been erected at the crash site, with an inscription saying that Toni Mrlak died for independent Slovenia. His widow however has been left with a bad taste in her mouth – because she fears people may think the allowance was granted to keep her quiet.

\* \* \*

---

195) Jelko Kacin during questioning by the investigative commission, Slovenian Parliament, Ljubljana, March 9, 2000.
196) " " (He was betrayed by his own country), interview with Emilija

The first attempt by the prosecutor's office in February 1992 to deal with the downing of the helicopter was a step backwards. Despite an expert's opinion to the contrary, the head of the Ljubljana public prosecutor's office, Tomaž Miklavčič, came to a conclusion that the helicopter crashed of its own accord and there was no relevant information proving that it was shot down.[197] Prosecutor Miklavčič however had done his work unprofessionally and incorrectly.

Five years later, the Mrlak case was addressed by prosecutors again. However, they decided it "was impossible to ascertain who shot down the helicopter". They also argued that at the time of the downing of the helicopter, war was raging and Slovenia was still part of Yugoslavia and thus not eligible to be considered under international law. They ruled that his death was therefore a case of murder and not a crime against humanity or international law. However the Slovenian Presidency never did declare war or a state of emergency, and in the last days of June and first days of July 1991, Slovenia underwent only low-intensity conflict.

As we know, Anton Krkovič, former commander of 30th Development group of TD, admitted in 2011 that he was behind this "heroic" deed. In 2014, Emilija Mrlak filed a criminal complaint at the Ljubljana Circuit Court against him and two others – accusing them of taking the lives or attempting to take the lives of Anton Mrlak and Bojanče Sibinovski.[198] However, the Ljubljana Circuit State Prosecutor Karmen Erčulj ruled on a technicality that the statute of limitation for the alleged offences expired on June 26, 2006, and threw out the case.

Toni Mrlak's widow meanwhile filed a further lawsuit requesting the Ljubljana Circuit Court to open an investigation of Krkovič. However, in October 2016, the Court turned the request down, and Krkovič himself never apologised to the relatives of the deceased pilot.

Although the Court failed to do its job, Emilija Mrlak remains the moral victor. With her relentless determination, supported by

---

197) Explanation of the Public Prosecutor Tomaž Miklavčič of the criminal proceeding, Ljubljana, February 27, 1992. Source: Draga Potočnjak, *Skrito povelje* (The Secret Order), Sanje, Ljubljana, 2013, pages 140 and 141.
198) Criminal complaint against Anton Krkovič and the unknown shooter and the unknown offender,

relatives and certain media, and the abundance of evidence presented in the book The Secret Order written by Toni Mrlak's sister-in-law, Draga Potočnjak, she got closer to the truth than any court did, and she unmasked several "heroes" of the independence.

"I will not appeal against the verdict because I do not see a chance of achieving anything more in the Slovenian courts than I already have. At the same time, I'm concerned for the safety of my family," she said.[199]

This is how Franz Kafka wrote about the frustrating inhumanity of such tortuous legal procedures:

> *In front of the law there is a doorkeeper. A man from the countryside comes up to the door and asks for entry. But the doorkeeper says he can't let him in to the law right now. The man thinks about this, and then he asks if he'll be able to go in later on. "That's possible," says the doorkeeper, "but not now." The gateway to the law is open as it always is, and the doorkeeper has stepped to one side, so the man bends over to try and see in. When the doorkeeper notices this he laughs and says: "If you're tempted give it a try, try and go in even though I say you can't. Careful though: I'm powerful. And I'm only the lowliest of all the doormen. But there's a doorkeeper for each of the rooms and each of them is more powerful than the last. It's more than I can stand just to look at the third one." The man from the country had not expected difficulties like this, the law was supposed to be accessible for anyone at any time, he thinks ...*[200]

---

199) Emilija Mrlak, at a press conference, Ljubljana, October 25, 2016.
200) Franz Kafka,          translator: David Wyllie, Posting Date: August 13, 2012 [EBook #7849]

- *Chapter 5* -

# "A MAN IN SHORTS WITH KALASHNIKOV DISARMS ARMY"

From June 27, official Slovenia took away the word "People's" when referring to the Yugoslav Army. Now it was an occupation force set on crushing the will of the Slovenian people. Right after the Slovenian declaration of independence, the YA started to transport reinforcements from other republics, as well as Federal police and customs officers from Belgrade. Helicopters with reinforcements landed at Cerklje ob Krki, the only permanent airbase of the Yugoslav Air Force in Slovenia.

On the morning of June 27, the Slovenian Territorial Defence Headquarters ordered an attack on the military airbase and in the evening two mines landed on the airport runway. Nobody was hurt, but the Yugoslav Air Force pilots took fright and flew their aircraft and helicopters away from the Slovenian base. Air Force fighter jets Jastreb (Hawk) and Orel (Eagle) found a temporary refuge at the military airport Željava near Bihać in Bosnia-Herzegovina, while most of the helicopters landed at Zagreb's Pleso Airport.

Janša and his supporters claim that the "attack" on Cerklje Airport was one of the most important events in the war for Slovenia. However, another reason for the demise of the Yugoslav Air Force in Slovenia was the desertion of 12 Slovenian pilots, whose courageous action disrupted the operation of the Air Force at Cerklje.

\* \* \*

The next day, Friday, June 28, was a very long and strenuous day, during which the Yugoslav Army's strategy started to unravel. The invading Army found itself encircled by its opponents, and planners of the military intervention took impulsive actions, based on the entirely wrong assumption that the Slovenian population would give

"If we look at it from the professional point of view, the Army should beforehand have ensured uninterrupted supplies, adequate communications systems, as well as media and police support," said YA General Marjan Vidmar, Commander of the Ljubljana Corps Headquarters at the time.[201]

Ordinary people were now openly hostile towards aggressors. Young soldiers stuck inside hot tanks were given no help, and without water barely survived on the rations they brought with them. They had no real idea why they were there. High-ranking Yugoslav officers brainwashed conscripts and non-commissioned officers into thinking that they were defending Yugoslavia against an attack by NATO.

Ranks of the Territorial Defence, whose main objective was to recapture the international border crossings, were in the meantime filling up with new recruits. Alongside police, they were throwing up barricades with the support of local residents. The Republic's Coordination Group was sending orders for offensive action against the YA wherever the defenders had a tactical advantage, but others tried to keep the conflict low-key.

Vinko Beznik emphasised that the police officers strictly followed instructions that they should first try to negotiate with their opponents: "The Territorial Defence functioned similarly, although not completely like our police officers. About one per cent of TD officers were trying to show their strength, while at the same time we were urging YA members to surrender. Because of this, they were causing fights and unnecessary casualties."[202]

Those mobilised into the Territorial Defence were for a long time wrongly convinced that they were taking part only in military exercises. They did not grasp that they were taking part in real military action and were shocked when YA tanks started shooting at them.

Members of the Slovenian TD were dressed in old-fashioned Army greatcoats and looked very untidy. Many of them did not even wear insignia of the Slovenian military, and the main difference between them and Yugoslav soldiers was that they wore ribbons on their sleeves. Another difference was that YA soldiers

---

201) "

were well-groomed and wore short hair, unlike the more slovenly Slovenians, many of whom were unshaven with disheveled hair.

On June 28, 1991, Yugoslav Air Force planes attacked Slovenian positions for the first time. Fighter jets machine-gunned the road between Novo mesto and Ljubljana on a slope called Medvedjek, repeatedly hitting a barricade of trucks that had stopped about 20 military armoured vehicles and other vehicles coming from the Croatian town of Karlovec. Six foreign drivers were killed next to their trucks, and another six were wounded. Among the dead was also a farmer living nearby.

Members of the Territorial Defence in the Dolenjska region had their baptism of fire, and for some it was too much. During the night of June 27 to 28, about 70 of 500 TD soldiers watching an Army armoured column deserted from their positions of ambush. Their commanders thus met their first Waterloo, even though they are loath to admit this today.

Members of the Dolenjska TD Command positioned on Medvedjek twice refused to obey orders by the Republic's Coordination Group; firstly, when it requested the blowing up of a nearby road viaduct, which could have endangered civilian lives, and secondly when the Coordination Group requested a night attack. The soldiers were psychologically and technically unprepared, and they also had no hedgehog barriers or anti-tank mines.

On June 28, Yugoslav Air Force jets targeted Slovenian Radio and TV transmitters on hilltops, as well as the main Brnik Airport, where they severely damaged Slovenian civilian aviation. Yugoslav Air Force Jastreb fighters shot at trucks in barricades which were blocking an Army tank column heading for Šentilj, the busiest international border crossing with Austria in the northeast of Slovenia. They attacked Štrihovec, the last village before the border, claiming the lives of three foreigners, among them a 10-year-old boy from Turkey.

One year later, the Slovenian leadership gave medals to individuals responsible for defending Šentilj. During the ceremony, security officer Bojan Babič remarked to a colleague: "Had you shot at tanks at the time, you would also have been among the recipients of the medals."

The man, a former non-commissioned YA officer in the Maribor barracks who does not wish to be named, took a different view: "Had I shot, I would have killed my friends and former colleagues." When it became clear that Yugoslavia was going to fall apart, he openly admitted to his soldiers that he was going to join the Slovenian armed forces. They did not try to prevent him from leaving and before his departure, he promised that he would not shoot at them.

"When I was manning the ambush at Štrihovec, I knew that in the tanks before the barricades were my comrades from the tank unit. I served almost 10 years with this unit … There was no force in the world that would at that time convince me to act differently. There's no medal that would make me break that promise."

The second day of conflict was marked by renewed efforts to take back control over the Slovenian borders. Slovenian military strategists tasked themselves with launching attacks on some of the border crossings to show the aggressor that his plans would fail.

A particularly terrifying battle took place at the international border crossing named Holmec, not far from the Austrian town of Bleiburg. Holmec and Vič near Dravograd are the largest border crossing in the Slovenian region of Koroška, and Holmec was easily accessible for members of the Slovenian TD, but less so for the YA tanks and armoured vehicles.

When on June 27, a colonel from the Maribor YA Corps and a lieutenant supposed to be in charge of the guardhouse at Holmec[203] saw a sign "Republic of Slovenia", and a pole with the Slovenian flag instead of the Yugoslav one, they gave an ultimatum to Slovenian policemen: if you don't pull out within 15 minutes we are going to start shooting at you.

Yugoslav soldiers encircled the border crossing, but members of Slovenian TD surrounded the Yugoslavs. When the ultimatum expired, the Yugoslav soldiers first fired a few warning shots, but the Slovenians made their presence felt, and took pressure off Slovenian policemen stationed at the border crossing only 200 metres away from the YA guard house.

---

203) Dragan Dragić, a lieutenant from Serbia who arrived at Holmec on June 26, 1991, took command

Meanwhile several dozen Federal police officers disembarked from helicopters at the nearby Slovenian town of Dravograd. They intended to take over Koroška's other border crossings with Austria, but were captured by a special unit of the Territorial Defence. The Yugoslavs carried written orders to "launch an attack on border crossings and seize facilities, by force if necessary, regardless of possible victims".[204]

The commanding officers of TD at Holmec decided to surprise the Yugoslav soldiers and attack first. During the night between June 27 and 28, they feverishly prepared for a decisive battle. In manpower, they were three times stronger. However, they did not have radio communications capabilities and their weapons were inadequate: alongside rifles and machine guns, they only had a few mortars and two anti-aircraft guns. They even used an MG-42 German machine gun from 1942 with an engraved swastika on the gun breech. Had the Yugoslav Army seized it as a war trophy, they would doubtless have displayed it as evidence that Yugoslavia was being destroyed by a born-again "Third Reich".[205]

The following day, guns in Holmec started to rattle before daybreak. In the darkness before dawn, the opposing sides almost ran into each other. When the YA soldiers at the guardhouse realised they were encircled by the TD, they started to shoot at the border crossing below them with a recoilless cannon. A sniper shot at Slovenian police hiding at the edge of the building, killing first one and then another, while a third was hit in the collarbone and was barely able to hide under a car.

On the other side, three soldiers of the YA were killed. However, despite these deaths and many wounded, the YA soldiers held their position. They hit the border crossing with an incendiary grenade and the building went up in flames. The Yugoslav commander of the guardhouse counted on help from the Army, but a large armoured car column that left Maribor towards Koroška encountered many

---

204) Damijan Guštin, Vladimir Prebilič, *Boj na Holmcu 27. in 28. junij 1991 in Koroška v vojni za obrambo neodvisnosti Republike Slovenije* (The Battle For Holmec, June 27 and 28, 1991, and Koroška in a War For Defending the Independence of the Republic of Slovenia), Narodna in univezitetna knjižnica, Ljubljana, 2006, page 40.
205) Serbian media often associated German support for Slovenia and Croatia with nazi Germany's

barricades on the road alongside the Drava River and finally got stuck in front of the Dravograd barracks.

The surrounded soldiers of the YA in the Holmec guardhouse were losing strength, however the commander and his deputy stubbornly refused to give up. When it became clear that the reinforcement would not come, the two officers and a group of soldiers escaped through their opponents' encirclement, and reached other border units of the YA. However, they were captured the following day.

Thanks to the initiative of a Slovenian farmer nearby, a group of the young Yugoslav recruits tried to surrender. They waved a white flag, but had to hit the ground when shots rang out. Were the Slovenians committing a war crime by firing on surrendered prisoners? The action was captured by a film crew of the Austrian TV ÖRF. Slovenian police did indeed shoot at the surrendering soldiers, but from where they were positioned they could not see that the Yugoslav soldiers waving their white flag. All nine of the Army soldiers, among them two Slovenians, survived this moment of drama.[206]

\* \* \*

On the same day, June 28, 1991, an officer of the Yugoslav Army by a cool deception achieved the surrender of YA troops occupying all the rest of the frontier guardhouses in the Koroška region. Captain Dragan Cvetković was among more than 50 captured members of the YA being detained by Slovenian police in a cultural centre in the town of Slovenj Gradec. By agreement with the police, Cvetković slipped out of the cultural centre, taking care that his fellow soldiers did not see him. He changed his uniform for civilian clothes and drove to a secret location, a house in a nearby village. From there, he telephoned his subordinate officers in the guardhouses and urged them to surrender peacefully to the Slovenian police.

The captain had to resort to this subterfuge, because his superior, Major Mladenović, commander of the nearby Dravograd garrison,

---

206) Damijan Guštin, Vladimir Prebilič, *Boj na Holmcu 27. in 28. junij 1991 in Koroška v vojni za obrambo neodvisnosti Republike Slovenije* (The Battle For Holmec, June 27 and 28, 1991, and Koroška

stubbornly insisted on executing orders from Belgrade and was demanding that all border crossing be held. After lengthy negotiations, Captain Cvetković convinced commanding officers in the guardhouses to lay down their arms and surrender to Slovenian police officers. After that, Cvetković put his uniform back on and returned to join the captured soldiers.

"Where have you been, comrade Captain?" asked an anxious non-commissioned officer.

"I went to see my commander and my boss."

The non-commissioned officer was surprised because he knew that Cvetković and Mladenović were like day and night: "And what were you doing over there?"

"We were drinking coffee," Captain Cvetković calmly answered.[207]

Meanwhile outside the Dravograd garrison, surrounding Territorial Defence soldiers took fright when they heard a military plane suddenly approaching. Eyewitnesses saw a Yugoslav Air Force Mig-21 jet swoop quickly over the nearby Drava River valley. Then there was silence. The TD and police were petrified, since they had no anti-aircraft weapons. They heard a hollow sound and a curtain of water rose over the Drava River. Then everything was quiet again.

The fighter-bomber was piloted by a Slovenian originating from nearby. He was tasked with destroying barricades that Slovenian defenders set up in front of the military barracks and a column of YA armoured vehicles. Had he launched a rocket into the barricades, he would almost certainly have killed his compatriots. On the other hand if he did not follow his orders, he would have been punished on return to base.

He had a few seconds to decide. He called the YA command centre in Zagreb, and got lucky. On the other end of the radio was yet another Slovenian national, who came from the village of Črneče right next to Dravograd. He was Major-General Marjan Rožič, Commander of the 5th Corps of the Yugoslav Air Force in Zagreb. This is approximately how the conversation went between the pilot and the General:

"What shall I do?"

---

207) Captain Cvetković and the non-commissioned officer were speaking in Serbo-Croatian.

"Drop the rocket. Drop it into the river. Be careful about the bridge!"
"I understand. It didn't work. I couldn't do it."
"Do it again! Into the Drava River, so you won't hit people."
"Yes, I was successful. Done! I'm returning to base."

General Rožič was walking a tightrope. At the time, he commanded 550 jet fighters, or two-thirds of the total number of the military fighter planes in the Yugoslav Air Force. In the war against Slovenia, he was torn between senseless orders from the YA supreme command in Belgrade and loyalty towards Slovenia. He chose the latter and took a great risk in deciding to stay as Commander of the Zagreb Air Force Corps as long as possible, because from this position he would be able to block, or at least lessen, the impact of the Yugoslav Air Force sorties against Slovenia.

On another occasion, when he was ordered to come to the aid of the Army soldiers in Holmec with an air attack, Rožič refused to do so, citing bad weather as an excuse. However, the pressure from Yugoslav and Serbian generals was getting greater, and the beginning of July saw the end of General Rožič's defiance and his career. He and General Kolšek, the overall commander of the YA in Zagreb, were replaced and placed under house arrest in Belgrade. The two were accused of disobeying orders to bomb Ljubljana. The Yugoslav military subsequently released General Rožič after Slovenian leaders came to an agreement with Belgrade about the withdrawal of the Yugoslav soldiers from Slovenia.

The retired Major-General does not want to talk about these events today. He wrote personal notes about the worst moments in his life, but his relatives convinced him to throw them away and leave the past behind. Now he lives alone – with a clean conscience – in Črneče next to Dravograd.

On July 2, only a day after Generals Rožič and Kolšek were replaced, members of the TD clashed again with soldiers from the YA motorised armoured unit stuck in Dravograd. This time, a Yugoslav Air Force plane did launch missiles at Slovenian positions, and a fierce battle erupted which caused injuries on both sides. But two days later, after exhausting negotiations, the TD was able to push the armoured unit back into the Dravograd barracks.

By July 5, all YA officers and soldiers in the Koroška region had surrendered. Slovenians seized many documents, military equipment and arms.[208]

\* \* \*

A few days earlier, on June 29, guardhouses in Prekmurje on the frontiers with Austria and Hungary started to fall like dominoes. But serious clashes had erupted the day before in the nearby town Gornja Radgona on the Mura River, which forms Slovenia's border with Austria. A motorised Yugoslav Army column of tanks, armoured vehicles and trucks left the Croatian city of Varaždin and headed towards the river border crossing in the Slovenian town. However here too the advantage laid on the Slovenian side: the unit was comprised of inexperienced conscripts, among them several Slovenians.

In the centre of Gornja Radgona, residents and volunteers acting on their own initiative attacked the military vehicles with Molotov cocktails and set fire to two trucks, a tanker truck and an all-terrain vehicle. In one of the trucks, ammunition caught on fire and started to crackle as in the middle of a fierce battle.

This is how Anton Orgolič, one of the 14 young men taking part, remembers it: "I lit a Molotov cocktail and handed it to another person who threw it. We were standing too close to the trucks and soldiers started to shoot at us with anti-aircraft guns and sparks were flying all around us. The following day, somebody came to us and said: you have to enlist in the Territorial Defence now, because as civilians you burned down those trucks."[209]

The commander of the military column, Colonel Berislav Popov, ordered his armoured vehicles to open fire. In a few minutes, 20 buildings in the town were destroyed, among them the steeple of Radgona's church, from where armed Slovenians were shooting at the YA soldiers. That day, two civilians observing the battle were killed, as well as five YA soldiers, while 17 people were wounded.

---

208) According to the Slovenian TD, they captured 7.9 mm sniper guns, machine-guns, automatic weapons, ordnance, Zolja rocket launchers and a large quantity of ammunition.
209) Anton Orgolič, in the collection of testimonies entitled: *Moški na položajih, ženske v strahu, otroci*

Colonel Popov and his superior, Major-General Vlado Trifunović, Commander of the Varaždin Corps, for several years faced charges of war crimes against civilians in a court in the nearby Slovenian city of Murska Sobota. They defended themselves by saying they were only executing orders of the Yugoslav government and the top military leadership to secure the Yugoslav borders. Colonel Popov also argued that he was defending his soldiers against attack by Slovenian volunteers.

Trifunović was acquitted in August 2013, while Popov was convicted in absentia and sentenced to five years in prison. He appealed against his conviction, and his sentence was overturned. At a retrial on October 22, 2015, witnesses and military experts testified that Popov did not issue an order to cause damage, and gave the opinion that his unit acted according to the Geneva Conventions. Regarding the two killed civilians, it was argued that the Slovenian defence forces were supposed to make sure that the civilian population was removed from the area of armed conflict. Based on these testimonies, the district state prosecutor withdrew the charges and the trial of Popov in Slovenia ended.[210]

However, he and General Trifunović were not out of the wood. They had later decided to move their unit from Varaždin in Croatia to Serbia, due to the desperate situation in Croatia. Even though they disabled all heavy weapons left behind, the two YA officers were convicted in Serbia for undermining the country's defence capability: Trifunović received 11 and Popov seven years in prison. However, while they were in prison, the Supreme Court of Serbia acquitted them of all charges and ruled that they should be awarded restitution.

But in December 2013, a court in Croatia found them guilty in absentia and sentenced each to 15 years imprisonment for war crimes against the civilian population in Varaždin in September 1991.

The legal cases of Vlado Trifunović and Berislav Popov have been keeping courts in all three states busy for more than two decades. General Trifunović once said he is probably the only general to be convicted of committing a war crime in one state, and to be sentenced in another state for not committing a war crime.

\* \* \*

After the first day of armed conflict, Slovenia's Defence Minister (Secretaries were renamed Ministers on June 27) Janez Janša, wanted to give encouragement to the defenders of the homeland and the Slovenian public, and at the same time make foreign countries wake up.

So on June 27, he announced on television that the conflict so far claimed "100 dead and wounded" on both sides, whilst in reality the whole 10-day war in Slovenia claimed less than 90 military and civilian casualties on both sides. He also claimed that Slovenian defenders shot down "six enemy helicopters",[211] even though members of the TD in fact shot down only the unarmed YA Gazelle flown by Toni Mrlak, and a Mi-8 transport helicopter over Barje on the outskirts of Ljubljana, killing three Yugoslav Air Force airmen.

It looked as if the Territorial Defence and police were not overly scared by the supposedly unbeatable Yugoslav Army with its heavy military gear. However, the luck of war was not on the side of defenders that day. There were too many mistakes, and too little experience on the part of commanders and soldiers. A few weeks before the outbreak of hostilities, Janša replaced several TD commanders with new ones, mostly his loyal followers. However they were not known to soldiers in the TD units, nor were they trusted by them. In the most critical moments, this caused much trouble and diminished their effectiveness in battle.

Some officers did not know how to motivate their soldiers or protect them. In the TD units, there were several cases of suicide, uncontrolled discharges of firearms and accidents associated with handling weapons. In the Ljubljana suburb of Šentvid, a TD member lost his life because he did not know how to handle an Armbrust rocket launcher, and another was wounded. Because of fear, many mobilised members of the TD drank too much alcohol. The police on the other hand drank less and had more discipline.

Many of the failings were covered up. Daily reports that contained – or should have contained – accounts of accidents and offences in the ranks of the Slovenian defenders disappeared from units and the headquarters of the TD. Some people looking for credit obviously tried hard to make the war look as clean as possible.

During the first days of war, the communications system did not function. The new Racal radio transmitters smuggled in a short time beforehand from Britain were excellent, but the users did not know how to operate them. Instructions were in English, which was beyond many members of the TD.

Fortunately, the defenders were communicating with their opponents – who until yesterday had been their friends, colleagues and neighbours. They found themselves on different sides of barricades and received orders from different leaders, but they still kept personal contacts amongst each other. They were communicating on telephones and exchanging messages, threats and ultimatums. When nothing else worked, the message was accompanied by a burst of fire from automatic weapons.

People in the field decisively influenced the course of events. At that time, most of them acted according to their own consciences, even when it was not in accordance with the demands of their superiors from the operations centre of the Republic's Coordination Group, located in basements of Ljubljana's Cankarjev dom concert hall.

On June 28, when the Slovenian defenders were facing collapse, the political leadership launched the so-called *Nabava* (Procurement) plan. Drawn up in autumn 1990, it envisaged removal of weapons and ammunition from the Yugoslav Army's 29 military depots in Slovenia. Usually, these depots were guarded only by small numbers of soldiers, mostly less than 100.

The first military depot to fall into Slovenian hands was in the town of Borovnica, about 30 kilometres southwest from Ljubljana. Early in the afternoon of June 28, the depot was seized by local residents and members of the 30th Development group, most likely without receiving any order from the Headquarters of the Territorial Defence. They also captured 30 soldiers of the Yugoslav Army.

"The commanding officer, who resisted arrest, was shot by one of our soldiers in his buttocks. We gave him medical help, and his wound was sewn up by a veterinary doctor because we had no medical doctor nearby," remembers Ludvik Zvonar, the

Advisor to the Government, who came to Borovnica soon after the takeover.[212]

The depot contained about 824 tons of weapons, ammunition and ordnance, including most of the artillery ammunition of the armoured brigade stationed in the town of Vrhnika.[213] They also seized several tons of food and about 2,000 uniforms. The Slovenians rejoiced most at capturing anti-aircraft and anti-armour weapons named Strela, Osa, Maljutka and Zolja.[214]

On the same evening, 10 fully loaded seven-ton trucks left the depot and drove to the railway station in the Ljubljana suburb of Moste, so that the weapons could be distributed as soon as possible to TD units throughout Slovenia. There was a constant danger that the YA might show up and retake the Borovnica military depot, or that the Yugoslav Air Force might attack it and destroy it.

However, late at night, something unusual happened. The commanding officer of the TD at the captured depot received information that Yugoslav tanks from Vrhnika were approaching Borovnica. The TD quickly abandoned it and left it without supervision for several hours during the night. Even today it is not known who was responsible for the false alarm which left the depot unguarded. When Zvonar learned about it, he flew into a rage and threatened to shoot the responsible officer.

At around 4 am, the TD returned and were surprised to find the military depot almost completely looted and trashed. The Ljubljana command of the TD estimated that in a few hours 90 machine guns, more than 100 different handguns, 32 sniper rifles, a large number of Kalashnikov automatic rifles and some valuable military equipment disappeared from the depot. So in only a few hours, between 15 and 20 tons of weapons were taken away from the depot.

---

212) Author's conversation with Ludvik Zvonar, Radovljica, June 28, 2010.
213) The list of seized facilities and assets by the Territorial Defence in Slovenia. The list was handed to the Deputy Slovenian Minister of Defence, Miran Bogataj, by a General of the Yugoslav Army, Andrija Rašeta on July 21, 1991.
214) On July 4, 1991, the Command of the Ljubljana YA Corps sent to the 5th Army Region Command in Zagreb a list of assets seized in the Borovnica military depot by the Slovenian TD. (no. 95-46/1). On that list, there were 500 pistols; 3,582 rifles; 277 machine guns; 366 mortars; 40 cannons; 4,000,000 rounds of ammunition; 308 Zolja rocket launchers; 1,400 anti-tank rockets; 150 Osa and

The threat of tank attack was a false alarm, but nobody knows who was responsible. During the night hours, this important military depot became a free self-service market. Local residents took weapons for hunting or as trophies. Some of the looted weapons later turned up in the buildings of the Borovnica municipal assembly and county administration, as well as at the police station at Vrhnika. Several years later, arms from Borovnica such as a sniper rifle, a Scorpion submachine gun or even a Zolja rocket launcher were found in the possession of "collectors" trying to sell them.[215]

Five weeks after the takeover of the military depot in Borovnica, the Ljubljana TD Command prepared a report stating that "the removal of material and technical assets, ammunition and weapons from the depot was done uncontrollably, in an unorganised and uncoordinated manner" and that securing of this strategically important facility during the first three days after capture was catastrophically inadequate.[216]

The report did not mention Krkovič's Moris brigade, though it soon became clear that members of the brigade stole the majority of the weapons. Milan Zorko, who in the past worked for the Yugoslav military and was from July onwards a security officer of the Ljubljana Region of TD, says: "About a third of weapons and military equipment from the military depot in Borovnica was taken by the TD members from the Ljubljana region, and the rest was moved to Kočevska Reka, where Krkovič's brigade was stationed."[217]

The military formation under the command of Colonel Anton Krkovič was until October 1992 officially named the 30th Development group of the TD. However once the fighting started in 1991, it started to be called the Moris brigade (Moris is an acronym for the Ministry of Defence of the Republic of Slovenia). Krkovič did not obey orders from the military leadership, and he consistently ignored the Headquarters Commander of the TD, Janez Slapar.

---

215) Document, showing that a member of the Moris Brigade, Darko Njavro, reported to the police names of private sellers of weapons. Informative conversation - official mark, code 3, Police station Ljubljana – center, February 22, 1991.
216) The report was prepared in August 1991 by the Deputy Commander of the 5th PŠTO Ljubljana (Regional Command of the Territorial Defence), Marino Medeot.

"The commander of the Dolenjska region of the TD threatened that he was going to consider the Moris brigade an enemy unit, because the brigade was appearing unannounced on his territory and in Ljubljana, performing activities without any prior notification," noted the military Security organ. It demanded that Commander Krkovič be "completely incorporated into the leadership and command system".[218]

Despite these structures however, Colonel Krkovič recognised only the authority of Janez Janša. He maintained this loyalty in the spring of 1994 after Janša's removal from the position of Defence Minister and also after 1998, when the Moris brigade was disbanded in a reorganisation of the Slovenian military. Janša however did not reward his loyalty in the same manner. When in 2004 Janša became Prime Minister, he did not appoint Krkovič to any important military post. The most Krkovič gained was promotion to the rank of Brigadier in 1993.

In the old system in Yugoslavia, the 30th Development group represented the professional core of the Territorial Defence force which guarded the top Slovenian Communist Party leadership, including Tito's ideologist and author of the socialist self-management system, Edvard Kardelj. For that reason, it was also called the Kardelj's brigade.

"Anton Krkovič, who at the time was a Captain in the reserve, was employed in the Kardelj's protective brigade ... Candidates for these kinds of jobs had to go through extensive background checks. The most important factor was total loyalty to the League of Communists," wrote Draga Potočnjak in her book.[219] So Krkovič first served the elites of the former regime, and later followed new leaders with the same enthusiasm, presumably so that he could stay near the top of the power structure.

The role of Krkovič and the Moris brigade in moving the weapons to Croatia was widely testified. According to statements by Ljubljana TD members, the Moris brigade entered the military depot in Borovnica without authorisation during the night of June 29, and this is corroborated by Jože Perko of the Criminal Police, based on information provided to him by two anonymous military informers.

---

218) Analysis of the Security organ of the Defence Ministry (VOMO), Ljubljana, July 22, 1991.

"The equipment of the entire Moris brigade and all the investment this unit made in Kočevska Reka and Škrilj[220] were paid with the money from the arms trade that was directed by Krkovič. He was personally selling weapons to Croatians. He took weapons seized from the military depot of the YA in Borovnica, removed them to the warehouse in Škrilj, and between the end of 1991 and the beginning of 1992 he gradually started to sell them to Croatians from Herzegovina. In August 1992, he sold the last lot of grenades ... Krkovič sent potential buyers to the director of the Orbis Company, Ivan Draušbaher, who wrote a formal permit. That way, Krkovič sold weapons and military equipment for about three million German marks. Allegedly, he also sold a tank for one million U.S. dollars. The money from these sales was at the disposal of Krkovič who was equipping the Moris brigade without any supervision," wrote the chief of the operational department of the Criminal Police.[221]

At the conclusion of an operation named *Ovira* (Obstruction) in June 1993, Ljubljana Criminal Police officers wrote similarly in their report that "large quantities of weapons and ammunition were taken to Croatia".[222]

During an internal investigation at the Defence Ministry in September 1994, Colonel Peter Zupan, Assistant Commander of the Territorial Defence in charge of the rear area, told investigators that he came to the following conclusion: "Already then, some weapons went to Croatia and some to Kočevska Reka, where they were under the control of Anton Krkovič. Even though I was Krkovič's superior officer, I was never able to enter the military depot in Kočevska Reka."[223]

The head of the counter-intelligence department of the VOMO, Bojan Babič told investigators that the sale of weapons to Croatia was directed by Ludvik Zvonar, his assistant Elo Rijavec, and after them, Anton Krkovič. They took some of the weapons directly

---

220) Škrilj, which is about 11 km east of Kočevska Reka, was the headquarters of the Moris brigade.
221) "Information" without a date, place or author's identification. Source: Matjaž Frangež, *Kaj nam pa morete?* (What Can You Do to Us!), Part 2, self-publishing, Radenci, 2008, page 536.
222) *Ovira* – report no. 22/1-31, Headquarters of the Criminal Police, Ljubljana, June 23, 1993.
223) Official minutes from the VOMO's commission interrogation of Peter Zupan, MO Ljubljana,

to Croatia and some to Kočevska Reka.²²⁴ In a hearing by the Parliamentary investigative commission in January 2000, this was further confirmed by Anton Peinkiher, the head of the intelligence department of the VOMO. He said members of Moris Brigade participated in removing weapons from Borovnica and other military depots of the Yugoslav Army.²²⁵

During the night of July 5 to 6, 1991, a convoy of about a dozen Croatian trucks, each capable of carrying loads between five and seven tons, entered the military depot in Borovnica, driven by members of the Croatian National Guard (Zbor narodne garde – ZNG).²²⁶ Members of Slovenian TD, who were present during loading of the trucks, remember that ZNG soldiers mostly took large-calibre tank and artillery ammunition.

That same night, Vid Drašček, assistant chief of police in Ljubljana's Vič district was on night-time duty. While patrolling a local road leading from Ljubljana to Borovnica, he and his colleague spotted eight trucks with Zagreb licence plates. They made a note of it and reported it to their superiors.

A year later, Drašček met a Slovenian officer who was present at the loading of Croatian military trucks at the military depot Borovnica. The captain told him the Croatian ZNG were brought to the depot by the assistant to the commander of the Slovenia's TD. Drašček realized these must have been the trucks he spotted at night in Ljubljana.

Drašček says he was also given information by a security officer who was working at the Defence Ministry on July 5.²²⁷ While he was manning the reception, two Croatians walked up to him and asked to see Ludvik Zvonar. One of them put a briefcase on the counter, opened it and exclaimed: "Here's the money, give us the weapons."²²⁸

---

224) Official minutes from the conversation between the Defence Ministry panel and Bojan Babič, no. 881-600, Ljubljana, September 19, 1994.
225) Minutes of testimony by Anton Peinkiher to the Slovenian Parliamentary investigative commission, no. 0610-12/99-00, Ljubljana, January 26, 2000.
226) Croatian National Guard (ZNG) was the name of the uniformed military defence formation in Croatia. ZNG was a predecessor of the Croatian military and was established by a decree of Croatian President Franjo Tuđman. Because at the time Croatia was formally still a part of Yugoslavia, units of ZNG were part of the Interior Ministry.
227) The captain is Janez Čerin and the security officer, Jože Petkovšek. Source: Vid Drašček, official

The security officer remembered that the briefcase was full of 1,000 German mark banknotes totalling at least three million. When Zvonar showed up, he escorted the Croatian buyers out of the building and took them to a nearby bar.[229]

In his 1994 weapons report to his successor Jelko Kacin, Defence Minister Janša wrote only that Croatia received from the Borovnica military depot 800 tons of ordnance for free, in order to stop Yugoslav army units advancing towards Slovenia.[230] He did not mention the sale of weapons for three million German marks brought by the two Croatians.

\* \* \*

While they were tussling over the booty of weapons in Borovnica, elsewhere in Slovenia real armed conflict was taking place. June 28 saw one dramatic event after another. Fingers on triggers became sweaty with nervous tension, but fortunately the encounters did not grow into a larger war, since men in uniform on both sides possessed composure and wisdom more than an inclination to warmongering.

The true heroes who resolved crises were often obscured later on by braggards boasting of exaggerated military escapades. A good example is the clash that took place that Friday at the international border crossing of Rožna Dolina between the town of Gorizia in Italy and Nova Gorica in Slovenia.

A large number of Yugoslav Army soldiers with tanks and armoured vehicles were occupying the border crossing that they had seized the previous day. The atmosphere was tense. Tired and irritable soldiers did not know what they were doing there, while the officers were convinced that they came to defend Yugoslavia from an Italian attack.

Local residents gathered at the army encampment and started to grumble more and more loudly at the military presence. Police advised people to move away since they had received information that the Slovenian Territorial Defence could attack at any moment. To

---

229) Statement of Vid Draščak, Security Organ of the 5th TD Regional Headquarters, Vrhnika, September 5, 1994.
230) Janša's weapons' report for the period between May 15, 1990 and September 1, 1993, no.

deal with the tense stand-off, the police brought up local resident Drago Kosmač, who knew the commander of the YA Unit, Captain Eduard Kovačić, a Croatian by birth. Kosmač used to be a member of the elite guard protecting Marshal Tito.[231] He thus had military experience and was versed in martial arts. He went up to Captain Kovačić, who was leaning against one of the tanks and urged him to surrender, but to no avail.

As day turned to night, members of the Slovenian TD hidden behind nearby houses started to shoot at the Yugoslav soldiers with infantry weapons. An anti-tank projectile hit a tank and a soldier sitting in the hatch was projected out of the vehicle in a large loop.

Kosmač took advantage of the confusion and ran from the customs building to the other side of the border crossing. He barged into a room where a group of 13 Federal police and YA officers were gathered, and yelled: "Throw away your arms! Surrender! You are encircled!" He repeated his order until one of the officers put his automatic weapon on the floor. Kosmač picked it up and cocked it, then leant against a wall and forcefully urged them to surrender several more times. Finally, they all threw down their arms and raised their arms.

At that time, a non-commissioned officer named Tomašević pointed out to Kosmač that Captain Kovačić was wounded. Tomašević bound up the captain's arm and, on Kosmač's orders, used a radio antenna and the rest of the bandage to make a white flag. Waving the white flag, they all came out of the building, except for the wounded captain.

However bursts of fire from the hidden Slovenian TD members started to pour down on them. Kosmač's prisoners dived for cover and only Kosmač himself and Tomašević kept on walking with the white flag.

At that moment another Yugoslav tank started to shoot. Kosmač threatened non-commissioned officer Tomašević with his weapon and told him to make the tank crew surrender. While the former Tito' guardsman stood with a finger on his trigger behind the tank out of sight of the crew, Tomašević banged on the tank's hatch and shouted at the crew to surrender. At first they paid no attention to him, but eventually they turned off the engine, jumped out of the tank and joined other prisoners.

Another tank, which was engulfed in flames after being hit by the Slovenian TD a few minutes earlier, looked as if it would explode, so Kosmač took all the captured Yugoslav soldiers, about 100 of them, some 50 metres away to a bus stop, where he ordered them to lie on the ground with their hands behind their necks.[232]

Just 10 minutes after the first shot came from behind an apartment building, Kosmač completed his action with a short burst of fire into the air to "convince" the captives not to try anything stupid and wait peacefully for the arrival of police officers and ambulances. Three YA soldiers were dead and a dozen more wounded. But the battle was not over: the TD soldiers were still shooting at the border crossing and wounding several civilian bystanders.

So when the TD showed up at the scene as "liberators", local residents greeted them with angry words. There were nine TD members, led by a Major Srečko Lisjak. As Kosmač was helping the wounded Captain Kovačić and escorting him to an ambulance, three armed Slovenians – Major Lisjak and two civilians – came up to him, and one ordered him: "Move away so we can shoot him." But Kosmač retorted that the wounded captain was a prisoner-of-war, and he turned him over to the ambulance personnel together with other wounded soldiers.[233]

The true story of the battle for Rožna Dolina was hidden for a number of years. Official accounts told only that Major Lisjak and his small band of soldiers defeated and captured a large unit of the Yugoslav military[234] positioned at the border crossing. Allegedly, he used a megaphone to urge YA soldiers to surrender, but none of the local residents heard that, and they saw him for the first time only after Kosmač finished his action.

Major Lisjak's superiors did not try to verify these inaccurate accounts. Instead they pinned so many medals onto his hero's chest that Lisjak became the most decorated officer in the Slovenian military. Praise for the outnumbered Territorial Defence for their "outstanding success,

---

232) Drago Kosmač, *Rožna Dolina 1991 - Spomini* (Memories), *Vojnozgodovinski zbornik, št. 39*, Logatec, 2010. Author's conversation with Drago Kosmač, April 14, 2014.
233) Ibid. and author's conversation with Drago Kosmač, Ljubljana, April 14, 2014.
234) Expressions like Yugo military or Yugo army were meant to be pejorative and were mostly used

determination, high level of preparedness and well-chosen battle positions",²³⁵ cast a more positive light on the military leadership than the more banal truth that a civilian in a white tee-shirt and chequered shorts disarmed more than 100 Yugoslav Army soldiers.

The official description of events of June 28 is contradicted by many police officers, emergency medical personnel and others who were able to see with their own eyes that the Slovenian flag was raised at the border crossing with Italy mostly because of the courage and composure of Drago Kosmač.²³⁶

Local resident Slavko Šuligoj was able to record the decisive moments at the border crossing with his amateur video camera.²³⁷ Julij Levpušček, a Criminal Police officer from Nova Gorica, says today that he is very disappointed that Kosmač was formally recognised as a veteran of the war for Slovenia only 17 years afterwards. Kosmač was awarded the Silver Medal of Slovenian Armed Forces, but that was a peace award that did not recognise participation in armed combat. The case of Drago Kosmač confirms the old proverb that bullets and medals never hit the right person.

Following the liberation of Rožna Dolina, Yugoslav Army soldiers from other border guardhouses around Nova Gorica surrendered to Slovenian police the following day, June 29, as did Yugoslav soldiers stationed at the major international border crossing of Vrtojba. Julij Levpušček, who was the first to arrive there with an automatic weapon, remembers how it happened: "YA soldiers commanded by Captain 1 Class Mile Strajnić were already lined up. I introduced myself to him as a member of the Special Police Unit, and he responded that they were surrendering. He ordered his troops to put down their arms. Some of the soldiers grumbled and did not want to follow the order, so I pointed my automatic weapon at them and threatened that I'd shoot if they didn't listen. When all the soldiers had put down their arms, they stepped back five paces, following my order."²³⁸

---

235) A passage about the battle in Rožna Dolina from the Janez J. Švajncer's book *Obranili domovino* (They Defended the Homeland), Viharnik, Ljubljana, 1993.
236) Among them were Dragan Maksimovič, Darij Simčič, Julij Levpušček, Franc Šumandl and Slavko Šuligoj.
237) Video is available at: https//www.youtube.com/watch?v+O_zPojyg9kc (May 21, 2015).

Levpušček then handed the captured soldiers over to Slovenian police officers who arrived on the spot. Like with Drago Kosmač at Rožna Dolina, most of the credit for the peaceful resolution of the tense situation at the strategically important Vrtojba crossing goes to the composure and decisiveness of Julij Levpušček. That however did not prevent Major Lisjak, who arrived on the scene afterwards, from claiming that the Yugoslavs surrendered to him. This was nonsense, since soldiers could not surrender first to Levpušček and then again to Lisjak. Some armies punish such false claimants with demotion or dishonourable discharge, but for 17 years there was only one truth about the liberation of border crossings Rožna Dolina and Vrtojba – the "truth" of Srečko Lisjak.

When Kosmač and Levpušček publicly revealed what really happened at the border crossings, Lisjak responded: "The battle for victory [requires] knowledge, a task, desire, courage and the necessary equipment. The two aforementioned gentlemen did not have any of it then, nor do they have it now."[239]

When Lisjak also accused Kosmač of links with the Counter-Intelligence Service of the Yugoslav Army (KOS),[240] Kosmač and Levpušček had had enough and filed a lawsuit against Lisjak, by then a retired Colonel. After five years of legal wrangling, Lisjak was found guilty of insults and slanders and given a suspended sentence. After a settlement in a civil lawsuit, Lisjak had to pay 5,300 euro.

In their verdict against Lisjak, the panel of three judges of the Ljubljana Superior Court stated: "Key claims of the plaintiffs were that the soldiers surrendered to them, and that their actions should receive adequate historical consideration."[241]

---

239) "War Operation Like a Sports Match", *Primorske novice*, Koper, October 10, 2008.
240) "Who conquered Rožna Dolina in 1991",

- *Chapter 6* -

# SURRENDER ULTIMATUM AS SLOVENIANS SEIZE YUGOSLAV ARMS DEPOTS

By the the first days of July 1991, all of the Yugoslav Army's major weapons depots in Slovenia had been captured. Because of the danger that the Army might counterattack and destroy seized weapons, they were quickly moved to secret locations. Some of them were distributed to Territorial Defence units, but a large portion of weapons, ammunition and ordnance was sent to Croatia.

Nobody was supposed to talk about these arms transfers. Even today, regional TD commanders do not know anything, or do not want to talk about it. However it has been possible to piece together a picture of what went on.

In the afternoon of Friday, June 28, members of the TD on Šentviška Gora overlooking the Slovenian town of Tolmin received an order to capture the commander of a depot where the YA was keeping the largest quantity of weapons in the northern Primorska region of Slovenia.

A member of a TD unit later wrote that perhaps they undertook the assignment "too cautiously and too timidly" because they planned to kidnap the commander when he and a group of his soldiers were visiting a local village pub. But they came to the realisation that such an attack would be very risky. One TD member says they "acted prudently and avoided possible casualties on both sides".[242] In other words, they did not seize him.

A member of the municipal headquarters of the Tolmin Territorial Defence says the TD unit missed the opportunity because it was not yet fully formed and its weapons were not yet "de-mothballed," and so were not ready to be used in combat. Less than half the people

---

242) Tonček Leban in: (Firm Network:

called up to join the TD showed up. Among the volunteers who did report, "18 were drunk on the following day", wrote a member of the regional TD Command.²⁴³

In the end, the disarray of the TD made no difference. After two days and brief negotiations, the Army commander, a bitter and disappointed second lieutenant from Bosnia-Herzegovina, decided he could no longer serve in the YA. He surrendered along with 25 of his soldiers.

Members of the TD regional headquarters in Tolmin seized about 900 tons of pure explosives and an enormous quantity of weapons, ammunition and ordnance. The list of these "spoils of war" shows that it was a real treasure trove of weapons.²⁴⁴ According to the commander of the TD regional headquarters at the time, Vito Berginc, they moved the captured arms into abandoned barns near his nearby hometown of Kobarid.

In the first half of July however, Croatians already began to take away anti-tank mines from these locations. Berginc remembers that they received an order from the TD Command to give weapons to the Croatians: "Nobody informed me or asked me about it, everything was done at the higher levels of command. Also, we don't have any lists of the weapons removed from Šentviška Gora. We had to hand over all the documents to the Ljubljana authorities."²⁴⁵

Božidar Horaček, an informant of the VOMO of the Primorska region present during the loading, stated that 40 trucks were filled with weapons for the Croatian Army and soon crossed the border from Slovenia into Croatia.²⁴⁶

This was confirmed by Rajko Velikonja, his colleague from Nova Gorica. He remembers that trucks from Croatia came at least twice to collect weapons from the former YA depot in Šentviška Gora. The first time it allegedly happened during the independence war and the second time in autumn 1991. He said the "merchandise was taken by

---

243) Roman Medved, Ibid., pages 110 and 111.
244) The list includes 96 Strela 2M anti-aircraft rockets; 5,600 rockets for the VMR multi-barrel rocket launcher; 383 Osa rockets; 920 Maljutka rockets; 932 Zolja rocket launchers; 13,275 mortar shells; 13,121 anti-personnel mines; 2,752,730 rounds of ammunition for infantry weapons; 28,630 hand grenades and 19 tons of TNT explosives. Ibid., page 111.
245) Conversation with Vito Berginc, Radovljica, April 26, 2014.

Croatians who paid for it in cash". Velikonja does not know where the money went, but he is convinced it did not go to the Territorial Defence regional headquarters in Tolmin.

Both transactions were organised by the Assistant Commander of the TD for the rear area, Zupan, and Intelligence chief Peinkiher. The first shipment contained mostly "assets for anti-tank combat and other assets with the calibre greater than 12.7 mm". The second shipment, containing 50 semi-automatic guns and 10,000 rounds of ammunition, ran into trouble when police stopped it on the way and saw parts of weapons. According to Velikonja, the incident was resolved after intervention by higher defence and police officials, and the shipment continued on its way to Croatia.[247]

On May 12, 1992, the regional TD Command in Tolmin registered another large transport of weapons from Šentviška Gora to Croatia. Two trucks with Croatian licence plates were loaded in the depot with 500 high-explosive fragmentation mines, 20 Strela 2M anti-aircraft missiles and 30 Maljutka anti-tank guided missiles.[248]

In the village of Ortnek in southern Slovenia, the Yugoslav Army kept huge quantities of propellants in reservoirs. On June 28, Sergeant Darko Njavro of the Moris brigade showed up at the depot with a comrade, both heavily armed. They shot at soldiers, broke a few windows and roof tiles, and then ran away.

The following day, a platoon from Krkovič's Moris brigade from Kočevska Reka arrived in Ortnek, and exchanged a few more shots with the YA soldiers. They retreated after the Yugoslavs threatened to respond with mortars and artillery, but a few days later the Army soldiers had gone, leaving behind several million litres of different kinds of fuel. These supplies also became "spoils of war".

According to a member of the VOMO who wants to stay anonymous, eight million litres of fuel from Ortnek were sold at the price of 18 Slovenian tolars per litre and then another 10 million litres for 43 tolars per litre.[249] The total proceeds were 574 million tolars or almost

---

247) From the official transcript of the VOMO Commission and the statement of Rajko Velikonja, no. 881-600, MO, Ljubljana September 3 and 5, 1994.
248) Official note of the Security organ of the Municipal TD, Marjan Dovžak, Nova Gorica, May 13, 1992.
249) Slovenian Tolar (SIT) was introduced in Slovenia on October 8, 1991 at the exchange rate of 1

2.4 million euro.²⁵⁰ Sale of fuel was under the control of the Defence Ministry, however Ministry officials say they have no information about the money nor the quantity of fuel sold.²⁵¹ It seems that documents about this undertaking disappeared, together with the money.

\* \* \*

There was more to come, much more. The weapons and ordnance depot of Zaloška Gorica is a 10-minute drive west of Celje in central Slovenia. It consists of five underground warehouses, a guardhouse and a building storing detonation devices kept separate from mines for security reasons.

On June 27, members of the Celje Territorial Defence planned to seize the depot, which was guarded by around 30 Yugoslav Army soldiers. However the lieutenant in command was at first unwilling to surrender, and the TD hesitated to attack because they were concerned about land mines planted around and inside the depot. Then on June 30, the soldiers were tempted by a promise that they would go home if they handed over the depot, and the lieutenant opened the door of the guardhouse and surrendered.

The Zaloška Gorica depot ranked second in terms of quantity of military supplies stored by the YA in Slovenia. According to the YA records, it held 920 tons of weapons and ammunition.²⁵² Viktor Kranjc, commander of the Western Štajerska Territorial Defence, estimated even more. His report said it contained about 1,100 tons of "spoils of war". Just over half the captured military supplies stayed in the depot, some of them were distributed to other areas and regions, and the Croatian National Guard (ZNG) from Zagreb on July 15, 1991 took delivery of 60 tons of weapons, ammunition and ordnance.²⁵³

---

250) Some of the documents about the sale of fuel to Petrol are kept by the author.
251) The answer of the Service for Strategic Communications, MO, Ljubljana, March 12, 2015.
252) Overview of facilities and assets of the YA seized by the TD in Slovenia. The document was handed over to representatives of Slovenia's Defence Ministry by YA General Andrija Rašeta on July 21, 1991.
253) Viktor Kranjc, Commander of the Western Štajerska TD Command, *Analiza bojnega delovanja 8. PŠTO* (Analyses of the functioning in the war by the 8ᵗʰ Regional Headquarters of the TD), from

It seems, however, that these were not the only shipments to Croatia. Defence Ministry advisor Ludvik Zvonar informed his group at the Ministry on July 22 that a military column authorised by the Commander of the Croatian ZNG, Tomislav Mesić, was arriving at Celje, passing through the Slovenian town of Ptuj. Colonel Mesić reported to General Špegelj, the Croatian Defence Minister, who in the summer of 1991 organised several truck-transports of weapons, ammunition and ordnance to Croatia from a number of Slovenian military depots, in particular Ložnica.

\* \* \*

When Viktor Kranjc, Commander of the Eastern Štajerska TD Command, was leaving his post on March 21, 1994, he "bequeathed" to his successor – with the knowledge of Defence Minister Janša – 10,000 German marks in cash which were found inside a safe in Celje. The Intelligence and Security Service (OVS) of the Defence Ministry established that the money probably originated from the sale of weapons from the YA depot in Zaloška Gorica to Croatia, and from the sale of trophy weapons to the Orbis Company in Velenje.

There should have been even more cash in the safe of the Celje TD Command, but in the meantime Colonel Kranjc had spent at least 16,000 German marks, 5,000 of them to pay for a trip by the Celje Paintball Club to the United States, even though the club was not associated with the Slovenian military. The OVS passed their findings to the Celje prosecutor's office.[254]

The biggest problem facing the investigators was incomplete documentation. Peter Mlakar, former head of the support area of Western Štajerska TD, admitted that documents relating to the sale of weapons to Croatia and Bosnia-Herzegovina from the Zaloška Gorica depot were destroyed: "I kept all these documents hidden at my home and burned them in a furnace on a verbal order by Viktor Kranjc. I did that in February 1994, about a month before the removal of Janez Janša from the post of Minister of Defence. Kranjc told me

---

254) Viktor Kranjc – Seized Cash, OVS MORS, no. 881-601/9K-DR-389/97, Ljubljana,

in confidence that Janša ordered him to destroy the documents, and he asked me several times if I had done it. He even threatened that he would send his people to my home to check if I have done it."[255]

Marjan Hočevar, another member of the support area sector who was handing over weapons at the depot, described to the investigators the flow of weapons sold to Croatia and Bosnia-Herzegovina between summer 1991 and the end of 1992. The order to hand over the weapons came from Ljubljana, but in the depot they had to draw up documents recording all the items. "At the end of 1992 or maybe the beginning of the next year, I was ordered to send all the documents to the Ministry of Defence," explained Hočevar.

At first he believed that by arming the neighbouring country, Slovenia was preventing the war that Serbia was waging against Croatia from spreading into Slovenia. But when he discovered that every piece of ammunition was overpaid with German marks, he realised that big business was going on. Once, a deputy of the Croatian Defence Minister offered Hočevar a wad of German marks for a crate of hand grenades from the depot in Zaloška Gorica.[256]

Colonel Ladislav Lipič, Assistant Commander of the Eastern Štajerska TD, was also caught up in these dubious activities. Criminal investigators determined that Kranjc and Lipič between them sold weapons for approximately 190,000 German marks to the Director of Orbis, Ivan Draušhaber, and that they illegally spent that money to cover the needs of regional headquarters.

All three were suspected of abusing their authority. However their case – as with others concerning the sale of weapons – stopped in the prosecutor's office.[257] Ladislav Lipič was promoted to the rank of Major-General and in December 2008 he became an advisor to the then-President of Slovenia, Danilo Türk. Viktor Kranjc was promoted to the rank of Brigadier-General and in 2002 he became the chief inspector of the Defence Ministry.

---

255) The official record of the discussion with Peter Mlakar, no. 881-82-11/6-95, OVS, MORS, Celje, June 27, 1995.
256) The official record of the discussion with Marjan Hočevar, no. 881-82-12/6-95, OVS MORS, Celje, June 28, 1995.

* * *

On the same day that they captured Zaloška Gorica, the Territorial Defence also seized another large Yugoslav Army depot at Vražji kamen, in the region of Bela Krajina in southeast Slovenia. Army guards in that depot were not particularly eager to fight and surrendered after just a single shot. According to the YA, this depot complex contained 400 tons of artillery and 18 tons of infantry ammunition.[258] They also found a few valuable anti-aircraft and anti-tank weapons which, together with ammunition for guns and handguns, were distributed to the nearby TD units. During the first three days of July, TD personnel emptied almost half the Vražji kamen supplies.

As they removed the weapons from the captured depots, TD officers often asked themselves where to take the seized weapons and material, because there was a danger that the YA might counter-attack. Therefore, they had to act swiftly and to improvise. As a result, military supplies were being kept in garages, stables and barns, and in administrative buildings, even in towns and close to schools.

In the next few days, guns in Slovenia were mostly silent, but a real war was erupting in Croatia. After Croatia declared independence on June 25, fighting intensified between Croatian Defence Forces and the Yugoslav Army backed by local Serbian rebels.

At the beginning of August 1991, the war quickly engulfed much of newly- independent Croatia. For the rest of the summer and the whole autumn, Serbs pounded the Croatian town of Vukovar, threatened Osijek, occupied Western Slavonia, and shelled the medieval port city of Dubrovnik. The Croatian regions of Lika, Kordun and Banija were burning, as was the town of Karlovec close to the Slovenian border. Not even Zagreb was safe.

At the height of the crisis on October 28, 1991, Josip Vukina, an officer of the Croatian Army about whom we will hear more, came to Slovenia.[259] The following morning, with the permission of the

---

258) Overview of facilities and assets of the YA seized by the TD in Slovenia. The document was handed over to representatives of Slovenia's Defence Ministry by YA General Andrija

Slovenian authorities, he accompanied 10 trucks loaded with 120 tons of artillery shells for the Croatian Army into Croatia.[260] Slovenian TD reservists from Bela Krajina had been loading the heavy crates until late the previous night. After leaving the Vražji kamen depot, the convoy of trucks drove over the Vahta pass to Novo mesto and entered Croatia through the Obrežje border crossing. The trucks were unloaded on the outskirts of Zagreb.

Among truck drivers, there were also employees of the Slovenian transport company Viator from Ljubljana, hired and paid for by the Slovenian Defence Ministry. The loaded vehicles were guarded by the head of security at the Dolenjska TD Command, Žarko Henigman, who that year also participated in night transports of weapons from Kočevska Reka to Croatia.

By the end of June, just a few days after Slovenia's declaration of independence, the Slovenian TD had captured all major YA weapons depots – except for the largest one: a gigantic underground complex under the Pohorje plateau in Zgornja Ložnica, located halfway between Celje and Maribor.

\* \* \*

In the meantime, the European Community was not overly interested in the armed conflict in Slovenia.[261] Only on June 28 in the evening did it send a diplomatic mission consisting of a troika of foreign ministers.[262] They went to Belgrade and demanded a ceasefire and election of the Croatian Stipe Mesić to head the collective Presidency of Federal Yugoslavia.

Ministers Jacques Poos, Hans van der Broek and Gianni De Michelis did not come to Ljubljana, where the war was underway. However they stopped in Zagreb where they met with the Croatian

---

260) Dispatch of the Operational-Communications Centre (OKC) at the Ministry of Internal Affairs about passing of 10 trucks through the border crossing Obrežje, no. 21/6-2-552, MNZ, Ljubljana, October 28, 1991.
261) In 1991, there were 12 countries in the European Community: Belgium, Denmark, France, Greece, Ireland, Italy, Luxembourg, Germany, Netherlands, Portugal, Spain and Britain.
262) The EC delegation was comprised of foreign ministers Jacques Poos (Luxembourg) who until June 30, 1991 presided over the Council of Ministers; Hans van den Broek (Netherlands), who

president Tuđman and would-be Yugoslav Federal President Stipe Mesić. After a long journey driving on side roads because of fighting and road blocks, Slovenian President Milan Kučan and Foreign Minister Dimitrij Rupel also joined the meeting. The European troika tried to convince the leaders of Slovenia and Croatia to postpone the implementation of their independence for three months.

By that time, Slovenia had already declared its independence, and the Slovenian president insisted that it was not possible to delay it. He pointed out that it was the Yugoslav Army that attacked Slovenia, and he astounded the European diplomats by handing them an English translation of of the Federal Yugoslav military plan named Trench (in Serbian *Bedem*). The plan showed that the socialist Yugoslav regime and its Army pretended they were being attacked by the European NATO forces with a help of Slovenian and Croatian "collaborationists". It was obviously based on a "training" document of the same name, drawn up by commanders of the 5$^{th}$ Army Region at the beginning of 1991 to prepare for defence of Yugoslavia against an attack by the western alliance. As it is now known, this document served the Yugoslav Army as a template for capturing border crossings in Slovenia.

Kučan defended Slovenian interests clearly and steadfastly, while Croatian president Franjo Tuđman put up a confused performance in front of the European troika. He accepted all the European conditions without reservations. Although Croatia declared independence on the same day as Slovenia, June 25, he showed he had little idea how to implement it.

In the following days, Tuđman came in for considerable criticism for his vacillation from Slovenian leaders and European diplomats. As for Slovenian President Milan Kučan, he ended the talks with Poos, van den Broek and De Michelis convinced that the wheel of history could not be rolled back to the time before June 25, 1991.

\* \* \*

During the night of June 29 to 30, the Slovenian Parliament gathered to decide an answer to an ultimatum sent by the generals in Belgrade that Slovenia should capitulate. The Parliamentarians endorsed the insistence of the Slovenian Presidency and government that the declaration of independence should be implemented. They demanded a halt to the armed conflict and a peaceful resolution of the conflict. They appealed to the European Community to send observers immediately to Slovenia.

During the session of the Parliament, Interior Minister Igor Bavčar came to the podium. In a dramatic tone, he told members: "I would like to inform you that a few minutes ago, a serious breach of ceasefire occurred on the part of a sabotage squad of the Yugoslav Army. At the crossing of Šubičeva and Titova Streets, the soldiers shot at the Slovenian parliament. The police intercepted them at the crossroad near the Smelt Company. Now, the police are enforcing a blockade. I presume this was one of the planned provocations meant to disturb even this extraordinary secret session, and if I may, I would suggest that Parliament shorten its work."[263] Members of Parliament retreated to the basement where they hurriedly completed their session and passed their resolutions.

Official versions of the event continued for a long time to allege "sabotage" or even a "terrorist" attack by Yugoslav Army soldiers. Eventually, the truth came out, that is, that our guys were shooting at our guys.[264] It happened that "the enemy" was only one person, a 20-year-old member of the Territorial Defence from nearby Kamnik, who most likely lost his nerve because of the war psychosis.

That same day, he and other members of his TD unit were distributing weapons from the Borovnica army depot. However, instead of then obeying an order to take a rest, he left his unit without permission at 11 pm, still wearing his TD uniform, and went on a murderous rampage with an automatic weapon in Ljubljana.

First, he shot at a car that would not stop for him. A few minutes later, he forced another driver to hand over his vehicle and drove at a barricade in front of the Ljubljana's Faculty of Philosophy. When two policemen tried to stop him, he killed one with a burst of fire from

---

263) Igor Bavčar at the extraordinary plenary session of the Slovenian Parliament, June 30, 1991.
264) The real background of the event on the night between June 29 and 30, 1991 was, it seems, first publicly revealed by the policeman Pavle Čelik, in his book        (Behind the barricades

his weapon and injured two drivers halted at the barricade. Then the attacker drove along Kardeljeva and Titova Streets,[265] heading towards the suburb of Bežigrad. In front of the Smelt Company building, police officers ambushed him, riddled his car with bullets and killed him. They shot so fiercely that they also wounded one of their own colleagues.[266]

The fake official version of the incident was unwittingly invoked the following morning by President Kučan and Prime Minister Peterle, who gave an answer to Yugoslav General Kadijević and the Prime Minister Marković, complaining of "the most drastic violation of the ceasefire, committed by a saboteur-terrorist squad of the Yugoslav Army" the night before in front of the National Parliament. The Slovenian political leadership had been manipulated into this combative stance.

\* \* \*

On Saturday, June 29, four days after Slovenian independence, there was another dramatic scene at the Pleso military airport in Zagreb, headquarters of the 5th Corps of the Yugoslav Air Force. In the morning, its Commander, Major-General Marjan Rožič, received a grim plan from the Supreme Command in Belgrade. Military airplanes were supposed to destroy strategic targets in Ljubljana: the main train station, the post office and the building of the Slovenian radio and TV. Rožič felt nauseous. He went to the centre of Zagreb to see his commander, General Kolšek, who informed him that he could not do anything about it. General Rožič returned to the airport and called the Commander of the Yugoslav Air Force, Lt-General Zvonko Jurjević,[267] informing him: "I will not fulfill this plan even if it costs me my life. I resign immediately."

"What? Why?"

---

265) Today, these are Slovenska and Dunajska Streets.
266) The wounded policeman was Franc Trbovšek, who later found employment at the Intelligence and Security Service (OVS) of the Defence Ministry and in September 2015 became its General Manager.
267) Pilot and General Zvonko Jurjević was a Croatian, from the Croatian city of Slavonski Brod. According to the information of the YA Security and Intelligence Service (KOS), he was advocating offensive actions and a bigger role for the Air Force – even when Air Force planes were bombing his

"You know that bombing the targets in that plan means that we're also killing families of our officers? Do you understand? At the post office, railway station … wives of our people work all over there."

Jurjević answered in a hoarse voice: "Oh my Marjan, you're forcing me again to go to Adžić and Kadijević at the Supreme Command."

Only in the afternoon, did General Jurjević call General Rožič back to tell him: "Marjan, you'll get a contingency plan!"[268]

However, neither Kolšek nor Rožič ever received that contingency plan, or knew whether it existed at all.[269] The crisis passed.

The following morning, June 30, the ultimatum set by the Yugoslav generals expired, and from early morning reports were coming in of imminent danger of air attacks. Yugoslav Air Force planes were constantly taking off and landing on Croatian airfields. From the Željava airfield, next to Bosnia-Herzegovina town of Bihać, MiG-21 and MiG-29 fighter jets were approaching, while Jastreb (Hawk) and Galeb (Seagull) strike aircraft reportedly took off from the Pula Air Force airfield. All over Slovenia, sirens sounded, and many residents of Slovenia spent their first Sunday in their newly independent country in cellars and shelters.

In the meantime, dramatic warnings were coming in to the Territorial Defence regional headquarters and police stations. It looked as if the Yugoslav Armed Forces were serious this time. Croatian observers sent a dispatch to the Slovenian military leadership whenever a military airplane took off, which only increased the tension. "Had a quarter of threats that we received in the Republic's Coordination Group come to fruition, it would have been really hard," admitted Janez Slapar, who at the time was in charge of the Territorial Defence Headquarters.[270]

Some Slovenian politicians later tried to take credit for convincing the Yugoslav generals at the last moment not to launch bombing raids.[271] The fact is that the General Staff in Belgrade was split between those who would yield and others who wanted to bomb Slovenia and "level

---

268) They were speaking Serbo-Croatian.
269) The event was described to the author by an unnamed officer of the 5th Corps of the Yugoslav Air Force Command. General Marjan Rožič does not want to talk about it.
270) Author's conversation with Janez Slapar, Radovljica, October 19, 2015.
271) Janez Drnovšek, in his book wrote that the bombing of Slovenia was cancelled when he (My truth), Mladinska knjiga,

it to the ground". According to officers present in Belgrade, Admiral Stane Brovet urged General Kadijević against bombing Ljubljana. Kadijević eventually relented, but if he had not, General Kolšek in Zagreb would have had to execute his order.

Admiral Brovet enjoyed the confidence of General Kadijević because he was married to a Serbian woman. However, that was not the case with General Kolšek, who was Slovenian himself. After the attack on Slovenia, the Security Service of the Yugoslav Army pretty much followed his every step. This is what he wrote later of the prospect of bombing the Slovenian capital: "To destroy key facilities and cause human casualties seemed indefensible to me. It seems that some people wanted that I, a Slovenian national, should make a decision about it, so that the Supreme Command could justify its actions in front of the world public and in history."[272]

General Kolšek was removed from his position on July 1, General Kadijević having already signed the order for his dismissal two days earlier. He was replaced by a Serbian "hawk", Života Avramović, nicknamed *Ledeni* (The Iceman). Also removed was Major-General Rožič, commander of the 5th Corps of the Yugoslav Air Force in Zagreb. His position was taken over by feisty Major-General Ljubomir Bajić from Serbia.

But this heavy-handed decision-making by the generals in Belgrade did not help the Yugoslav Army's soldiers in Slovenia. More and more officers, voluntarily or by force, ended up in the Slovenian captivity, while fewer and fewer of them were obeying the confusing and often contradictory orders of their superiors.

On July 1, an agreement was reached for the withdrawal of the Yugoslav Army units from the border crossings of Rateče, Korensko sedlo and Karavanke, adjoining Austria. All border crossings and military guardhouses in this northern Gorenjska area thus came under the control of the Slovenian police and the Territorial Defence without casualties and with only a few shots fired. "Besides weapons, we also fought with words," wrote Rina Klinar, who reached agreement with the YA officers in her capacity as president of the municipal executive council of Jesenice.

\* \* \*

They fought not only with words, but with procrastination, passivity and common sense. On one of those days, a small unit of the Territorial Defence in Gorenjska walked up a hill, tasked with capturing a small Yugoslav Army guardhouse in the Karavanke mountain range. They reached the enemy and the two sides eyed each other. The two commanders, a non-commissioned officer on the TD side and a sergeant on the Yugoslav side, knew each other well and were almost drinking buddies. Moreover they were resourceful. They decided that they did not want to shed blood, neither their own nor anybody else's. They agreed that the "butchering"[273] evoked in Prešeren's epic poem could wait a little longer, and in the meantime they would wait for further orders.

The two military men each contacted their respective superiors down in the valley to inform them that the situation was under control, but they did not tell them who was controlling whom. Weary members of the TD settled under a spruce tree in the shade in front of the fence around the guardhouse, while some of the YA guards made themselves comfortable on a bench under a maple tree by the entrance to the guardhouse.

The Slovenian TD non-commissioned officer and the YA sergeant sat around a table, and to prevent unforeseen problems ordered their respective subordinates, a TD corporal a YA private first class, to sit next to them. After that, they started playing tarot cards, concentrating very seriously. Day started to turn to night and a red sunset appeared behind the peaks of the Julian Alps.

All of a sudden, there was a rustling sound and more members of the Slovenian Territorial Defence rushed out of the woods. There were many of them, with guns ready to shoot, and their commander shouted: "Hands up! Useless to resist!"

The YA sergeant sternly looked at his soldiers sitting scared on their rickety bench. Then he stood up and turned towards the Slovenian non-commissioned officer with whom he had been playing cards, and exclaimed: "Comrade Slovene, in our region we have a saying: he who is sentenced to be fucked, his pants come down by themselves."[274]

The Territorial Defence seized weapons and equipment and sealed the guardhouse. They took border guards and their sergeant into captivity for a few days and then sent them home to the various parts of Yugoslavia from which they came.

\* \* \*

Slovenians defending their homeland fought guerrilla warfare against a Yugoslav Army which was trained for frontal warfare. Local residents were of incredible value to the Territorial Defence, just as they had been to the Partisans who defeated foreign occupiers in World War II. Improvised road blocks brought advancing YA tanks and other armoured vehicles to a halt, and they were unable to deploy their firepower.

At first the TD had trouble with calling up and motivating recruits, different commands were badly coordinated, there were too few weapons and equipment, and as often among Slovenians too much drinking. The TD soldiers only gradually became aware that this time they were not just training. It was for real, and they had to defend Slovenia's independence. The most difficult tasks during the first days of fighting therefore fell on better-trained professional police units. After a time however the TD stiffened its ranks.

In the words of Ciril Kosmač, a Slovenian novelist and World War II Partisan:

> *First you need love, which at every hour is willing to raise its arm in defence. Only then come forth a prudent thought and a weapon.*[275]

The Slovenian defenders were influenced by the bellicose messages put across by the government's media specialists. The time of holidays was approaching, when they wanted to be free. Instead they had to

---

274) He was speaking in Serbo-Croatian.

endure a nerve-wracking wait behind their barricades, anticipating a devastating counter-attack by an insulted and irritated Yugoslav Army.

This psychosis led to some deadly incidents. On the last day of June 1991, a nervous police patrol from the Vič[276] police station spotted a car driving at high speed through the suburb of Ljubljansko barje. The car artfully dodged road barricades and the police patrols that tried to stop it. The driver gathered speed as he headed toward the nearby town of Ig, where a Territorial Defence unit was stationed. The police took no chances. At a roadblock at the entrance to Ig, they started shooting at the car with all available weapons …

They killed three youths, who turned out to be members of a Ljubljana sports club returning home from training. When they realised what they had done, the police officers were stunned. A young police officer holding a smoking gun buried his head in his hands, wishing he could undo what he had just done. Then a colleague shouted: "This one is still alive. Get an ambulance! Quick!" One of the four people in the car was seriously wounded but survived.

A few hours later, on the other side of Slovenia, the ground shook and a boom could be heard for dozens of miles around. A YA depot storing ammunition and weapons in Črni Vrh above the town of Idrija was blown up by a YA officer, an ethnic Albanian from Kosovo, after failue of negotiations the previous day. A gigantic blast wave broke the glass windows of almost all the houses in the village and shingles fell from roofs. Local residents said it was as if the gates of hell had opened. Where the military depot once stood was a crater six metres deep and 18 metres wide.

Negotiations also failed at the Puščava YA depot next to the town Mokronog in the Dolenjska region, where YA was keeping seven million litres of diesel fuel and 10,000 litres of engine oil and lubricants. Army Sergeant Dragomir Grujović from Serbia got into a vehement dispute with his own commander, Slovenian Captain Branimir Furlan, after the latter reached an agreement with the Slovenian TD to hand over his unit and the depot. Grujović vehemently objected and opened fire on the Captain. "From my service pistol, I shot three times towards him, and I was aiming below his waist," he recounted later.[277]

Grujović took control over the military installation, and allowed paramedics to take away the wounded commander. He told the soldiers that it was up to them whether they wanted to leave the unit. Soldiers who were ethnic Slovenians and Albanians all left the unit, while Serbs and a Croatian sergeant stayed.

Residents of Mokronog and the surrounding area were terrified that the "madman in Puščava" intended to blow up the barracks and the warehouse with diesel fuel and lubricants. An ecological disaster seemed imminent. A Special Police Unit besieged the depot for several weeks and only after Slovenians and the Yugoslav Army agreed that the YA should leave Puščava did Grujović come out of the depot. He returned to his hometown, Čačak in Serbia, where he was promoted to lieutenant and continued his military career.

His former commander Furlan was in the meantime promoted to Brigadier and embarked on a stellar career, becoming Assistant to the Commander of NATO's Allied Joint Force Command in Naples, Italy.

The story from Mokronog would have been forgotten had the justice ministers of Slovenia and Serbia not signed an agreement on mutual extradition of citizens. Interpol issued a warrant for the arrest of Dragomir Grujović, and in May 2015, a circuit court in Novo mesto sent him an indictment, charging him with a crime against the civilian population and attempted murder.

Grujović swears he will never surrender alive to the Slovenians. "What's important is not what happens to me, but what all this tells professional soldiers, who pledged their allegiance to their homeland. To defend it to the end or desert at the first moment of danger?"[278]

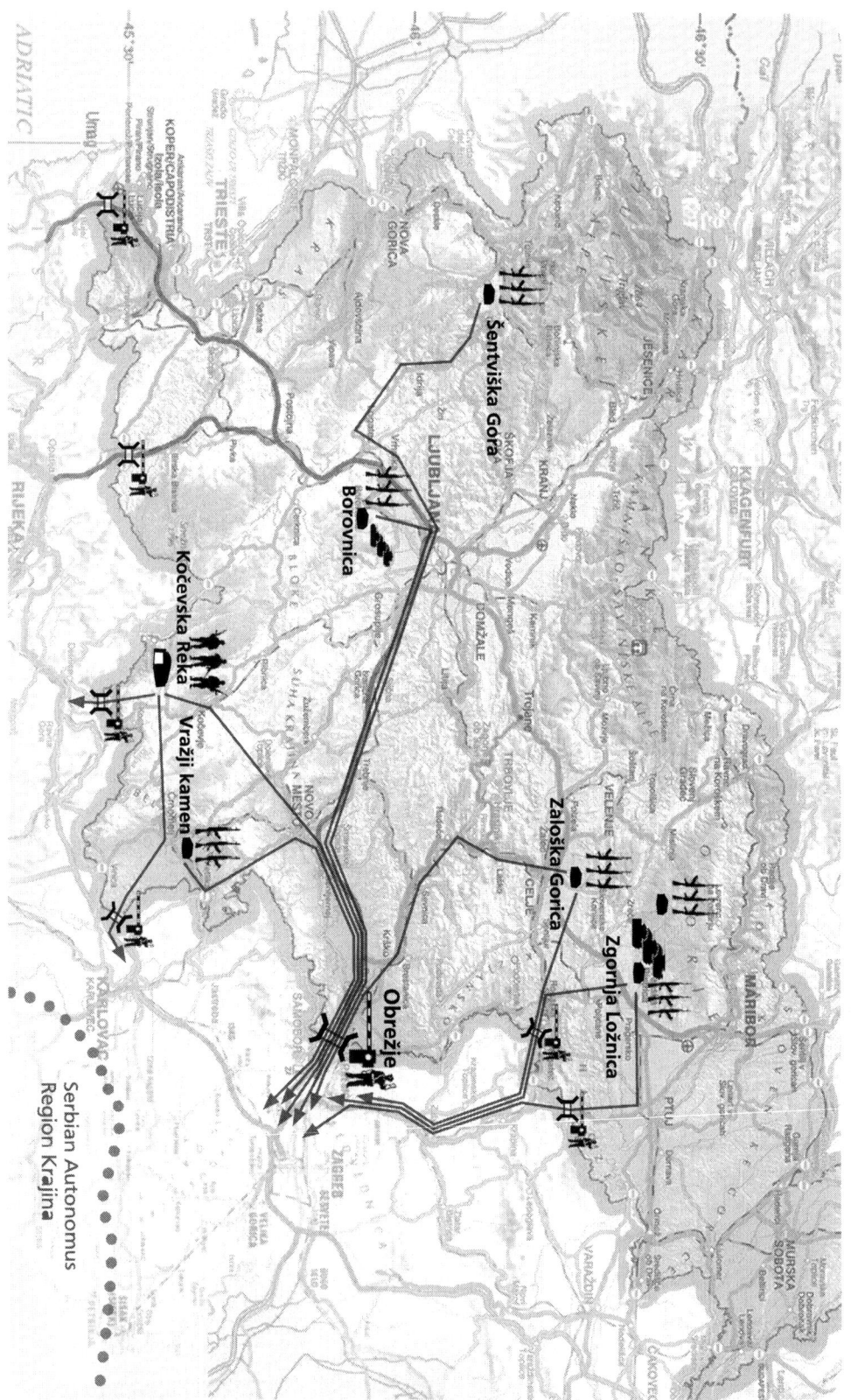

*- Chapter 7 -*

# SAVING LJUBLJANA FROM A BLOODBATH

*As long as we nurture the myth about a military victory over the YA, there will never be peace in Slovenia.*

Neven Borak

At the beginning of the 1990's, a joke circulated around Europe that in the capital of some small Balkan country a meeting of the Communist Party and the state leadership is taking place. The Secretary General of the Communist Party announces that they will declare war on the United States of America.

A meaningful murmuring follows, and the party boss continues with a stern voice: "The Americans will attack us and destroy our country and then we'll sign a peace agreement with Washington and will enter into an alliance with them …"

"The Americans will bring us democracy," adds the Prime Minister.

"And with a help of the American investments, our economy will flourish," continues the Foreign Minister.

Everybody nods approvingly, except for the Defence Minister, who frowns and mumbles to himself. After a while, he asks: "What if we win?"

\* \* \*

The first few days after the declaration of Slovenian independence felt as if they would never end. As dramatic events followed one after the other, many people lost a sense of time because of the constant insecurity and the stress of waiting to see what would happen. Unbeknown to most of its inhabitants, Ljubljana found itself on the edge the abyss of total war.

The commander of the Ljubljana Territorial Defence, Lt-Colonel Miha Butara maintains that on June 28 the commander of the Republic's Territorial Defence Headquarters, Colonel Janez Slapar, ordered over the telephone that his troops attack Yugoslav Army military barracks in Ljubljana and Vrhnika. Butara later wrote that he rejected the order on the grounds that military actions in large towns would not be appropriate, but that Slapar countered that "there were gunshots and battles everywhere in Slovenia except in Ljubljana".

Butara later wrote that he lost his cool and that he forgot he was talking to his superior officer at the other end of the line. He remembers that he replied sternly: "Janez, I have enough of this! Here are 300,000 civilians, we have the situation completely under control, there's no need to expose innocent people to danger." When Slapar insisted that his order be executed, Butara says he angrily exclaimed: "Janez, without a written order I will not do anything. Go fuck yourself and stop playing war." People present who heard the phone discussion on the loudspeaker were stunned and speechless.[279]

Lt-Colonel Butara believes this conversation took place on June 28,[280] but Slapar told the author in an interview it was early on June 27: "I called him a few minutes after we received information that YA tanks left their barracks in Vrhnika. I ordered him to make all necessary preparations to attack the aggressor's military targets. This was only an order to get ready which would be followed by an executive order to attack."

"Butara demanded a written order. Did he receive it?"

"Of course! During its extended morning session, the Slovenian Presidency approved the use of firearms. Based on that approval, the operational group of the Republic's Coordination Group issued the following order: – Enable protection of facilities, borders and communications, and prevent manoeuvering by the YA by decisive battle actions, focusing on fighting armoured units and other technical assets. –[281] At around 10 am I sent this order to the

---

279) Extract from the notes of Lt-Colonel Miha Butara and Dr. Martin Premk, *"Za resnico se je treba boriti"* (You have to fight for the truth). Available on: http://www.veterani ljubljane.si/e_files/content/Aktualno/Miha%20Butara%20Izvlecek.pdf (September 22, 2014).
280) Ibid. Author's conversation with Miha Butara, Vrhnika, January 22, 2014.

regional TD Command and therefore also to the commander of Ljubljana TD, Lt-Colonel Butara."

"The Presidency did not issue an order actually to attack YA barracks and also it did not declare a state of emergency."

"Wait a minute. I was present at that session in the morning of June 27. When the YA tanks left the barracks, President Kučan and other members of Presidency became aware that Slovenia was under attack and that it should defend itself. The Yugoslav Army became an aggressor and therefore, every part of it became a legitimate target."

"How did you take those vulgar words that Butara uttered at you over the phone?"

"Butara never said these words to me. Had he told me to go fuck myself, he would have been fired on the spot from his position as commander of the Ljubljana TD …"

"So he was fired 'only' two days later. What were the reasons for his dismissal?"

"It wasn't only that he resisted obeying orders. Nothing in the Ljubljana region was functioning the way it was supposed to … Also, the weapons from the Borovnica depot were taken away in a completely disorganised way. Butara was more a politician than a soldier," answered Slapar.[282]

Miha Butara wrote that approximately 20 minutes after his telephone conversation with Slapar, one of his subordinate officers, Captain Neven Borak, called the Ljubljana Territorial Defence Command wanting to find out whether an order he received to attack the YA barracks in the suburb of Moste was valid. It had been given to him by someone with the code-name *Sonce* (Sun), which would normally mean it came from Butara. Butara replied to that he did not issue the order and forbade its execution. Captain Borak breathed a sigh of relief.

In the meantime, similar orders were received by other Territorial Defence units in Ljubljana. When the commander of the Second Partisan Detachment learned that he had to attack the largest YA barracks in Ljubljana, in the suburb of Šentvid, he reportedly had

---

Decree of the Slovenian Presidency, no. SZ 800-030-1991, Ljubljana, June 27, 1991.
282) Author's conversation with Janez Slapar, Radovljica, October 19, 2015. After the end of conflict,

to go to the lavatory. As luck would have it, this "call of nature" prevented his unit from attacking the barracks before it received an order from Lt-Colonel Butara calling it off.

That evening, Butara and Borak asked each other who in the name of the Ljubljana command could have decided on such a senseless action: "Gradually but fairly quickly we realised that it was a coded order which I did not issue. The order could have come only from the Republic's Headquarters of the TD or the Republic's Coordination Group.[283]

Captain Borak confirmed this: "Before I would transfer an important order my unit, I had to be completely sure that it came from the right quarter – from the commander of the regional TD, and not from the Republic's Coordination Group, which was … only an advisory group."[284]

The dismissed Ljubljana regional commander Butara is convinced he acted correctly. Several years later, he wrote to Borak: "As soldiers, we know that those foolish things from the Republic's Coordination Group were not legitimate military orders but a product of something else, most likely a panic and misunderstanding of the conditions on the field."[285]

Borak says none of this took place on June 27, as claimed by Slapar: "All of this was happening on Friday June 28, which is the same day Butara and I were trying to figure out where these orders were coming from."[286]

\* \* \*

Milan Zorko of the Territorial Defence Command in Ljubljana remembers that Andrej Lovšin, director of Security organ of the Defence Ministry (VOMO), on June 29 demanded that they should enter the Yugoslav Army barracks situated on Metelkova Street which was the headquarters of the Army's Ljubljana Corps: "Lovšin was determined. He ordered that we get in through sewer

---

283) Extract from the notes of Lt-Colonel Miha Butara and Dr. Martin Premk, "*Za resnico se je treba boriti*" (You have to fight for the truth) – quoted work.
284) Author's conversation with Neven Borak, Ljubljana, October 13, 2015.
285) From the letter of Miha Butara to Neven Borak, Ljubljana, February 24, 2003.

shafts, capture it and disarm the soldiers. At the Ljubljana VOMO headquarters, we were appalled. I served long enough in these barracks to know all its spaces. I tried to explain to Lovšin that the YA put concrete plates on the bars of the sewers. The barracks had sharpshooters and military police always ready to act, as well as anti-aircraft guns. The revenge of the Yugoslav Army against the inhabitants of Ljubljana could have been terrible …"[287]

Fortunately, in the last days of June, there were no attacks on Yugoslav Army barracks in Ljubljana, though the Territorial Defence did block them and monitored that the YA soldiers were staying inside. Most of the TD officers were aware that they were not ready for such an all-out attack. They lacked large-calibre weapons and also soldiers, since less than half the TD members answered the summons to mobilise.[288] During the first days of the conflict, confusion, unclear orders and lack of coordination meant that occasionally battlefield units did not even know whether they were allowed to shoot.

However the pressure for outright attacks only increased. On July 2, the Ljubljana TD command again received orders to attack the YA barracks in Ljubljana's Šentvid, Moste and Metelkova Street.

Lt-Colonel Janez Lesjak, successor of the dismissed Butara as commander of the Ljubljana TD region, was known as a trusted Janša man, and Janša kept ordering him by phone to launch an outright attack on the Ljubljana Army barracks. But Lesjak was in a quandary and full of doubts.

The barracks in Moste, the second largest in Ljubljana and ironically named the Barracks of Brotherhood and Unity,[289] were blocked on their east and northeast sides by the Litija TD under the command of Captain Avgust Cvetežar. This unit with its 420 soldiers was the strongest in numbers in the Ljubljana region, but they were not very well armed at the time. According to Cvetežar, they were able to

---
287) Author's conversation with Milan Zorko, Ljubljana, October 28, 2014.
288) According to the Miha Butara, the TD units in Ljubljana region were only at 42 to 48 per cent of full strength.
289) After independence, the barracks was renamed as the Barracks of Franc Rozman – Stane, a Partisan commander of World War II. However, in 2012, following the suggestion of a Slovenian right-wing politician, it was renamed again, this time as the Barracks of Edvard Peperko, a Territorial

stop a breakout of Army tanks from the barracks with hand-held rocket launchers and Partisan-style fighting. However, he doubted the armoured units would succeed in making a break for it because they had no logistical support.

In the west, the Barracks of Brotherhood and Unity were blocked by the First Partisan Detachment of the Territorial Defence, commanded by Neven Borak, with about 150 TD members positioned inside warehouses of the nearby BTC shopping centre.[290]

"I received an order to attack the barracks in Moste by telephone at around 3 pm when the media was already reporting cease-fire negotiations," said Neven Borak in an interview a year later. "Four hours earlier, I received an order to remove barricades so that employees could enter the BTC. My first reaction was to ask the man on the other end of the telephone line if he was serious, and whether he was aware of what he was demanding."[291]

The person who talked to Borak was Lt-Colonel Lesjak. Borak asked him for a three-hour delay so he could consult his officers to find the best options for an attack. At about 3 pm, the order also reached Cvetežar, without any advice or instructions for preparation, only a rebuke asking, "Why is there no sound of shooting at your place?" Cvetežar's Litija Headquarters requested a written order, which never arrived.[292]

Soon afterwards, Cvetežar met Borak and the two agreed that the order was "lunacy". The phone rang, and it was their commander Janez Lesjak. He rudely reprimanded Borak and asked him why his unit did not attack the Army barracks. After that, he also wanted to talk to Cvetežar. This is how the commander of the Litija detachment remembers that phone conversation: "Avgust, are you going to attack the barracks or what?"

"Are you completely crazy? Where's your head?"

"What shall I do? Janša is on my back constantly!"

"What does Janša have to do with you? Tell him to get off somewhere! Where's Slapar? Does the Presidency know about this crazy order?"

---

290) Today, the same area houses the BTC City, the biggest shopping centre in Slovenia.
291) "Veno Karbone", interview with Neven Borak, *Mladina*, Ljubljana, June 30, 1992.
292) According to the testimony of the officer in the Litija TD detachment, Boštjan Grošelj, Litija,

"I don't know about that, but Janša is calling me every 10 minutes requesting an attack."

"Janez, are you aware at all where you're sending us? There are civilians in their homes. First, evacuate them! When it starts, the bombs will hit the Ljubljana suburbs of Novo Polje, Zadobrova, Črnuče, Zalog, Trzin, Domžale, Fužine, Nove Jarše and also the centre of Ljubljana … Janez, I will not issue an order to attack the barracks, and you can do with me whatever you want! I will not leave half the Litija detachment on this field."[293]

With that, the conversation ended. After it was all over, a Yugoslav Army officer told Borak they were intercepting the Territorial Defence's phone calls. So in the barracks, they knew about the order to attack, and also Borak's refusal to follow it …[294]

Today, Lt-Colonel Janez Lesjak says of the two subordinate commanders: "The blockade of the barracks was the basic task of their units in the decisive moments, but that task also included a possible attack and disabling of the barracks."[295] He also admits: "Yes, that day was quite chaotic because we received many orders to assume highest readiness before an attack which was later called off. I'm aware that because of all that, there was a lot of confusion in the TD units."[296]

At 4:30 pm, Lt-Colonel Lesjak received another phone call from Janša, who wanted to know whether the Territorial Defence was ready to attack Army barracks, and told him Yugoslav Army officers had received an order to attack important facilities in Ljubljana. Janša ordered that the TD should act ahead of the YA, which meant immediately.

"Now it's for real. An outright attack on a well-armed and well-staffed YA barracks in the middle of Ljubljana. My mind was racing: –

---

293) Extracts from the memoirs of Avgust Cvetežar, Litija, December 2012. Available on: http://www.veterani-ljubljane.si/e_files/content/Razprave/Gusti,%spomini.pdf (September 23, 2015). Author's discussion with Avgust Cvetežar, Litija, October 20, 2015.
294) Author's conversation with Neven Borak, Ljubljana, November 5, 2015.
295) Officers of the Litija detachment state that the commander of the Ljubljana TD should have summoned units' commanders to discuss with them possible consequences of an attack. In this case, the preparations for the attack would have lasted two days. However, it seems that the Republic's Coordination Group was in a hurry, maybe because it wanted the TD to hit the YA barracks before the ceasefire agreement was reached.
296) Janez Lesjak, "*Osamosvojeni, izigrani in razočarani*" (Independent, double crossed and disappointed). An internal record, Grosuplje, April 27, 2014. Janez Lesjak's answers to author's e-mail

Can we do it? What will the consequences be? Do they know about it at the Republic's Coordination Group? There's no time to check it out. Surely, they agreed to it!? I was aware of Janša's headstrong character, and I was trying to convince myself that he was not acting on his own. I was torn between responsibility and fear: – Will we be able to handle the situation so that it doesn't get out of control? How far will the Yugoslav Army go? –"[297]

This is how Lesjak, in his book On the Edge described probably the most dramatic moments of his life. He and Janša had known each other from an early age, and he was aware that it was not good to get into quarrels with Janša, especially during those feverish hours. On top of that, Janša at the time was Defence Minister – and therefore should be obeyed.

Twenty-five years later, Janez Lesjak said in reply to questions by the author: "Let me be forgiven but I'm no pathological militarist. However when it is a question of your nation and country, a man has to get over many things. Don't ask me if in today's circumstances and knowledge if I'd do it again. I would be in a big quandary."

"Defence Minister Janša was not able directly to order you, the commander of the Ljubljana TD, to attack YA barracks in Ljubljana?"

"No, the Defence Minister only expressed his initiative. At that time, I could have flatly refused to follow his initiative or order. Without any consequences! My superior commanding officer was the head of the Republic's TD Command, Colonel Janez Slapar. Soon after, he also called me and demanded an attack, as did Igor Bavčar, the head of the Republic's Coordination Group. All three called me within several minutes, and all of them confirmed that the decision to attack YA barracks was made by the Republic's Coordination Group … Today it's hard to describe those moments. I don't wish to anybody to be in a situation like this."

"What about the Presidency of the Republic of Slovenia, as the Supreme Commander of the Slovenian Armed Forces?"

"A collective body like the Presidency is not capable of discussing war operations of such vast and rapid dynamics and even less to make

---

297) Janez Lesjak, (On the Edge

decisions about them. That's why the Republic's Coordination body was created and given the authority from the highest state organs to act."

"Did you also receive a written order to attack, or did you at least demand it?"

"For me, Slapar's confirmation of Janša's demand was valid, but I cannot confirm whether or not a written order followed. I no longer have access to the archives of the Ljubljana TD Command and it's questionable whether or not those documents were preserved. In any case, I'm completely aware of the responsibility that I received and carried at that time."[298]

Those are the words of Janez Lesjak, while Janez Slapar has the following opinion: "Words on the phone don't mean anything. Had the attacks on YA barracks caused fiercer fighting, chaos and civilian casualties, the greatest responsibility for all that would have fallen exactly on the commander of the Ljubljana TD. In a legal sense, he would have been the biggest culprit … But let's leave the hypothesis of what if …"[299]

After trying to figure out with Cvetežar whether the Presidency of Slovenia knew about all this, Captain Neven Borak, whose detachment was blocking the western side of the Army barracks in Moste, checked the authenticity of the order to attack at the highest level: "I called my colleague, Bojan Ušeničnik, who was an advisor for defence matters to the Presidency of Slovenia. He promised me that he would check on it, but I never heard back from him. Later, I checked with the former YA Major-General Milovan Zorc, who was President Kučan's advisor but he didn't know anything about it. So I presume that the Slovenian Presidency didn't have a clue about it."[300]

Even though the order to attack was not specific, Borak and his unit prepared a provisional attack plan. In the beginning, they would blow up two very large reservoirs for lubricants inside the barracks, even though nobody knew for sure what was really inside them. Maybe they would ignite, but most likely it would cause much smoke which would impair the defensive strength of the Yugoslav Army.

---

298) Janez Lesjak, "*Osamosvojeni, izigrani in razočarani*" (Independent, double-crossed and disappointed). An internal record, Grosuplje, April 27, 2014. Janez Lesjak's answers to author's e-mail questions, Ljubljana, October 21, 2015.

Captain Borak was surprised that there were no ambulances on standby to help with possible casualties resulting from such a risky operation. Also, nobody ordered the evacuation of residents living close to the barracks. In those hours, the BTC shopping centre was full of people. After the firing of first shots, curious people would most likely gather on the streets or watch fighting from the windows and balconies of their apartments.

Cvetežar and Borak were still inside the warehouses of the BTC shopping centre when about an hour after the first order another one arrived from Lesjak which was much more comforting: "Avgust, we're lucky. Most likely, a cease-fire will be declared, but you must remain vigilant."

"I told you my decision," curtly replied the commander of the Litija detachment.[301]

Lesjak summoned the two commanders to report to him.

"The Lt-Colonel first castigated me, asking me where I was during the plebiscite for Slovenian independence, what kind of a TD member I was, what kind of a Slovene ..." remembers Neven Borak.[302]

Cvetežar showed Lesjak an assessment of the YA barracks in Moste prepared by his unit. Inside were 450 soldiers, 110 of them Slovenians, dug in behind barbed wire. Besides guns and ammunition they had two crates of hand grenades, and they had trained large-calibre artillery guns on civilian dwellings in the suburbs of Fužine and Nove Jarše. They had nine tanks ready for action and on the east side of the barracks they placed triple-barrelled anti-aircraft guns. On the top of all that, they also had surface-to-air missiles with a range of four kilometres.

"Thirty six barrels with 20 mm grenades were trained on us. Had we attacked, the YA would have massacred us on an open field. This would look like one of the suicidal attempts to break through the front in World War I," recalls Cvetežar.

Lesjak looked into the papers and asked: "Avgust, who is screwing with me, you, or my people?"

---

301) Extracts from the memoirs of Avgust Cvetežar, Litija, December 2012. Author's discussion with Avgust Cvetežar, Litija, October 20, 2015.

"Janez, check it out."[303]

Meanwhile Lesjak was receiving new orders: "I called Borak and Cvetežar to report to me at the Ljubljana Headquarters after they refused to follow my order ... However, I had to interrupt the discussion or, if you want, their excuses, because I had just received a message that a tank column from the Vrhnika barracks was heading towards Ljubljana. This saved them, because otherwise I would have replaced them immediately due to their insubordination. However after that, Janša informed me of the decision to stop the attack ... Then every second counted because I immediately had to inform all the subordinate commands about the cancellation of the attack. In this atmosphere, I simply forgot about Borak and Cvetežar," remembers Lt-Colonel Lesjak.

Andrej Grošelj, deputy commander of the Litija detachment, says that he checked with the local communities of Polje and Zadobrova and police whether they had evacuated residents, but nobody knew anything. So the officers of the detachment supported Cvetežar's decision to refrain from offensive action and "wrote down their overall opinion that they would not attack the Moste barracks for this reason".[304]

After the order to report to Headquarters of the Ljubljana TD, the Litija commander said farewell to his subordinates. He was convinced that he would have to go to jail. At the same time, he was relieved because he was envisioning scenes from the Nuremberg Trial.[305] Had he followed the order and ordered his soldiers to attack, he would have risked revenge by the Yugoslav Army against the residents of Ljubljana. As he put it, most war criminals tried to excuse themselves by saying that they were only following orders.

---

303) Extracts from the memoirs of Avgust Cvetežar, Litija, December 2012. Author's discussion with Avgust Cvetežar, Litija, October 20, 2015.
304) From the e-mail of Andrej Grošelj to Avgust Cvetežar, Škofja Loka, November 20, 2012. There is a copy of a decision of the officers of the Litija's detachment that they were unanimously against any attack on the barracks until all the negotiation options were exhausted. However they were ready to act had the YA units moved out of the barracks.
305) On November 1945, in the German city of Nuremberg, an International Military Tribunal began a trial of top war criminals of Hitler's Germany. The tribunal was established on the initiative of Truman, Churchill and Stalin and it sentenced several nazi leaders to either to

Cvetežar's deputy, Andrej Grošelj, accompanied him to Ljubljana and was with him during the questioning. Grošelj's son, Boštjan, said that the rest of the Litija detachment prepared a plan to seize the Ljubljana Territorial Defence Command and free Cvetežar and Grošelj if they were punished and put in jail.[306] However, Lesjak "forgot" about the two captains. He figured out that it took one hour and 15 minutes between Janša's request to attack barracks and the revocation of the order.[307] He recalls a dramatic discussion with Janša: "He quickly asked me: – How far are you with your attacks? –"

"Another 30 minutes."

"Stop immediately! Cease all battle activities! Our side has decided on a unilateral cease-fire!!!"

"How can I cancel and stop all units in only a few minutes? That's impossible!"

"Do everything you can to stop them!"

"Janez, that's impossible, we'll not be able to prevent some units from attacking. The attack was coordinated and is already taking place."

"You have to stop. Wherever you cannot stop, go all the way …"[308]

By some miracle and luck, the phone calls and couriers reached all commands and units of the Territorial Defence in the Ljubljana region. Cancellation of the order to attack the three Yugoslav Army barracks in Moste, Šentvid and on Metelkova came at the last moment when the "boys already had shells in the tubes".

In front of the YA barracks in Šentvid, Janez Kušar, Commander of the Second TD Assault Unit, had received an order for a mortar attack on the barracks: "A soldier was already holding a shell on the top of the mortar barrel, when we received the order to stop. They only thing that I could do was to shout: – Stop! –"[309]

Franc Boldin, commander of a reserve TD unit stationed in the village of Hrušica near Ljubljana, still has a beaten-up piece of paper dated July 2, on which somebody hand-wrote: "At 4 pm received

---

306) Testimony of Boštjan Grošelj, Litija, November 6, 2015.
307) Cvetežar said that the time-gap between the order to attack and its cancellation was at least five hours. However, he allows for a possibility that Lesjak announced the cancellation when he and Borak were together inside the BTC. Had the cancellation happened then, the time difference would have been only two hours.
308) Janez Lesjak, *Na robu – Ljubljanska pokrajina med osamosvajanjem* (On the Edge – Ljubljana Region During the Battle For Independence), Modrijan, Ljubljana, 2011, page 94.

Lesjak's order to attack YA barracks, named July 4th, on Metelkova in Ljubljana at 4:30 pm."

The Territorial Defence troops started preparations to move towards the barracks five kilometres away. Suddenly, somebody exclaimed: "Oh shit guys, we have the same uniforms as the Yugoslav Army. With these Titovkas[310] on our heads, they're going to shoot us dead."

When they were ready and the day almost turned to the night, somebody said: "We better hurry up guys or we will get caught by darkness. After that, somebody else remarked: God dammit, in that case, we better wait a little bit longer." This is how Boldin's unit awaited the cancellation of the order to attack.[311]

The only attack on Yugoslav Army barracks in Slovenia was carried out by TD units close to Defence Minister Janša, which attacked a barracks in Ribnica on June 28 with mortars.[312] They killed one recruit, and there were a few injuries on both sides. YA officers threatened they would bombard targets all around the town with artillery if the TD did not cease. However the attack came to nothing and calmer spirits prevailed.

\* \* \*

After a ceasefire agreement was reached on July 2, the commander of the Litija Detachment met the commander of the YA Moste garrison, Lt-Colonel Rudi Vrabič, a Slovenian by birth. Vrabič told Cvetežar that he had received clear orders that in the event of an attack on the barracks, they should defend it and also destroy civilian targets, especially energy plants in Ljubljana.[313] Both officers exchanged phone numbers and they never saw or heard from each other again.

In 2008, Lt-Colonel Butara wrote that the commander of the Yugoslav Army barracks in Šentvid, Colonel Tomislav Šipčić,

---

310) *Titovka* or *Partizanka* was a triangular hat with a red star in the front, worn by the Yugoslav Partisans during WWII. After the war, the YA adopted the hat and renamed it Titovka, after their leader Tito.
311) Testimony by Franc Boldin, Litija, November 6, 2015.
312) The town of Ribnica in the Suha Krajina region of Slovenia is about 40 km south of Ljubljana.

threatened the municipal leadership in Ljubljana that if there was even a small provocation by the Slovenian TD, he would use all available resources to start destroying the Slovenian capital. The Colonel said his heavy artillery would bombard the buildings of the Slovenian Presidency, the Government, Parliament, the post office, the telephone centre, the radio and TV centre, the gas company, the electrical grid and industrial plants.[314]

Twenty years later, a former head of the YA Zagreb region, General Andrija Rašeta, coldly stated: "The command of a unit that was attacked had the order to strike any part of Ljubljana with its most powerful weapons [...]. Not far from Šentvid, there was a large transformer station. Had the YA damaged it, half Ljubljana would have been without power for at least half a year."[315]

\* \* \*

Janez Lesjak, Avgust Cvetežar and Neven Borak were caught up in the whirlpool of events. In decisive moments, they reacted differently and made contradictory decisions. Despite that, they had many things in common.

The most important thing is that none of these commanders lost any of their men. When the situation got serious, men in the Litija detachment asked themselves who would find the courage to notify relatives of the loss of loved ones. In the end, all the parents hugged their sons, and all the wives hugged their husbands ... even though some of them returned from the "military exercises", as they called the conflict with the Yugoslav Army, dirty, with no sleep and some of them reeking of alcohol.

All three TD officers attended the famous academy for YA reserve infantry officers in Bileća in southeast Bosnia-Herzegovina. Even today, they are proud of their several months of military training in an institution they considered a European "West Point",[316] experiencing the hot sun and the freezing cold of Herzegovina's rocks.

---

314) Extract from the notes of Colonel Miha Butara and Dr. Martin Premk, *"Za resnico se je treba boriti"* (You have to fight for the truth) – quoted work.
315) Andrija Rašeta in a news programme 24 hours on

In Bileća, Borak learned a golden rule: never shoot at civilians. Cvetežar said the trainers warned future reserve officers: when you receive an order you try to follow it but never do it blindly, especially if its execution would cause unnecessary casualties of men under your command. Use your mind, take into account the circumstances you find yourself in, and inform your superior.

All three TD officers – maybe because of their experiences in Bileća – were respectful towards their adversaries, even though some were looking at them through the "barrels of their guns". They considered the YA officers to be professionals, "normal" people who knew what they were doing. In conscript soldiers on both sides, they saw scared young men, mostly teenagers, whom higher authorities wanted to use as cannon fodder.

Soon after the end of the conflict with the YA, the three Slovenian *Bilećanci* took off their military uniforms. They were no longer active in the TD or the Slovenian military and for several years they did not talk about those dangerous hours of confusion over attacking the Army barracks.

However when General Slapar was later going through a ream of documents, he found an analysis of the combat operations prepared on July 18, 1991 by the Republic's Coordination Group. In it, he was puzzled that he could not find any information about orders to attack Yugoslav Army military barracks.[317]

In December 2011, during protests throughout Slovenia against the second government headed by Janša,[318] Cvetežar's memoirs circulating on the internet asked: "Why were there several years of silence and why even today, do we still not have a clear picture of events that happened at that time in Ljubljana?"

"We were scared. We were afraid of the consequences. All of us were pushing aside the unpleasant experiences from the war. However, we talk about it now because in the past we were hoping that we would live better in an independent Slovenia," explains Dušan Drnovšek, commander of the first Litija Territorial Defence company.

---

317) Conversation with Janez Slapar, Radovljica, October 19, 2015.
318) The second mandate of Janez Janša's government lasted from February 10, 2012 to March 20, 2013. Janša had to resign because of a report by the Anti-Corruption Commission which

However Janez Lesjak did about the events in a book published in 2011, and the former TD commander Neven Borak did not remain silent either. Borak is a respected economic expert, who advised former Presidents Janez Stanovnik and Milan Kučan, as well as Prime Minister Janez Drnovšek. In the 1990s, Borak wrote resounding columns for the Slovenian daily *Dnevnik* under the pseudonym Veno Karbone. In them, he criticised the impetuous, reckless and violent actions of some Slovenian politicians, and accused Janša and Bavčar of continuously threatening people with secret documents.

"Janez Janša was waving in his hand an intelligence document that originated before Slovenian independence. The document alleges that I was bringing information to my father, a Serb from the Croatia's Krajina region, and that he was forwarding it to the Serbian leadership in Knin. The intelligence service that Janša was referring to also claimed that my father was such a fanatical Serb that he requested that his grandson speak to him only Serbian. What kind of an intelligence service is it that would overlook the fact that my father has no grandson?" asks Neven Borak.[319] He was one of the first intellectuals in Slovenia to become the target of hostile attacks from Janez Janša.

Just before independence, Avgust Cvetežar of the Litija detachment became general manager of Elma, a state-owned company making small household appliances in the Ljubljana suburb of Črnuče. Like many Slovenian companies, Elma suffered from the loss of Yugoslav market after the the break-up. In order to keep the company afloat, Cvetežar decided that everybody from general manager to cleaning staff would receive a minimum wage.[320] The company was drowning in debt, but in two years Cvetežar brought it back into the black. However, when the subsequent government programme of privatising state property reached Elma, Cvetežar did not want to participate and in 1993 he left the company.

Cvetežar is proud that he was considered a "red" manager. He says that during the time he was managing Elma, he benefited not financially but morally. Even today he puts on a tee-shirt sporting a red

---

319) Author's conversation with Neven Borak, Ljubljana, November 5, 2015.
320) For comparison: Cvetežar's monthly salary was 14,000 tolars, while his wife, who was a secretary

star. Ideologically, he is on the opposite side to twice-Prime Minister Janša, who advocates neoliberalism and comprehensive privatisation.

Of Janša at the time of independence, former Captain Cvetežar says: "He did not care whether sons were fighting their fathers, or brothers fighting brothers. Even Janša's new uniform sporting a patch with number one on the sleeve was inappropriate for a minister. It showed his great desire for power and importance."[321]

Like Borak and Cvetežar, Janez Lesjak in his soul is a leftist, who swears by a welfare state and social justice. He is angry at "leeches and plunderers" who are putting the last vestiges of state property into their pockets. He is almost ashamed that he participated in the struggle for independence, and he is angry with himself that he naïvely "held a bag" for various thieves. Contrary to Borak and Cvetežar, who did not have personal contacts with Janša, Lesjak knew him from the time they were teenagers in the town of Grosuplje.

Lesjak grudgingly concedes that Janša then was the right man in the right place, because he would not give in, and his stubbornness "kept everything up". But during the independence conflict, he also saw the problems this caused. When Lesjak as commander of Ljubljana Territorial Defence once publicly drew attention to worsening discipline in its ranks and warned that its soldiers were irritated at the cushy life of their superiors, Janša quickly forced him off the podium. At that point Lesjak realised: Janša does not allow criticism.

"When I see today how Janez is trampling on his former beliefs and how he's twisting historical truths, I get more and more angry, not so much at him than at the fact that he's forcing other Slovenians and me into ideological radicalisation and mutual hatred," wrote Lesjak recently.[322]

---

321) Extracts from the memoirs of Avgust Cvetežar, Litija, December, 2012. Author's conversation with Avgust Cvetežar, Litija, November 6, 2015.
322) Taken from: " " (Independent, double-crossed and disappointed),

Janez Janša testing a new type of anti-armour weapon. Photo taken at the army polygon Bač, August 13, 1991.

Photo: Marjan Ciglič (photo courtesy of the Museum Contemporary History of Slovenia - MCHS)

From left to right: Colonel Janez Slapar, Commander of the Headquarters of the Territorial Defence; Milan Kučan, President of the Presidency of the Republic of Slovenia and Janez Janša, Defence Minister.

Photo: Tone Stojko (photo courtesy of the MCHS)

Members of the Slovenian Presidency after their inauguration in May, 1991. From left to right: Matjaž Kmecl, Ivan Oman, Milan Kučan (President), Ciril Zlobec, Dušan Plut.

Colonel Vladimir Miloševič (left), with Maribor municipal officials, moments before his kidnapping by the Yugoslav Army special unit, May 23, 1991.

Photo: Joco Žnidaršič

Presidents of the six Yugoslav republics during the meeting in Ohrid, Macedonia, April 18, 1991. From left to right: Milan Kučan, Slovenia; Alija Izetbegović, Bosnia and Herzegovina; Franjo Tuđman, Croatia; Kiro Gligorov, Macedonia; Slobodan Milošević, Serbia; Momir Bulatović, Montenegro.

Photo: Joco Žnidaršič

On June 27, 1991, at around 2:30 am, tanks of the Yugoslav Army military garrison at Vrhnika moved toward the main Slovenian Airport of Brnik, near Ljubljana. There were no steel roadblocks or hedgehog barriers on the road to stop them, only fragile cars. The tanks crushed them like

Aftermath of the first battle between the Slovenian Territorial Defence forces and the Yugoslav Army in Trzin near Ljubljana, June 27, 1991.

Photo: Joco Žnidaršič

From left to right: aviation technician Bogomir Šuštar and pilots Jože Kalan and Toni Mrlak, all Slovenes, serving in the Yugoslav Army. The helicopter crew Kalan - Šuštar escaped to the Slovenian side together with their helicopter, Gazelle, on June 28, 1991, whereas Mrlak was shot down by Slovenian Territorial defence only a day earlier.

Photo from the archive of the Mrlak family.

From left to right: Jelko Kacin, Colonel Janez Slapar (with binoculars), Milan Kučan, Igor Bavčar and Janez Bogataj. Photo taken at the army polygon Bač, August 13, 1991.

June 28, 1991: Yugoslav Army armoured column on a slope called Medvedjek on the strategically important road to Slovenia's capital Ljubljana. Slovenian defenders received an order to stop them at any cost. Picture taken a few hours before the battle.

hoto: Igor Mali

Borovnica, the first Yugoslav Army depot to fall into the hands of the Slovenian Territorial Defence. This photograph shows the aftermath of looting and trashing committed a military unit under command of Col. Krkovič.

Photo: BOBO

The attacks of the Yugoslav Army in Slovenia were often observed by crowds of angry civilians. This picture was taken near Bregana, close to the border between Slovenia and Croatia, July 3, 1991.

Igor Bavčar, Interior Minister, at press conference, July, 1991. In public, he appeared mostly focused and self-confident.

Photo: Borut Krajnc, Mladina

Janez Janša, Defence Minister, frequently appeared frightened and confused.

Photo: Borut Krajnc, Mladina

Scene after a battle near the Slovenian-Croatian border, July 3, 1991. Coffins with young Yugoslav Army soldiers killed in Slovenia infuriated their relatives, particularly in Serbia. They demanded that President Milošević take decisive measures against Slovenia.

Photo: Joco Žnidaršič

Worried, uncertain and tired. The Slovenian delegation at the crucial negotiation in Brioni, July 7, 1991. From left to right: Foreign Minister Dimitrij Rupel, President of the Parliament France Bučar, Prime Minister Lojze Peterle, President Milan Kučan and member of the Yugoslav Presidency Janez Drnovšek.

Dubrovnik, historic city on the Adriatic coast, in ruins. Croatian old town was shelled mostly by the Yugoslav Army units stationed in Montenegro. May 1992.

Photo: Igor Mali

Weapons distributed to Croatian fighters in Posavina, northern Bosnia-Herzegovina, June, 1992.

- *Chapter 8* -

# CEASEFIRE AS YUGOSLAV ARMY DISINTEGRATES

On Tuesday July 2, the Yugoslav Army tried to turn the fortunes of war one last time, and the fiercest battles so far broke out. Battles raged at the frontier crossings of Šentilj and Gornja Radgona, and at some YA border guardhouses. By now the only people dying in the battles were YA soldiers, mostly officers who consciously chose Yugoslavia or death.

From Zagreb and Varaždin in Croatia, additional armoured units headed towards Slovenia to help their surrounded comrades. Yugoslav Air Force jets shot up barricades and destroyed television transmitters on the Nanos and Kum mountains, and in the town of Domžale.

That same day, the Territorial Defence attacked a column of armoured vehicles outside Dravograd near Austria. In retaliation Yugoslav military stationed in the barracks of nearby Maribor shelled the Pohorje plateau,[323] cutting the wire of the winter sports cable car. The Yugoslav commander, Major-General Mićo Delić, claims today that he shelled Pohorje's meadows in response to angry orders by his superiors in Zagreb. As he did not want to shell Maribor and its inhabitants, the wire of the cable car had to suffer collateral damage instead.[324]

The fiercest battle of the day took place at Krakovski gozd in the Dolenjska region near Croatia.[325] In the early hours of the morning, a YA column of 12 armoured vehicles was heading back towards its garrison in Karlovec, Croatia, when it came up against barricades of mined trucks which the TD had pulled across a road running through a forest, with another one a few kilometres further on near Novo mesto.

---

323) In northeast Slovenia.
324) A statement of Vladimir Milošević who visited Delić in Belgrade. Milošević stayed in touch with Delić even though they were on opposite sides during the fighting in Slovenia. Author's conversation with Milošević, Murska Sobota, May 20, 2014.

The TD, which had anti-tank weapons and twice as many soldiers as the Yugoslav unit, tried to convince the YA commander to surrender but to no avail. So the TD attacked, disabled several armoured vehicles and caused heavy casualties among the enemy.[326] After that they quickly withdrew in buses, and were well clear by the time Yugoslav Air Force jets arrived to shoot up their vacated positions.

The surviving Yugoslav soldiers retreated with the remaining armoured vehicles to a nearby petrol station and waited for reinforcement, but none came. So next night, the commander and most of his men set off on foot towards Croatia, but ended up in Slovenian captivity.

In Krakovski gozd, the TD proved that it was capable of destroying a well-equipped unit of armoured vehicles. About four hours later, the TD attacked a column of armoured vehicles coming to help their compatriots at Krakovski gozd, and destroyed two tanks and two other vehicles near the border crossing of Prilipe. Four YA soldiers were killed while the Territorial Defence had no casualties, again evading a Yugoslav Air Force counter-attack. The Slovenians hurriedly switched off the nearby Krško nuclear power plant.

As negotiations for a ceasefire neared conclusion, both sides were on the offensive and weapons were never again as loud as they were on that July 2. Major-General Ljubomir Bajić, a hardline Serb who advocated destruction of the Slovenian TD, took command of the Yugoslav Air Force in Zagreb,[327] and around 2 pm a Yugoslav Air Force MiG-21 jet broke the sound barrier over Ljubljana.[328] Sirens sent inhabitants of the capital scurrying to their shelters, and old folks grimly recalled the horrors of World War II.

In the area of Kočevska Reka, Yugoslav Air Force jets piloted by overzealous Serbs who had been drinking dropped cluster bombs, which are forbidden by international conventions. However they only hit a transformer station and caused no casualties.

In response to the Slovenian destruction of armoured columns at Krakovski gozd and Prilipe, the enraged Chief of the General Staff of the Yugoslav Armed Forces, General Adžić, accused Slovenia and

---

326) According to unverified information, eight YA soldiers were killed, while the YA claimed that only two were killed.
327) Information was confirmed also by Colonel Alojz Ferlinc.

its leadership of "starting the most merciless and dirty attacks on anybody who was wearing a YA uniform". Appearing on Belgrade TV, he swore: "We will force our opponents to respect the ceasefire and to act less arrogantly. We will find those who are hiding in their dens. You cannot win with manipulations and hatred. We will take control and bring things to an end."[329]

One reason for this crescendo of fury was that the tide was turning against the Yugoslavs. German Foreign Minister Hans-Dietrich Genscher had just paid a visit to Belgrade and now wanted to meet Slovenian President Milan Kučan. He could not come to Ljubljana because of the fighting, so they met at Beljak (Villach) in Austria just over the border. There Genscher dropped his insistence that Yugoslavia's territorial integrity should be maintained.[330] In doing so, he distanced himself from the European troika. The door for international recognition of independent Slovenia swung half-open.

\* \* \*

On that same critical day, July 2, a dispute which had been simmering for several months erupted in the Slovenian leadership. In the morning, Ministers Janša and Bavčar failed to show up for a meeting of the Slovenian Presidency called by Ciril Zlobec, who was filling in for Kučan. As hours passed, Zlobec called Kučan in Beljak and told him it smelled like a *coup d'état*. Kučan told him to make sure that Janša and Bavčar respected the ceasefire.

The members of the collective Presidency went over to the underground centre where Janša as Defence Minister and Bavčar as Interior Minister were directing operations. There was a sharp exchange of words, and Zlobec told the two Ministers they could take over if they wanted, but they would not be doing it behind the back of the Presidency, or in its name.[331]

---

329) From a public appearance of General Blagoje Adžić. The whole speech was published in the Serbian newspaper *Borba*, Belgrade, July 3, 1991.
330) Viktor Meier, *Zakaj je razpadla Jugoslavija* (Why Yugoslavia Disintegrated), Znanstveno in publicistično središče, Ljubljana, 1996, page 314.
331) From a book by Ciril Zlobec, (It's Nice to Be a Slovene But It's

Zlobec feels that that after the military success in Krakovski gozd Janša became obsessed with victory: "He was telling journalists that they killed dozens of enemy soldiers. Based on the information we received in the Presidency, I warned him not to exaggerate this victory. However, he harshly rebuked me, telling me that it was going to be as he said. I told him again that the Presidency was telling foreign countries that Slovenia's strength was in its defence, so it was not appropriate to announce conquests and spoils of war or indulge in triumphalism. However Janša could not be persuaded and stuck to his opinion. After that, Bavčar tried to calm him down saying: Janez stop fooling around, you don't need to exaggerate."[332]

Zlobec was right to be concerned about foreign perceptions. Officials at the Presidency intercepted a sharp statement by the Italian Foreign Minister De Michelis wondering why Slovenians could not control their Territorial Defence. The President of Bosnia-Herzegovina, Alija Izetbegović, called up to check on stories about alleged Slovenian atrocities against Yugoslav Army soldiers, which had been spread by their mothers.

Zlobec told Italian diplomats that there was a division in the Slovenian leadership between "hawks" and "doves". In his later book, Janša accused "some second-rate politicians" of wavering: "This is why information was sent to the world that in the Slovenian leadership there were hawks and doves, and that especially Igor Bavčar and I were too unyielding towards the Yugoslav military. This kind of explanation of things was very detrimental to us ..."[333]

In fact foreign intelligence services knew perfectly well about the divisions in the Slovenian leadership. A Slovenian security officer, Milan Zorko, says they distinguished which leaders followed the "hard" and "soft" lines. Some of their observations took the view that Janša would have caused even greater problems had there not been more rational politicians above him. The Defence Minister was frequently seen as acting on his own, not wanting to account to anybody, says Zorko, and adds: "I had access to several reports from foreign intelligence services and could not find any positive opinion about Janša. He never

---

332) Author's conversation with Ciril Zlobec, Ljubljana, June 6, 2014.

let anybody give him orders. He accepted into his circle only those people who blindly followed his orders, did not ask any questions and were bloodthirsty. Among them were Anton Krkovič, Srečko Lisjak, and some other TD commanders."[334]

Kučan admits that there were disagreements how to manage the war. "Janša's logic was frequently different from that of the Presidency. He ordered attacks on barracks, he wanted a total war, and when he didn't get what he wanted, he was accusing us all of capitulation. I often told him that his methods were not compatible with democracy. I also remember that the first day of the Trzin battle, Bavčar yelled at Beznik, calling him names, and demanded him to order shooting. Yes, Bavčar too was among those who wanted to create a larger war. Fortunately, we had France Bučar as the President of Parliament, who in the most difficult times knew, how put things in their right place."[335]

\* \* \*

As soon as Kučan returned from Beljak, he prepared the text of a ceasefire together with the "new" President of Federal Yugoslavia, Stipe Mesić, who was in Ljubljana and from there coordinated with Yugoslav Defence Minister, General Kadijević. According to the Federal constitution, Mesić was the Commander-in-Chief of the Yugoslav Armed Forces, but during this visit to Slovenia he had to hide from attacks of his own army.

As a Croatian, Mesić's role by now was totally incongruous. He had accepted to take up his new post as President of Federal Yugoslavia – and the European troika insisted on a functioning Federal Presidency – but it had become too dangerous for him to travel to Belgrade. In Croatia, and now on his visit to Slovenia, Mesić contacted leading European diplomats to convince them that the YA was no longer under civilian control and that Yugoslav Prime Minister Marković was a prisoner of the Army.

On the evening of July 2, still formally in his role as Yugoslav Federal President, Mesić joined Slovenian President Milan Kučan in

---

334) Author's conversation with Milan Zorko, Ljubljana, October 28, 2014.

presenting a four-point ceasefire agreement to domestic and foreign journalists. The first point stipulated an immediate cessation of all hostilities; the second separation of the YA and the Slovenian TD and return to their respective barracks or initial positions; the third the unconditional release of all prisoners; while the fourth stipulated that Slovenian police take control of borders and that Slovenia freeze all further steps towards its independence.

President Kučan sensed that the freeze was of little consequence. He declared that Slovenia would sooner or later achieve independence and confidently stated: "When this war ends, nothing will be as it was before. Not anymore. The new era will arrive, and we'll be talking about new things."[336]

\* \* \*

Yugoslav Army commanders meanwhile received orders to avoid armed conflicts. They retained "limited" goals – to cut Slovenia's connections with the outside world, to disarm the Territorial Defence and police, to maintain Yugoslavia's sovereignty and to install a marionette regime in Slovenia.

However, these plans concocted in Belgrade were illusory and naïve. Once it became clear that Slovenians would resist the Army's occupation of Slovenia's border crossings, the General Staff in Belgrade sent so many confused and contradictory orders, that the much-vaunted Yugoslav military seemed close to collapse.

When first clashes broke out in Slovenia, Slovenian men started to leave the YA in droves. Croatians and Albanians followed, as did Macedonians and Muslims during the later war in Bosnia-Herzegovina. In the YA, the only professional military personnel were officers and staff members in military institutions, all the regular soldiers were recruits.

News about the humiliating defeats of the Army quickly spread to other Yugoslav republics. Serbian politicians and media channeled their ire not only against Slovenian and Croatian "separatists", but also

---

336) President of the Slovenian Presidency at the news conference in Cankarjev dom, July 2, 1991.

against the Federal President Mesić and Prime Minister Marković. Also in Serbian sights were Austria and Germany, because of their support for Slovenia's and Croatia's independence. Serbian media spoke darkly of a resurgent "Fourth Reich".

According to figures of the Yugoslav Army, 48 soldiers were killed in clashes in Slovenia, most of them Serb nationals. Before the battle at Krakovski gozd, a Serbian soldier swore that he would not surrender because his father would kill him if he left the YA.[337] However, when coffins of dead soldiers started to arrive in Serbia, mothers of recruits rebelled and on July 2 angrily broke into the Serbian parliament, where they sharply criticised parliamentarians. Fathers joined the mothers and a long campaign began for the return of young recruits from capture in Slovenia and from the YA.

At the General Staff Headquarters, General Adžić brushed them aside by saying: "Peace can only be achieved by war, and in a war, casualties are inevitable."[338] Hundreds of relatives of recruits then set off for Croatia and Slovenia to look for their young men. After the Army's defeats at the hands of Slovenia's Territorial Defence, this mission to rescue the recruits merely undermined morale further.

The Yugoslav Army was humiliated and beaten, and its combat manpower was depleted, but still it would not surrender. The military leadership no longer had clearly defined goals regarding what it wanted to achieve in Slovenia, but it remained driven by a lust for revenge.

The following day, July 3, early in the morning, the Yugoslav command made one last attempt to regain the initiative, by sending an elite armoured-mechanised guard division from Belgrade towards Slovenia. It was a vast column of about 100 tanks and armoured vehicles. However, before long some of the tanks ruptured their tracks and others had engine problems. So the Brotherhood and Unity road from Belgrade to Zagreb was that day full of helplessly stuck military vehicles.[339] They were the last relics of Yugoslavia's "real socialism".

---

337) Summed up from: Janez Švajcer, *Obranili domovino* (They Defended the Homeland), Viharnik, Ljubljana, 1993.
338) Summed up from the newspaper *Borba*, Belgrade, July 3, 1991.
339) The Brotherhood and Unity road stretched from Jesenice, near the Austrian border, to the border between Macedonia and Greece. The construction of the road after the WWII was partly carried out by youth brigades and the Yugoslav Army. The road bec

\* \* \*

The key role of the Yugoslav Army was to support Yugoslavia's positioning between NATO in the west and the Warsaw Pact in the east. As Communism showed signs of crumbling in the 1980s, YA generals warned that the Yugoslav Federation was in peril and they demanded more money to operate and arm the Army.

By 1986, the YA cost more than 2.5 billion U.S. dollars, representing more than 70 per cent of the Federal budget. The standard of living in Yugoslavia suffered accordingly. In the end the Army never had to encounter a foreign enemy. Instead it fell apart under internal contradictions.

After the breakup of Yugoslavia, politicians had to calculate the value of its property to be divided up among the new independent republics. Experts working under the auspices of an International Conference on Former Yugoslavia estimated that the total net assets of the former Yugoslav Federation amounted to the equivalent of almost 94 billion U.S. dollars.[340] The largest share, 70 billion dollars, was accounted for by the YA.[341] As calculated for the Conference by the Crest Institute of France's *École Polytechnique*,[342] the value of military bases, infrastructure and military industry was almost 25 billion dollars,[343] stocks and supplies approximately 12.5 billion and military equipment more than 32.5 billion.

Infantry armament valued at 28 billion dollars consisted of 1.2 million rifles, handguns and machine guns, 145,000 mortars, a large quantity of rocket launchers and anti-tank equipment, and almost 6,000 howitzers and other artillery pieces. The Army had 1,000 tanks and other armoured vehicles, as well as electronic warfare equipment made in the United States, in particular eavesdropping systems.[344]

---

340) On December 31, 1990, one U.S. dollar was worth 10.655 Yugoslav dinars. Source: the exchange rate list of the National Bank of Yugoslavia (NBJ).
341) A report on assets and debts of Yugoslavia. Source: the International Peace Conference on Former Yugoslavia, Brussels, January 12, 1993 (inflation calculator: www.bls.gov).
342) Estimate of assets and debts regarding the military capabilities of former Yugoslavia. Source: *École Politechnique* (final report), Paris, December 28, 1992.
343) Because of time constraints, the Crest estimation undervalued most of the YA real estate, especially its research institutions.
344) The YA possessed: 820,000 pieces of semi-automatic and automatic weapons (mostly AK-47

There were also more than 100 naval ships, more than 500 fighter planes – among them modern MiG-29s – and 240 helicopters estimated to be worth 4.5 billion dollars.

The Yugoslav Army accumulated strategic supplies to last for at least three months,[345] so its warehouses contained several billion rounds of infantry ammunition, several million hand grenades, mortar shells and huge amounts of artillery grenades. According to Crest estimates, the value of all these ammunition supplies was more than 11.5 billion dollars.[346]

In all, the valuation of the YA was just two billion dollars less than Yugoslavia's total external debt at the breakup of the country.

Slovenian Defence Minister Janez Janša soon realised that these huge amounts of military materials in the depots abandoned by the YA were a gold mine. That is why, even before and during Slovenia's 10-day war, numerous convoys with weapons and ammunition made their way to Croatia. From the end of June 1991 to the fall of 1992, about 150 arms transports were registered crossing into Croatia from Slovenia.

Under Tito Yugoslavia was respected in the world for its non-aligned and peaceful politics, in particular in the decolonised "Third World" countries of Africa and southeast Asia. But over time, the non-aligned movement lost its significance, and the Yugoslav public realised its government was supporting some repressive regimes.

At the beginning of 1988, Slovenian media revealed that the Yugoslav Secretary of People's Defence, Admiral Branko Mamula, traveled to Ethiopia to facilitate the sale of weapons to the government troops fighting to prevent the secession of Eritrea. That fomented discontent among the Slovenian public towards the Yugoslav Army.

The 77 billion dollars-worth of Army property transformed multinational Yugoslavia into a powder keg. Its officers bragged how the YA was the "fourth strongest army in Europe" – which however

---

and 90,000 hand guns. Source: Crest, *École Polytechnique* (final report), Paris, December 28, 1992.

345) On December 31, 1990, the Yugoslav Army kept in its depots throughout Yugoslavia 227,000 tons of ammunition, mines and ordnance; 293,000 tons of fuel; 126,000 tons of food; 900 tons of medicine and approximately 1,800,000 of uniforms. Source: *École Polytechnique* (final report), Paris, December 28, 1992.

346) The estimated value of infantry ammunition was 7.9 billion dollars, tanks and armoured vehicles

was untrue, since most of the arms were obsolete and arsenals were scattered throughout the country.

This military extravagance forced citizens of Yugoslavia for decades to tighten their belts on behalf of a military machine which in the end pointed its barrels towards the people that were feeding it. Once the Yugoslav Army attacked Slovenia, the YA officers living there overnight became birds without nests, or clay pigeons, while the Belgrade generals who threatened to level Slovenia to the ground crumpled into grotesque shadows of a dying giant.

In those fateful days for Yugoslavia, General Blagoje Adžić, Chief of the General Staff of the Yugoslav Army, was probably thinking about his childhood trauma in a village in eastern Bosnia-Herzegovina. During World War II, when he was 10 years old, from a hiding place he witnessed collaborationist Croatian ustashes savagely kill his parents. He was the only survivor of a massacre that altogether killed 15 of his relatives.

One of his former colleagues said of him: "Yugoslavia became his second mother, and when he saw how Slovenian and Croatian 'separatists' were demolishing it, he probably had a feeling of a *déjà vu*. This might have been the reason for his animosity and unyieldingness."

When it became clear that the YA would not be able to break the resistance of Slovenia's defenders and when more and more YA officers were defecting to the Slovenian side, Adžić on July 5 issued the following statement: "Officers who are not ready to side with Yugoslavia and who do not measure up to the situation should leave the Yugoslav People's Army, while traitors should be killed on the spot without mercy or consideration. Units that will be under your command shall carry out their tasks in their entirety, even if everybody in the unit gets killed ... Huge losses are not necessary, we have technical superiority. War does not choose victims. Losses in Slovenia are minimal and completely negligible. You have as many casualties in a crash between two buses. From now on, fear has to force our adversary into capitulation, which means that we have to use fire against all that are resisting our actions."[347]

---

347) YA Colonel-General Blagoje Adžić in a speech to high-level officers of the YA, in the centre of

Soon after, Adžić visited the 5th Army Region Command in Zagreb. He angrily criticised officers, accusing them of sloppiness, indecisiveness, incompetence and mismanagement. At the same time, he branded officers from the Pula Air Force base as traitors, because their pilots refused to carry on sorties. In Pula, almost 80 per cent of pilots were Slovenians, and none of them wanted to fly to Slovenia. Adžić threatened that he would mobilise reserve pilots from Niš and Skopje to carry out the YA plans. However he did not know that those pilots had already run away.

Adžić shouted so hard that his face turned red, but nobody present dared to make a move. Embarrassed officers stared at the floor when Colonel Slobodan Stajčić, a Serb working in the finance department of the YA Ljubljana Corps, stood up and looked General Adžić into the eyes and said: "Comrade General, for 30 years you were teaching me to defend this nation. What's this idea now that I have to attack it? I don't want any part of it!"[348]

---

Colonels, who were of Serbian and Montenegrin nationality, to command units in Slovenia and Croatia.
348) "The fact is that we were waging war in Bosnia," interview with the Croatian General of the YA,

*– Chapter 9 –*

# SLOVENIA SETTLES AT BRIONI ... BUT A MAJOR KNOWS TOO MUCH

*We have to resist oppression and refuse to lie about what we know.*
Albert Camus[349]

Marjan Strehar died in unexplained circumstances on Friday, May 25, 2012, on a gravel forest road leading from the Muc cliff to Marolt farm in the area of Smolnik, on the Ruše side of the Pohorje plateau.

After a curve on the road, a black car approached Strehar's car without slowing down. Strehar quickly moved to the side of the road in his silver-grey car Škoda Octavia, and when it looked as if he was going to crash into the gorge of the Lobnica stream on the right, he instinctively hit the brakes. The black vehicle also stopped. The next moment, two unknown men appeared in front of him. A strong blow to Strehar's head followed and it was all over.

Did the two men push the body down the hill into the gorge of the stream or did they carry it 100 metres up the road to the bridge over the creek, and drag it on a path that led to the deep part of the stream in the gorge? That is not known for sure, but they probably placed the dead man face down in the stream where it was 30 centimetres deep. From there, about 80 metres further up, one could see Marjan's car on the side of the road with the engine still running.

The unknown men obviously knew the area and assumed that nobody would see them. The same day a wood-cutter noticed the car and notified the nearby farm, but nothing happened. When the two men finished their job and perhaps received payment, they immediately left the country. This is the most likely reconstruction of the event, based on accounts by eyewitnesses who wish to stay anonymous.

Marjan Strehar most probably died because he would not accept what was offered to him: a good job or money. The fact was that the Pohorje forests were full of valuable items, and buyers were coming there with millions of German marks or even gold. Marjan however did not put a single mark or gramme of gold into his pocket. He was clean and publicly chastised those who were doing just the opposite.

Whom did he disturb so much that he had to die?

\* \* \*

The story goes back to the night of June 28, 1991. As rain came pouring down, a shout went up: a soldier is dead, a Slovenian! The news spread around Territorial Defence troops hiding in front of a Yugoslav Army depot in Zgornja Ložnica located under the Pohorje plateau. They were waiting for an order to attack, and the news badly affected their spirits.

Two shots rang out behind a double barbed-wire fence protecting this largest YA depot in Slovenia. Some of the 24 soldiers on guard called desperately on the TD troops to help, but when an ambulance arrived a young man who had been shot was dead. Even today it is not clear whether he was killed because of a misunderstanding during changing of the guard. Or was he killed during an attempt to escape to the Slovenian side by a Serbian corporal who was unable to utter a word after the capture of the depot?

When a shell hit the guardhouse, Yugoslav soldiers became scared and started to heed calls for their surrender.

In response, the commander of the YA barracks in nearby Slovenska Bistrica threatened to shell the depot with his artillery. However he had reasons not to be acting in earnest. Colonel Rade Turović, a Serb, had been living in Slovenia for many years and had started a family in Slovenska Bistrica. In the middle of the independence crisis, his daughter proposed to marry a local man. Despite his threat, the colonel insisted that nobody was going to force him to shoot on Slovenians. A few days earlier, he had offered to surrender all the artillery in the barracks in exchange for the safety of himself and his soldiers.[350]

---

350) The artillery brigade under command of Colonel Rade Turović was equipped with 18 multi-barrel

The arms depot, which sprawled over about 27 hectares, finally fell to the Territorial Defence on July 2, and the Army retreated from the Slovenska Bistrica barracks. The Yugoslav soldiers took heavy weapons with them, but after crossing the border Croatian armed forces intercepted them. As a result Croatia laid hands on artillery pieces, but Slovenia had ammunition for them – and this gave arms merchants an opportunity for great business deals.

The commander of the TD of that region, Colonel Vladimir Milošević, remembers: "The YA loaded artillery pieces and Ogenj (Fire) multi-barrel rocket launchers on to a train … But it was diverted on to a siding, and since only one soldier guarded each wagon, it was not hard to seize the whole train. From the Ložnica depot, we captured about 2,500 missiles for rocket launchers. Each was worth several thousand German marks and they were sold to Croatia for more than 10,000 marks apiece.[351] I only learned several months later that everything was being sold at such high prices. When I add up everything that we had in those warehouses beneath Pohorje, I get dizzy. Tank grenades, anti-aircraft guns … Ten trains could not take all this stuff away. However, everything was gone, they took everything. Slovenian officials told us it was to help Croatia. When I ask people about it now, they are all quiet and look at me with fear in their eyes …"[352]

After the surrender of Colonel Turović, Janša verbally ordered that the depot must be emptied immediately, and local inhabitants started to remove the huge arsenal of about 5,500 tons of weapons, ammunition and ordnance.[353] According to Colonel Milošević the amount captured was sufficient for the whole of Slovenia, and included enough anti-tank weapons to destroy all YA tanks.[354]

---

20 mm-calibre cannons. Source: Janez Švajcer, *Obranili domovino* (They Defended the Homeland), Viharnik, Ljubljana, 1993, page 171.
351) According to the estimate by the French Institute Crest, each rocket for a multi-barrel rocket launcher was worth 3,927 DEM. Source: Crest, *École polytechnique,* Paris, December 28, 1992.
352) Author's conversation with Vladimir Milošević, Murska Sobota, May 20, 2014.
353) The list of weapons and locations of temporary storages was prepared on July 12, 1991 on Pohorje. By then, some of the arms were already sold to Croatia. The list also does not include assets transported to the town Kidričevo, or given to units of the TD and police.

The Colonel told his subordinate officers and everybody else who knew about the capture of the depot to be quiet about it. He was concerned that the YA might attack again, and destroy the depot and maybe also nearby houses. In particular they feared warplanes that were at that very time destroying radio and TV transmitters in Slovenia.

Women and children joined in loading the trucks that were transporting the dangerous cargoes into the forests of Pohorje. The first loaded truck drove off on July 2 and the last at 1 am on July 9 – over seven days and nights in all.

The main depot hall was so big that even the largest articulated trucks were able to drive in, and weapons were also stored in underground shafts on both sides. As there was no electricity due to safety concerns, people had to work by flashlight at night.

Inhabitants of nearby communities and farms alongside the road between Zgornja Ložnica and the Pohorje forests did not sleep well those days. Heavy trucks loaded with weapons continuously rumbled under their windows. About 100 vehicles were involved in the transports, many of them taking two of three cargoes a night. Over in Pohorje, the dark forests, which were normally so quiet, were filled with military crates full of grenades and mines of different calibres, as well as ammunition, anti-tanks missiles and detonators.

On July 1, the Interior Ministry prohibited members of the Slovenian Armed Forces from delivering, distributing or using alcohol. However in this traditional wine-producing area the ban was not followed very strictly. Weapons and ordnance were temporarily stored in outhouses of Pohorje farms, hunting lodges, log cabins and so on. Major Marjan Strehar, the local TD commander in charge of the relocation of the weapons on behalf of the Defence Minister, found about 100 temporary storage places altogether.

However they were not enough: most of the crates were left under the clear sky covered only by tarpaulins and spruce tree branches to protect against the elements. Once, a truck dumped crates of hand grenades on the forest ground in front of surprised workers. When lightning struck near a lakeside location where Plamen (Flame) missiles were stored, they installed lightning rods on the crates stacked several metres high. TD troops guarded the temporary storage places day and night, sometime assisted by hunters, but they still could not prevent thefts by locals.

"So today, at least half of Pohorje residents possess one, two, three or even more automatic weapons from that time, or even something else," said Rafael Mohorko, a lawyer and former anti-terrorism official.[355]

Ascertaining the exact quantities of weapons and their eventual destinations has not been easy. In autumn 1999, an investigative commission of the Slovenian Parliament led by parliamentarian Rudolf Moge requested that list from the Ministry of Defence, but to no avail. The committee estimated that around 300 tons of weapons and ammunition were handed over to units of Territorial Defence and police,[356] and between 500 and 700 tons found its way into an abandoned military warehouse in Kidričevo. The regional TD Command in Maribor, in a report to the Slovenian leadership, meanwhile submitted its own partial list of infantry arms[357] and other weapons taken from Ložnica for storage on Pohorje.[358]

Milošević remembers that in the first days of July Slovenian TD troops needed the anti-tank weapons most of all. Among weapons taken to Kidričevo were about 1,100 missiles for the Ogenj multi-barrel rocket launcher, but they disappeared from there too. The Colonel is convinced that they were sold to Croatia rather than given to the Slovenian defenders.

Croatian armed forces started to drive away the weapons from Pohorje immediately after the Ložnica depot was emptied. Trucks with Croatian licence plates were first seen on Pohorje roads when the Slovenians were still moving weapons into the area. Guards at the temporary storage areas said the whole Pohorje area in that damp

---

355) Author's conversation with lawyer Rafael Mohorko, Celje, December 1, 2011.
356) Units of TD and police received Strela 2M anti-air rockets, Maljutka anti-tank rockets, Zolja hand held anti-tank weapons and ammunition of various calibres. Source: Major Marjan Strehar, *Oprema in orožje v skladišču Ložnica*, (Equipment and weapons in the Ložnica depot), a letter to the Slovenian Parliament, Slovenska Bistrica, April 26, 1999.
357) On the list of captured weapons, there were five machine guns, one mortar, twelve 82 mm-calibre howitzers, 585 automatic rifles Kalashnikov, 32 sniper rifles and 461 other rifles. Source: The analysis of battle operations of the 7th PŠTO (Regional Command of TD) from June 26 to July 15, 1991, Maribor, July 20, 1991.
358) The report encompasses 19,000 hand grenades; 105 tons of TNT explosive charges; 13,000 artillery and tank shells; 45,000 mortar shells; 8,000 small-arms rifle grenades; 3.8 million rounds

summer looked like a true arms bazaar. They saw how Croatian buyers and Slovenian sellers were making deals and counting money on top of hot car hoods.

After a few days of a wild trading, the business activities settled down, and Croatian buyers needed a permit from commanding Slovenian officers in the area before they were able to make a purchase. Slovenians only provided papers about the quantities of sold weapons, and weapons traders made sure that no prices were listed on them.

By July 1991, Croatia found itself in a terrible war and very short of weapons. "They were offering cash, Austrian radio transmitters, Colts ... I remember that they offered gold for mortar tubes," said Marjan Strehar, who insisted he always had permission from his superiors and never took anything for himself.[359]

During the uncontrolled selling of Ložnica's weapons which lasted to the end of November, local residents of Pohorje often saw Anton Krkovič, commander of the Moris Brigade, Ludvik Zvonar from the Defence Ministry and members of the military Security organ (VOMO). They tried to make sure that there were as few documents as possible, to remove all traces of trading in cash. Nevertheless a few documents have survived with indisputable evidence of price gouging during these sales of weapons to Croatia.

The Croatian who removed the largest amount of weapons from Pohorje was Major Ivan Bećir, who on July 22 took away ammunition and mines,[360] returning a few days later to pick up 8,000 mine detonators. In the middle of October, he also took possession of 70 Strela 1M anti-air missiles and several thousand mines and grenades.

Another Croatian visitor to Pohorje was Josip Vukina, who took possession of ammunition at the beginning of September and grenades at the end of October. Among the buyers was also the famous Croatian general, Janko Bobetko. After a car accident in the middle of August 1991, he left two documents about weapons purchases in his Porsche sports car.[361]

---

359) Author's conversation with Marjan Strehar, Maribor, November 28, 2011.
360) The handover of ordnance was signed on 13 certificates for items received by the Croatians. These certificates are usually issued when items are simply borrowed. However, there is no evidence that the

Vladimir Šeks, commander of Croatia's Slavonia defence, received 8,000 obsolete M-48 rifles and half a million rounds of ammunition, while Defence Minister General Martin Špegelj received 70 tons of various military assets. Both Croatians took receipt of the armaments without "any purchasing contract", according to Ludvik Zvonar in his report about the weapons sale to the Croatian Defence Ministry.[362]

As there are no documents about the sale of weapons from Ložnica, there are also almost no records of police dispatches about border crossings of convoys into Croatia. The only exception is the Krško police station, and according to their documents 15 convoys with 38 trucks loaded with weapons from the Pohorje forests crossed into Croatia from Slovenia.

The quest for quick money by certain Slovenian politicians taking advantage of Croatia's dire need for weapons caused much bad blood, recrimination and ill-will in Croatia. Some people in Slovenia were also becoming uneasy about the huge deliveries of weapons to Croatia and were asking awkward questions to the government and parliament.

As an exchange for the weapons, the Croatian government also offered oil, which gave Janša an excellent opportunity to cover the illegal arms trade. "Weapons for Oil" made it easier to hide dealings in cash and the enormous amounts of foreign exchange that Croatians were bringing to Slovenia from the first days of July 1991.

Zvonar's correspondence with the Croatian Defence Ministry in Zagreb indicates that the Croatian buyers wired part of the money for purchasing weapons to Lloyds Bank in Zurich, where another account-holder was Cranex AG,[363] a Swiss subsidiary of the Slovenian Iskra company, which was the largest producer of electronic equipment in former Yugoslavia.[364] Zvonar's invoices show that Croatians deposited at least 16.6 million German marks into Cranex's account in Switzerland.

---

no. 2208/11, Novo mesto, August 19, 1991.
362) Information about the handing over of material-technical assets MORH, Uprava za logistiko, August 26, 1991.
363) Account no. 155.365.01, Lloyds Bank, Zurich. Source: invoices that the Republic of Slovenia advisor Ludvik Zvonar, submitted to the Croatian Defence Ministry.
364) Dušan Šešok, General Director of Iskra, was from 1993 Finance Minister in the first Slovenian government. In 2006, Šešok and a group of his colleagues carried out a so-called management buyout

According to sources who wish to remain anonymous, the money was exchanged for gold bullion that was taken for safekeeping to one of the Swiss banks. As a result, owners of this bullion could take out loans from this bank secured by gold. The bank made sure that the identity of the loan-taker, that is to say the owner of the gold bullion, remained secret.

For a long time, Major Strehar believed that the whole operation was selfless help by Slovenia to Croatia, which was under attack. Only later did he realise that it was thoughtless exploitation of the Croatia's distress. He was convinced that Defence Minister Janez Janša and the people around him were profiteering from the outrageous prices that they set for selling the weapons: "The first time I became suspicious was when I saw the weapons price list. During the loading of weapons, a Croatian lost a paper showing the prices. I remember very well that the Croatians had to pay 150 German marks for a hand grenade that was not worth more than 50 marks ... Everything was three times more expensive than usual. 60 mm mortars shells were sold for 300 German marks apiece, 120 mm mines for 800 marks, small-arms rifle grenades for 250 marks ... All of these prices were three times higher than normal ... I could not believe my eyes."[365]

\* \* \*

Marjan Strehar's body was found on May 26, 2012, one day after his alleged murder. When he went missing, a search party comprising his relatives, friends, firemen and hunters set out from the town Šmartno na Pohorju to find him. News of his disappearance also reached the Ruše side of Pohorje, where people knew him well.

At about twenty past one in the afternoon, a local resident called the police station in Ruše and informed them that he found an unlocked Škoda Octavia car on the forest road in the area of Smolnik, and that he saw a body in the Lobnica stream down the hill.

Two police officers arrived, but without an examining magistrate or a prosecutor. No official authority launched an investigation, even though Strehar's son told the two police officers that his father would never voluntarily go down the slope towards the stream. He said his father was a retired major of the Slovenian Armed Forces and had been receiving threats, but the police officers took no notice.

A journalist who knew Marjan Strehar well also came to the site of the incident. She called two police commanders in the area to inform them about the odd behavior of the two police officers, who were acting as if the whole thing was just an ordinary traffic accident. However, the commanders told her that based on the findings of the police who attended, there was no need for forensic technicians to visit the site.

Firemen and paramedics pulled the unfortunate Major Strehar from the stream, after which a doctor from the Maribor's community health centre checked the body but could not find anything that would point to foul play. He determined that cause of death was drowning, but nonetheless ordered a so-called sanitary autopsy to determine the exact cause of death. Besides stating that there were traces on the body showing that the deceased was sliding down the slope leading from his car to the stream, the police report says: "The body shows no clear signs of violence, only a small scratch on the lumbar region of his back and buttocks, while the backside of his pants is muddy from sliding down the slope."[366]

In the University Medical Centre of Maribor a doctor-pathologist performed a sanitary autopsy and confirmed that Strehar died because of drowning.[367] He also told the man's relatives that his heart was so weak that he would most likely have died even if he survived the fall.

However, had Major Strehar all of a sudden felt nauseous and fallen into the ravine under the road, he would have landed with his face down in a shallow part of the stream. The laws of gravity would have to be changed to explain how he ended up in a deeper part of the stream located a bit higher up. The slope beneath the road is covered with soil, grass, small plants, leaves and small beech trees but no big rocks. And contrary to the sanitary autopsy report, witnesses who saw the body said there was a large gaping wound above the deceased's right ear.

---

366) Ruše police station's report to the Maribor District Prosecutor's Office, no. 2212-34/2012/3, Ruše, July 24, 2012.
367) University Medical Centre Maribor's report on the cause of death of Marjan Strehar, sent to the

Strehar received threats from unknown people, and his friends warned him to be careful, because he knew too much about the weapons from Ložnica. He was asking increasingly angry questions about where the money from the sale of weapons went. He exposed himself too much. He did not want to cause problems for his family, so he handed over the documentation that he kept at home to various individuals and institutions.[368]

Major Strehar was a professional officer of the Territorial Defence. "He was one of those honourable men who are hard to find," said Branko Petan, former commander of the TD Command in Slovenska Bistrica.[369] Colonel Milošević, who knew Major Strehar well, remembers him as an honest man, not afraid to tell the truth to somebody's face. He is convinced that Strehar angered people who were selling weapons from Ložnica.[370]

Marjan Strehar loved the Pohorje forests and was an avid beekeeper. He died under suspicious circumstances while looking for a place for his beehive in the Ruše area.

\* \* \*

Events in the last days of June and first days of July 1991 were very unsettling. Fears rose that Slovenian and Croatian independence would trigger another terrible convulsion in the Balkans, just as the assassination of Archduke Franz Ferdinand in Sarajevo set off World War I. Among older Yugoslavs, memories were still vivid of the appalling cruelty of World War II in their country. Once again a growing number of innocent people were dying at the hands of armed people in various uniforms.

Croatia, now threatened by the brunt of the violence, had little sympathy for its newly-independent neighbour. Croatian President Tuđman mockingly described the armed conflict in Slovenia as an "operetta war".

---

368) In April and May 2008, the author of this book, and a journalist, Matjaž Frangež, (separately) requested from the Slovenian Parliament and the Defence Ministry documentation about the weapons trade. After a long procedure, the Information Commissioner removed the classification designation from a large portion of requested documents. As result, documents about weapons and ammunition from Ložnica depot became publicly available.

Meanwhile Serbia started to shift its focus from the crumbling Federal Yugoslavia to its new vision of a "Greater Serbia". After the first fighting broke out in Slovenia, Serbian leader Slobodan Milošević and his fellow-Serb Borislav Jović, who was a member of the Federal Presidency, held a number of meetings with the Federal Defence Minister, General Kadijević. Milošević demanded of Kadijević that the Yugoslav Army should defend borders of the "future" state: why should we defend Slovenian borders if they are only temporary? We should defend what will be lasting.[371]

This ominous demand by the Serbian leader made the death of the Yugoslav Federation even more inevitable.

When the mediation troika from Brussels first visited Belgrade, Jacques Poos, President of the Council of the European Community, proudly exclaimed: "The hour of Europe has dawned." His Italian colleague De Michelis asserted that the Community could be united when a situation became critical.[372] However, the European troika got burned, and their efforts to solve the Yugoslav crisis only demonstrated the absence of European unity.

In those days, foreign powers generally closed ranks against Slovenia. The three European ministers had received a mandate from the European Community effectively to tranquillise the quarreling Balkan "tribes". They were put out that little Slovenia spoiled their plans and wanted the status of an equal negotiating partner.

At a second meeting on June 30 in Zagreb between the European troika and the Slovenian delegation, led by President Kučan, there were disagreements whether further steps towards Slovenian independence should only be put on hold, or should Slovenia be pushed back into Yugoslavia? If the latter had happened, the Slovenian Territorial Defence and police would have had to leave captured border crossings immediately, the Yugoslav Army would have returned to Slovenian borders, Federal officials would have collected customs dues again and the TD would have had to return all the assets it seized from the YA.

---

371) Borislav Jović, *Poslednji dani SFRJ* (The Last Days of the SFRY), Politika, Belgrade, 1995, page 343.
372) Viktor Maier: (Why Yugoslavia Disintegrated),

After the meeting, the nonplussed Slovenian leaders tried to figure out whether the presiding Poos from Luxemburg was so artful that he was intentionally ambiguous about the troika intentions, or was he ambiguous only because he was not very artful?

The most unyielding towards Slovenia was the Dutch Foreign Minister, Hans van den Broek, who on July 1, 1991 took over as head of the European Council, and fought against mounting evidence to show that Yugoslavia still existed. Domestic concerns played a role in the opposition of most European states to the independence of Slovenia and Croatia. They were afraid that this could trigger an avalanche of secessions – such as the Basque country and Catalonia from Spain, and South Tyrol from Italy – and that it could worsen relations between England and Northern Ireland. So the 12-member European Community advocated adhering to the 1975 Helsinki Declaration that European borders cannot be changed by force. A subtle but noticeable deviation from these principles was advocated by Germany, which at the time was presiding over the Conference on European Security and Cooperation. Chancellor Helmut Kohl warned on July 1 that Yugoslavia could not be kept together by force.[373]

The United States followed the beginning of the Balkans crisis from a distance and left the initiative to the European Community. U.S. Secretary of State James Baker explained: "We just finished the war in the Persian Gulf. We were much preoccupied with the preparation for the Madrid peace conference to bring Arabs and Israelis to the negotiating table again. We just integrated unified Germany into NATO's Partnership for Peace, we fought a war in Panama. The United States was really busy ..."[374]

Baker chided Slovenia that its seizure of international border crossings violated the Helsinki Charter. Washington was clearly concerned that Slovenia's secession and a violent disintegration of Yugoslavia could cause an unsupervised breakup of the Soviet Union. However, after the unsuccessful coup attempt in Moscow on August 20, 1991, the Soviet Union started falling apart anyway.

---

373) Božo Repe, *Jutri je nov dan* (Tomorrow is Another Day), Modrijan, Ljubljana, 2002, page 349.
374) "I said that if you declare your independence with use of force – and that's the message I sent to all of the republics – you will cause a devilish civil war. If you violate the Helsinki Accords, if you declare your

After the Slovenian leadership declared a unilateral ceasefire on the evening of July 2, minor skirmishes continued between the Territorial Defence and the Yugoslav Army, mostly in the east of Slovenia. By July 4, Slovenian armed forces controlled all Slovenian border crossings, and the YA started to move back into its barracks, leaving 2,000 YA soldiers in Slovenian captivity. On July 5, nearly all weapons on the territory of Slovenia fell silent.

That same day, Milošević, Jović and Kadijević held another meeting in Belgrade, which confirmed their change in tack. The Serbian leader demanded that General Kadijević hit back with all the might of the Yugoslav Army, including the Air Force, to prevent Slovenia from maltreating their soldiers and officers. After that, according to Milošević, the Yugoslav forces should withdraw from Slovenia. He called for Slovenians and Croats to be removed from the YA, and that most of the Army forces be positioned on a line connecting Karlovec – Plitvice in the west, Baranja – Osijek – Vinkovci – Sava River in the east, and the Neretva River in the south. With this repositioning, they would protect areas in Croatia with sizeable Serbian populations and at the same time scare Croatia and calm spirits at home in Serbia.[375]

During the war in Slovenia, Milošević mostly stayed on the sidelines – so why did he suddenly get an urge to hit back forcefully? Very likely, the reason was that parents of young YA recruits were putting pressure on him and demanding the return of their children from Slovenia. For several days, there was no reliable news about their fate. The parents could only listen to Serbian broadcasts full of anti-Slovenian propaganda telling them that the Slovenian Territorial Defence was treacherously killing YA soldiers. Ironically, Slovenian "hawks" were themselves indirectly responsible for the Serbian leader's fury, since at the time they were vigorously fanning the war.

On that afternoon of July 5, Yugoslav Army General Blagoje Adžić seized his chance. At a military college in Belgrade he addressed 150 majors and colonels – all of them Serbs or Montenegrins – selected by General Staff Headquarters to command units in Slovenia and Croatia. Most of the Serbs recruited for this military action lived outside Serbia proper.

He told the officers immediately to take command of garrisons in Slovenia and prepare their units for fighting. "We have to force Slovenia to stay in Yugoslavia until such an agreement is reached that will be dictated by us," threatened General Adžić. By this he meant a recent decision by the Federal Yugoslav Presidency to re-establish conditions in place before the war, and to obtain the release of all captured YA soldiers and the return of seized military equipment.

A summary of the speech was forwarded to the Slovenia intelligence service the next day by the Slovenian YA Colonel Karlo Gorinšek, who at the time was in Belgrade to prepare an exam to become a general. He informed Slovenian intelligence officers that the YA leadership was convinced that the forthcoming attack on Slovenia enjoyed Milošević's support.[376]

General Adžić remained a dogmatic Yugoslav who wanted to keep the Yugoslav Federation together at any price, even if it meant use of brutal military force. At the same time, Defence Minister General Kadijević was more pragmatic and concerned to preserve the core of the YA.[377] The decision to punish Slovenia reached in Belgrade on July 5 by Milošević and Kadijević shows that Tuđman's advisors were wrong in accusing Slovenia of making a "secret agreement" with the Serbian leadership at the expense of Croatia.

Jović later wrote that Kadijević would have needed at least six to 10 days to prepare a plan to punish Slovenia. However, he was granted only two to three days. Milošević thought that Germany and Austria might recognise the independence of Slovenia and Croatia at any time – and then the military action would no longer be viable.

As it turned out, even two days were too many, as developments elsewhere were rapidly moving Slovenia's way. The first signs of a break in the deadlock already came at the ill-fated meeting with the European troika in Zagreb on June 30. At a session of the Slovenian Presidency the day after, Foreign Minister Dimitrij Rupel reported

---

376) Božo Repe, *Viri o demokratizaciji in osamosvojitvi Slovenije, III del: Osamosvojitev in mednarodno priznanje*, (Sources about Democratisation and Independence of Slovenia, Part III, Independence and International Recognition), Arhivsko društvo Slovenije, Ljubljana, 2004, page 103.
377) In 2001, Veljko Kadijević escaped to Russia. A year later, Croatia prosecuted him on three counts

that on the sidelines of the meeting, De Michelis secretly hinted to him that "a three month-long return of Slovenia into Yugoslavia" was the best option, because in that time a solution would be found and Slovenia would be independent, but not Croatia.[378]

Rupel's conversation with De Michelis proved invaluable.

The European troika, led by the van den Broek – in the meantime De Michelis was replaced by the Portuguese João de Pinheiro – arrived on Croatia's Brioni Islands[379] on Sunday, July 7 and immediately started negotiations. The delegation from Slovenia was led by President Milan Kučan.[380] Croatia was represented by Tuđman, while the Federal side consisted of Prime Minister Marković, President Mesić and three ministers. From Serbia, only Jović showed up, because Milošević insisted that Serbia was not involved in the conflict in Slovenia.

After painstaking negotiations, they agreed on the Brioni Declaration, which the European troika insisted all parties sign by midnight that same day, July 7.[381] The Declaration brought Slovenia a fragile peace and assurances by the European Community that it would immediately recognise Slovenian independence if the Yugoslav Army attacked.

However the Slovenian leaders were nervous about reactions back home. The Declaration demanded that Slovenia and Croatia were to suspend all activities stemming from their declarations of independence for three months. Slovenia had to establish on its borders conditions that were in effect before its declaration of independence on June 25, 1991. Slovenian police would control the borders, but they had to operate according to the Yugoslav regulations. Air traffic control stayed under the control of the Federal authorities,

---

378) Božo Repe, *Viri o demokratizaciji in osamosvojitvi Slovenije, III del: Osamosvojitev in mednarodno priznanje*, (Sources about Democratisation and Independence of Slovenia, Part III, Independence and International Recognition), Arhivsko društvo Slovenije, Ljubljana, 2004, page 124.

379) Brioni is an archipelago of 14 islands off the southwest coast of the Croatian Istria peninsula. Former Yugoslav leader Tito had a summer residence on the islands. Later the Brioni Islands were declared a Croatian national park.

380) Besides President Milan Kučan, members of the Slovenian delegation included: President of the Slovenian Parliament, France Bučar; Prime Minister Lojze Peterle; member of the Yugoslav Presidency Janez Drnovšek and Foreign Minister Dimitrij Rupel.

381) The Brioni Declaration sets forth the following principles: only nations of Yugoslavia can decide about their own future; new dimensions in Yugoslavia demand more detailed oversight and unconditional negotiations have to begin by August 1 and have to be based on the principles of the Helsinki Declaration on Security and the Cooperation in Europe (1975), as well as the Paris Charter on the New Europe (1990); the Yugoslav Presidency must have total control over armed forces; all

while Slovenian customs officers had to pay customs dues into an account controlled by Slovenian and Yugoslav authorities, as well as international observers. The Slovenian Armed Forces had to remove blockades and return seized buildings and equipment to the Yugoslav Army, which had to move back to its barracks.

The most problems were caused by the European intermediaries' demand that Slovenia release prisoners. The YA affirmed that 91 officers and 110 soldiers were detained in Slovenia's Dob prison at the time. According to President Kučan, "the Serbs told van den Broek that Slovenes more or less eat prisoners alive, that maggots are coming out of their eyes, that their wounds are festering and that they don't have any medical help or food".[382] The Slovenian delegation denied all this as Yugoslav military propaganda.

Back in Slovenia, Defence Minister Janša hesitated to release the prisoners and tried to set additional conditions for their liberation. Meanwhile the midnight deadline for accepting the Brioni Declaration was quickly approaching. So the prisoners-of-war became hostages of a confrontation within the Slovenian leadership. At the last moment, Janša decided to accept the proposals but only if the Slovenian Presidency ordered him to do it.

Kučan then called a member of Slovenia's collective Presidency, Ciril Zlobec, and asked him to call a session of the Presidency. Zlobec woke up the other three members of the Presidency, and in about 15 minutes called back: "I was able to inform Kučan in Brioni of the unanimous decision of the Slovenian Presidency that the Defence Minister should heed the demands of the mediators in Brioni and sign the memorandum documents in the agreed time, which was until midnight, set by the European troika ultimatum."[383]

The Brioni Declaration, which sealed the eventual breakup of Yugoslavia, was finally signed on July 7 at 11:55 pm, five minutes before expiration of the deadline. Even though in the following days some members of the Slovenian parliament criticised the Brioni agreement as a capitulation, Slovenia used the three-month moratorium very well to prepare itself for becoming an independent state.

---

382) President Kučan at the extended session of the Slovenian Presidency in Ljubljana, July 8, 1991.

On July 10, the Slovenian Parliament endorsed the Brioni Declaration by 89 votes to 11, with seven abstentions. In a statement, Parliament declared that it wanted to prevent further shedding of blood, and agreed for three months to suspend implementation of its Constitutional amendment declaring independence.

However, in the following weeks and months, nothing was the same as before, and the rest of Yugoslavia had to come to terms with losing its borders with Austria and Italy. Few of the customs dues collected by Slovenia made their way to Federal Yugoslavia, and thanks to skillful negotiation Slovenia also returned none of its spoils of war.

Slovenia would have suffered much worse had it rejected the Brioni agreement. The generals in Belgrade were hungry for revenge, and the Yugoslav Army was becoming uncontrollable, as events soon showed in Croatia and Bosnia-Herzegovina. Had the YA attacked Slovenia with all its might and with new Serbian commanders, the triumphalism of the Slovenian Territorial Defence would have been crushed, or maybe Slovenia would have won a Pyrrhic victory with numerous casualties and terrible destruction.

When European negotiator Hans van den Broek became hostile towards Slovenian proposals during the negotiations on the Brioni Islands, the Slovenians showed him the transcript they had obtained of General Adžić's warlike speech of July 5, in which he threatened Slovenians that they had won a battle but not the war. His belligerence in the end had the opposite effect to what he intended.

Croatia was the most negatively impacted by the Brioni agreement. Slovenia accelerated its efforts to cement its statehood during the three-month moratorium, but Croatia lost a valuable ally, and President Tuđman somewhat hypocritically accused the international community of allowing the fire of war to spread into Croatia.[384]

In Serbia, the Brioni Declaration was used to denounce the Federal authorities, "traitors" in the Yugoslav Army and the Slovenian leadership. Nationalists set about creating a new "Yugoslav" state with a pro-Serbian orientation. Milošević's followers rejected the view of Federal Prime Minister Marković that Croatia should decide

---

384) Summarised from several authors, (Creating the

about Serbs living in Croatia. They called on Federal authorities to ensure the presence of the YA in those Serb-populated areas of Croatia – Kninska Krajina, Slavonia, Western Srem and Baranja.

On the sidelines of the Brioni Island discussions, a meeting took place between two colleagues in the collective Yugoslav Presidency, Borislav Jović and the Slovenian Janez Drnovšek. Jović promised Drnovšek he would ensure that the majority of the Yugoslav Presidency members would vote for the withdrawal the Yugoslav Army from Slovenia. And indeed on July 18, the Presidency of Yugoslavia with all its members accepted the historically important decision that the YA should leave Slovenia in three months. Brioni had turned into a victory for Slovenia.

\* \* \*

While in Slovenia heated discussions about the pluses and minuses of the Brioni agreement were taking place, and while Drnovšek was waiting for a message from Serbia on withdrawal of the YA from Slovenia, convoys of trucks loaded with weapons, ammunition and ordnance from the Pohorje arms depots were heading for Croatia.

This was arranged by Slovenian Defence Minister Janša and Croatia's General Martin Špegelj, who had requested the supplies after hearing that Slovenia had seized a large quantity of "spoils of war". He ordered Croatian heavy trucks to drive to Slovenia. In his report to President Tuđman on July 31, 1991, he wrote that he brought back to Croatia about 500 tons of ordnance, but "on July 24, Minister Janša informed me that we can continue to move away ordnance and other military assets only if we pay the regular price. I didn't have money, so the transports stopped".[385]

The following morning, Janša ordered that a column of trucks that came to Slovenia to pick up ordnance be stopped because he had not yet reached an agreement with Špegelj about their handover. About an hour later, Janša gave the Croatians permission to go ahead "on condition that the payment was made in western currency", as was stated in a daily report of the Republic's Coordination Group. The column of trucks was to be stopped near Ljubljana if the delivery had not been paid for.

In about half an hour, Colonel Tomislav Mesić informed the Slovenians that on that same day they were going to send a letter of guarantee from Zagreb and the following day they would transfer the western currency. In the afternoon of July 13, Slovenian Interior Minister Igor Bavčar stated that the payment for the ordnance should be made in cash.[386] Criminal Police investigators later established that Tomislav Mesić was "carrying money to pay for weapons. Mesić is now retired and lives in Ljubljana".[387]

This ordnance was most likely carried away from Ložnica, Pohorje, or maybe from some other depot in Slovenia that was captured by the Territorial Defence. Slovenia did not give these armaments to Croatia for free. It did however donate some military equipment: old YA uniforms, blankets and suchlike, and for a good measure possibly also a crate of shells and one or two anti-tank weapons. Janša sticks to his claim that in return for weapons, Slovenia received Croatian oil. However, he is belied by reports of the Republic's Coordination Group, which show Croatians were mostly buying weapons for cash. By selling off these weapons even before the beginning of the YA withdrawal, the Defence and Interior Ministries were undermining the defence capabilities of Slovenia and needlessly exposing Slovenia to danger.

In May 2000, Janša testified before the investigative commission of the Slovenian Parliament. In order to intimidate the head of the committee, Rudolf Moge, he informed him beforehand that he had filed a criminal complaint against him for possible violation of his authority and misuse of his position.

After interrogating various witnesses, the committee members came to a conclusion that they learned the most from the officers of local and regional TD commands, while those at Republic's headquarters and the Defence Ministry, who were in charge of selling the weapons from Ložnica depot, did not reveal anything. In his report, Moge sarcastically stated: "The scale of arms trade was obviously so large

---

386) Ibid., pages 499-501.

that an individual could not even remember several thousand tons of merchandise sold from Ložnica. The investigative commission could not obtain a single piece of usable information."[388]

Estimates of the value of the Ložnica military assets given to the Parliamentary investigators by TD officers and officials of the Defence Ministry varied enormously: from 100 million to one billion German marks.

"The most realistic estimate is somewhere in between. With a high degree of certainty, I can say that the value of weapons from Ložnica was at least half a billion German marks," stated the commander of the Maribor TD, Colonel Vladimir Milošević.[389]

Yugoslav Army Lt-General Andrija Rašeta, who after the Brioni agreement led the commission on the withdrawal of the YA from Slovenia, on July 21 1991 handed a list of YA facilities and assets captured by the Slovenian TD to Miran Bogataj, Assistant Slovenian Defence Minister. However, the list included only the largest YA depots: Borovnica, Šentviška Gora, Zaloška Gorica, Vražji kamen and Ložnica, even though they also kept military equipment at approximately 20 smaller warehouses and barracks. In the five aforementioned depots, 3,800 tons of infantry, artillery ammunition and ordnance were documented,[390] while the Slovenian side estimated that in Ložnica alone there were 5,500 tons of infantry and artillery ammunition and ordnance. Several hundred tons were taken by the Slovenian TD and police, while the rest was probably sold to Croatia and the proceeds "evaporated".

On August 15, 1991, Colonel Peter Zupan took over the support function, because his predecessor refused to deal with weapons. Zupan says that during the distribution of weapons to Croatia and Bosnia-Herzegovina there was anarchy, "mostly because of interference by the leadership of the VOMO of the Ministry of Defence".[391]

---

388) The interim report about the course of investigation of the Slovenian Parliament on the involvement and responsibilities of public officials with regard to the discovery of weapons at Maribor Airport and the equipment and weapons in the Ložnica depot, Ljubljana, May 17, 2000.
389) Author's conversation with Vladimir Milošević, Murska Sobota, May 20, 2014.
390) Overview of facilities and assets of the YA captured by the TD in Slovenia. On July 21, 1991, this document was handed to members of MORS in Ljubljana by YA General Andrija Rašeta.
391) Official note of the VOMO investigators about the interrogation of Peter Zupan, MORS,

If we take into account the estimate of the French Institute Crest, Slovenians took possession of about 10,000 tons of military assets, mostly ammunition and ordnance. The report of a later London Conference states that on December 1990, one ton of Yugoslav Army ammunition was worth about 54,000 U.S. dollars. So the core of Slovenian "spoils of war" was worth more than half a billion dollars.[392]

Most of the proceeds from the sale of these armaments disappeared into the pockets of war profiteers. This is even more painful considering the fact that for almost 50 years, Slovenians contributed high percentages of their wages and pensions to arm the YA. This was denationalisation with a vengeance.

When Slovenia asked the international community to be recognised and accepted into the European family of nations, some people in Slovenia had their own plans: to obtain a lot of money and then buy power and influence the media. They did not care about friends, because if their plan succeeded, friends would come to them.

---

392) Estimate of assets and debts regarding military capabilities of former SFRY (Yugoslavia), final

- Chapter 10 -

# SHOWDOWN IN BELGRADE: SLOVENIA TRAINS CROATIAN USTASHES

During the armed conflict with the Yugoslav Army, there were only a few extremists on the Slovenian side, and they did not belong to the Slovenian military or police. One was Zmago Jelinčič, a provocateur hostile to non-Slovenian nationals, and a frequent lawbreaker. In the first days of fighting, he sported military fatigues and paraded armed on the streets of Ljubljana. When he asked a passerby for an ID without proper authorisation, the Slovenian security forces detained him for a few weeks. Later he became president of the Slovenian National Party and from 1992 spent 20 years as a member of the Slovenian Parliament, always in opposition.

During the fighting with the YA, a sort of "Slovenian Guard" operated in the Štajerska region. It was linked to a former YA non-commissioned officer and future VOMO member, Ivan Borštner.[393] Its aim was to recruit volunteers to fight against the YA.

In the battle for the Šentilj border crossing, the "guardists" were assigned to the Eastern Štajerska Regional Command of the TD. This is what the commander, Vladimir Milošević, said of them: "I ordered the security officer to monitor those guys closely, because I found out that they were members of a paramilitary formation. Fortunately, nothing unusual happened, except that on some of them we found double-edged daggers and garrot wires. I get goosebumps when I think what they could have done had the situation worsened. When fighting subsided, we sent them home …"[394]

\* \* \*

---

393) He was linked to the JBTZ affair explained in chapter 1, footnote 49.

In the critical days for Slovenia, other extremists also showed up – ustashes from neighbouring Croatia. Their name derives from the collaborationist marionette state set up during the World War II by Hitler's ally, Ante Pavelić.

Dobroslav Paraga, president of the Croatian Party of Rights (Hrvatska stranka prava – HSP) aligned with the ustashes had a close relationship with the Slovenian Social-Democratic Party (SDSS). Leaders of both parties met several times during preparations for independence. At the beginning of June 1991, Paraga and his party vice-president Ante Paradžik[395] offered 500 volunteers as help during the Yugoslav Army aggression against Slovenia. The offer was made to Jože Pučnik, then vice-president of SDSS, and Janez Janša, Defence Minister and an influential member of SDSS. Paraga said Minister Janša accepted the proposal and they agreed that Slovenia would train, equip and arm the volunteers.[396]

In the 1980s, Paraga had been a political prisoner in Yugoslavia. When in 1988, the Yugoslav People's Army detained Janez Janša for stealing a secret document, Paraga organised a demonstration in support of Janša in Canada, and reported about the detention to the U.S. Congress. He also claims to have convinced the Albanian government to move troops to the border with Yugoslavia.

After the end of the 10-day war in Slovenia, a group of Croatian Defence Forces (HOS) volunteers came to Slovenia's Kočevje forests for anti-sabotage training in the second half of July 1991. Training took place in deep secrecy and the political leaderships of Slovenia and Croatia were not informed. The Slovenian Defence Ministry was responsible for logistics, while a former YA officer, Frane Radobuljac, originally from Croatian Dalmatia but living in Slovenia in Grosuplje, was charged with carrying out the training.

The offer to take over the training was made on July 20 on behalf of Janša by Andrej Lovšin, director of the Ministry's VOMO. After a short reflection, Radobuljac accepted, on condition that he would not have to go with volunteers to the Croatian battlefields. According to

---

395) Ante Paradžik became the first chief of staff of Croatian Defence Forces (Hrvatske obrambene snage – HOS), the party's militia that only HSP members could join. HOS was formed on June 25,

Radobuljac, training of the first group of about 80 Croatian Defence Forces (HOS) personnel began on July 22, 1991.[397] It took place on a mountain plateau named Medvedjak, overgrown with forest, with a clearing in the middle, about 1,000 metres above sea level in one of most remote areas of the vast Kočevje forests.

Radobuljac remembers an incident after he finished working with the first group. One of the HOS volunteers completing his training course began shouting during his farewell address: "We don't want live witnesses, we don't want prisoners!" The outburst caused a stir and led to a delay in the training of the second group, consisting of a further 80 ustashes.

Paraga and Lovšin came to Medvedjak to check on the preparations of the units for battle. In recompense for training the HOS members, Radobuljac said he received an advance of 32,000 Slovenian tolars from the Slovenian Defence Ministry, and a subsequent award of 1,000 German marks for his hard work, delivered to him by Lovšin in the name of Paraga.[398]

In September 1993, a Croatian named Mladen admitted that he trained for eight days in Medvedjak. He remembered that his unit was visited during training by Paraga and Janša, accompanied by Lovšin.

"In his welcome speech, our president pointed out that the training was very expensive because the training of each volunteer cost the HSP party 5,000 German marks. At the same time, he gave us an order that if somebody asked us who we were, we had to say that we were volunteers in the Slovenian Territorial Defence, because our training was illegal."[399]

Military instructors and volunteers remember that the course was particularly rigorous. They were shooting without a break and learning how to kill from close range. They trained with several kinds of weapons, and mastered how to stab or strangle an enemy. The commander of the first group of the HOS members, Ante Šiljak,

---

397) Official record of the conversation of Stane Smolej, Brane Lukač and Damjan Režek with Frane Radobuljac, no.881-600, VOMO, Ljubljana, October 24, 1994.
398) Ibid.
399) Interview with Mladen (last name is not mentioned), a member of HOS,

pushed them mercilessly.[400] Šiljak, from Foča in Bosnia-Herzegovina, proudly wore on his uniform insignia saying *Allahu Akbar*, right next to his HOS emblem.[401]

The Croatian Party of Rights chose the slogan *Bog i Hrvati* (God and Croatians) while its military wing put the ustashe greeting *Za dom spremni* (For Homeland Ready) on its flag. Members of the HSP openly praised the World War II ustashe "Independent State of Croatia", and hailed the collaborationist ustashes as true patriots fighting Communism. With his militia, Paraga wanted to "clean" Croatia of Serbs and re-establish an "Independent State of Croatia", which would also include Bosnia-Herzegovina.

The HOS fighters had the opportunity to prove their heroism on the Croatian battlefield. They showed willingness to continuing fighting in the surrounded Croatian town of Vukovar when the Croatian government gave up defending the town. They were feared by Serbian irregulars and were brutal in their treatment of Serb civilians. Tuđman's regime soon started to treat them as terrorists who were harming Croatia's reputation at home and abroad, and made plans to get rid of them.

Dobroslav Paraga is convinced that the ruling Croatian party HDZ conducted "state terror" against his HSP because of the HSP's objection to dividing Bosnia-Herzegovina, as envisaged by Tuđman and Milošević. "Because of the secret deal with the Serbian president about dividing Bosnia-Herzegovina, the Tuđman's regime abandoned the defence alliance with Slovenia," he said.[402]

The HOS paramilitary units under training in Slovenia sported distinctive black uniforms, sewn for them somewhere in the Kočevje region. According to the HSP president, they bought 3,000 black uniforms and 3,000 pairs of boots in Slovenia. "For every uniform, Janša charged us 300 German marks. For Shpagina and Kalashnikov automatic rifles, Lovšin was charging us from 800 to 1,200 German marks apiece. The price of Makarov pistols was 500 marks and of hand grenades 400 marks each. We also paid very high prices for mortars, anti-tank mines and anti-tank grenades. We received

---

400) Brane Praznik, *Trgovci s smrtjo* (Merchants of Death), self-publishing, Ljubljana 2007, page 42.

no discounts even though we offered them volunteers to defend Slovenia," complained Paraga.

He also said that for training, weapons and uniform purchases, money was collected in Janša's name by Lovšin, in the presence of Moris commander Anton Krkovič and "some accountant".[403] Lovšin requested cash, which he took to a building close to Kočevska Reka. However, he did not issue Croatian buyers receipts or any other proof of purchase. The sellers also destroyed all technical manuals, probably because they wanted to hide the source. According to Paraga's estimates, they handed to Lovšin cash payments totaling at least five million German marks.[404]

The fact that they received five million marks from HSP was also confirmed by the head of the VOMO intelligence department, Anton Peinkiher. In his statement to an investigative panel of the Defence Ministry, he said that this was the first weapons deal made by the military intelligence. He said five truck deliveries were made to the Croatian border for ustashe paramilitary units.[405]

A military informant, Ciril Andoljšek from the nearby town of Ribnica, gathered information about the training and arming of the HOS members, but was told by VOMO headquarters that "what was going on in Kočevska Reka was none of his business and that he should worry only about issues within his jurisdiction".[406]

By giving this training, Slovenia violated international laws and conventions that forbid training of paramilitary units on the territory of another country. The event was briefly described in the 2006 report of a Parliamentary Commission in charge of supervising intelligence and security services (KNOVS). The report stated that members of a foreign state's military formation were trained in July 1991 in the region of Dolenjska, and that the payment for their training and armament was "made in cash".[407] There is no indication that the five million German marks earned from this undertaking ever ended up in Slovenia's budget.

---

403) According to VOMO employees, the accountant was Franci Cimerman.
404) Author's conversation with Dobroslav Paraga, Zagreb, May 9, 2009.
405) Official note of the VOMO investigators about the conversation with Anton Peinkiher, no. 881-600, Ljubljana, September 21, 1994.
406) Statement of Ciril Andoljšek, no. 881-600, Ljubljana, September 6, 1994.

After the first training course, Croatian members of the HOS met in the Majolka restaurant in the middle of Ljubljana and displayed their ideology and iconography. People present heard about the necessity to connect with Slovenian patriots – but especially about the need to liberate all Croatian regions from the Serb irregulars and communists. Among ustashe symbols they showed off wrapped in the red-and-white chequered Croatian flag were portraits of the WWII ustashe leader Ante Pavelić and other war criminals.

By September 1991, the HOS had approximately 12,000 registered fighters. Despite pogroms against their leaders, their strength and influence in Croatia increased dramatically, and at a certain moment Paraga's para-military movement had more than 30,000 sympathisers, most of them members of the Croatian National Guard (ZNG) and police officers.[408] Paraga received most of the money for training and arming his volunteers in Slovenia from Croatian emigrants in the United States and Canada, while Croatian municipalities also contributed.

Paraga's black-shirt followers may have been fighting in the front lines, but they became embroiled in a vicious battle with Croatian President Tuđman and his government. With both sides competing for power and funds from the rich Croatian diaspora, Tuđman decided to get rid of his rival's party and its ustashe militia.

According to Paraga, before and during the homeland war, at least 30 important members of his party were liquidated.[409] In September 1991, Croatian police killed Ante Paradžik, vice-president of the HSP, in suspicious circumstances. In early November, the Croatian Interior Ministry prevented about 200 armed HOS members from going to the Vukovar battlefield – and two weeks later the town fell into Serbian hands. One of the commanders of the Vukovar defence, HOS officer Mile Dedaković – Jastreb, escaped when there was no more hope of saving the town. Although his bravery and self-sacrifice were all too evident, he was accused of terrorism and armed uprising against the Croatian government.

---

408) From the report of the Minister of Defence, Janez Janša, to the Presidency of the Republic of

During winter 1992, Paraga escaped an assassination attempt largely by luck, while General Blaž Kraljević, commander of the HOS forces in Bosnia-Herzegovina, was killed near Mostar in August of the same year. That caused disintegration of the ustashe formations in Bosnia-Herzegovina, who were fighting against the Serbians alongside the Army of Bosnia-Herzegovina. In Croatia, the HOS disappeared to the political margins after proliferation of competing political factions, and its military wing was incorporated into the regular Croatian military in 1993.

Without receipts for his weapons purchases from the Slovenian side, Paraga found it hard to convince patriotic Croatian donors that their money went to the right hands. The saddened ustashe politician was furious at the Croatian regime and also at Janša, his former comrade from dissident days. He reproached Janša with hiding their friendship and taking up a relationship with the Croatian HDZ governing party.

"Later, when Janša was head of the Slovenian government, I tried to re-establish my contact with him and his SDS, but I didn't even receive a polite answer," Dobroslav Paraga complains today.[410] He promises that he will tell a court or a Parliamentary investigative commission the truth about what really went on in the Kočevje forests in the summer of 1991.

\* \* \*

Once the moratorium was agreed at Brioni, Slovenian politicians set to working out how to continue the process of building an independent state while observing the freeze on Slovenian independence imposed by the European Community. Slovenian officials traveled around the country to assess the damage caused by the Yugoslav military, and also listened to the views of the population.

Members of the Slovenian Presidency met again on July 15, together with Prime Minister Peterle and France Bučar, President of the Parliament, to review their observations. They noted that the 10-day war had damaged houses, outbuildings and other property. The economy was hurt by loss of industrial production, the tourist season was ruined, and damage done to planes of the Slovenian airline could not be recouped because insurance companies exclude war from their coverage.

The politicians praised the Territorial Defence, police, volunteers and others for standing up to defend their homeland. But they also picked up many critical comments, mostly about poor communication between Headquarters and local TD units. Commanders of smaller forces were often left to fend for themselves, but in most instances they acted thoughtfully and responsibly.

The survey showed there had been tensions between the Territorial Defence, the civilian authorities and the police, while attempts to mobilise recruits for TD units in Ljubljana and Maribor had not been successful. They heard complaints of authoritarian behaviour by some TD commanders and officials of the Security and Information Service (VIS). Also, it was evident that the Slovenian justice system was completely unprepared for conditions of war.[411]

The Slovenian Presidency noted that many accidents happened because of carelessness and negligent handling of weapons,[412] as well as "problems with maintaining discipline and excessive use of alcohol". As for the police and other law enforcement agencies, the Presidency stated that their activities were just as valuable as those of the Territorial Defence, but that "the media did not present them as such. Media mostly paid attention to the TD".[413]

On July 15, the first group of the European Community observers for monitoring implementation of the Brioni agreement arrived in Slovenia, and on the same day police headquarters in Ljubljana announced that the last prisoner-of-war had been released.

That was true then, but not three months later. On October 17, soldiers of Krkovič's Moris brigade captured a group of Yugoslav Army officers and non-commissioned officers and locked them up for 10 days inside the Medvedjak Training Centre on the same mountain plateau where the ustashe units were trained three months earlier.

---

411) Taken from: Findings, conclusions and recommendations of the Presidency of the Republic of Slovenia, no. 060-58/91, Ljubljana, July 15, 1991.
412) Janša in his report pointed out that in the ranks of the TD, more people were injured because of accidents than because of armed conflict. For that reason, he recommended that conscripted TD members be sent home as soon as possible.
413) Findings, conclusions and recommendations of the Presidency of the Republic of Slovenia, no.

The Moris brigade soldiers did not tell their captives why they were detained and they did not allow them any contact with relatives or lawyers. In one incident, a guard shot a prisoner in his hand when he wanted to light a cigarette.[414]

In view of this illicit act, it is perhaps lucky that the Hague International Criminal Court for War Crimes in former Yugoslavia did not indict any politician, soldier or a civilian from Slovenia for war crimes in Slovenia.

One Slovenian was indicted however, but for activities outside his home country. Franc Kos, nicknamed *Slovenac* (the Slovene), at first fought for the Croatians, but in July 1995 escaped to the Serbian side, where he joined the Army of the self-styled Republika srpska in Bosnia-Herzegovina under the command of General Mladić. Since December 2011, he has been on trial in Sarajevo accused of war crimes, including taking part in killings of Muslims from the Bosnian town of Srebrenica.[415]

As the Yugoslav Army slid to defeat in Slovenia, Pavle Ivić, a Serbian nationalist and academician, fiercely criticised Slovenia in an interview with the Belgrade newspaper Politika. He accused Slovenians of attacking columns of powerless young soldiers, demanded reparations for relatives of the dead, and concluded that Yugoslavia could not be saved: "The explosion of hatred that we experienced destroyed the last vestiges of confidence in brotherhood and unity. In a house that is dominated by hatred and the division, there is no happiness and progress – the only solution is splitting up."[416]

Just after the article was published on July 17, 1991, the collective Presidency of Federal Yugoslavia met in Belgrade in its full composition for one last time. The question before them was whether the YA should withdraw from Slovenia. Exceptionally, Slovenia's

---

414) Taken from the graduate paper of Marko Prešeren, *Tožilska in sodna praksa v RS glede kršitev mednarodnega prava oboroženih spopadov leta 1991*, (Prosecutorial Practices and the Case-Law in Republic of Slovenia Regarding Violations of the International Law During the Armed Conflict in 1991), Fakulteta za družbene vede, Ljubljana, 2009.
415) The court in Sarajevo includes judges from Bosnia-Herzegovina, as well as foreign judges, and deals with war crimes which the Hague Criminal Court could not take on because of time constraints.
416) The quote from the interview that Pavle Ivić gave to the Belgrade newspaper Politika on July 18, (Vukovar Tragedy

Janez Drnovšek also attended, and his quiet, uncharismatic character helped determine the outcome.

By decision of the Slovenian parliament, Drnovšek was supposed to be out of the collective Yugoslav Presidency, but his fellow Presidency member Borislav Jović said he had secured four votes of the so-called "Serbian bloc" for the withdrawal of the YA from Slovenia, so Drnovšek should come to Belgrade to provide the decisive fifth vote.

The men talked long into the night in the Palace of the Federation. From time to time, the powerful voice of General Kadijević echoed through empty corridors in a grim rumble, which to Slovenian journalists waiting outside the meeting room sounded like a voice from hell. Every two hours or so, a confused Prime Minister Marković staggered out of the conference room, saying he did not know what was going on, that nobody kept him informed, and nobody asked him anything.

Finally at around 4:30 am, the conference room door opened also for journalists. The result in favour of withdrawal of the YA was clear. The Serbian bloc and Slovenia's Drnovšek voted in favour of withdrawing the Army, as did the Macedonian member, Vasil Tupurkovski. Bogić Bogićević from Bosnia-Hercegovina abstained, leaving only the Croatian member, Stipe Mesić, against.

Sporting his trademark thin moustache and with a timid grin on his face, Drnovšek explained the technical details of the withdrawal to the journalists and was then asked if this represented a historic turnaround.

"Yes, we could say that. Without tonight's decision – Well it's already morning – all the declarations, agreements, accords, decisions and other documents don't mean anything. Slovenia only now really became a sovereign and independent country."

"Is this the greatest diplomatic achievement of your career?"

"Maybe it is. Even before I talked to my Serbian and other colleagues about the possibilities of a YA withdrawal from Slovenia, General Kadijević and Admiral Brovet showed a constructive attitude ..."[417]

Technically, the Yugoslav Presidency only confirmed a decision of the Supreme Command of the YA to withdraw its units from

---

417) Author's conversation with Janez Drnovšek right after the end of the session of the Presidency

the territory of Slovenia. In order to cover up the real reason for withdrawal – the military and psychological breakdown of the Army – the Command stated that they were pulling out because Slovenia did not follow the provisions of the Brioni agreement. They set a three-month deadline for their withdrawal, and decided that the Ljubljana Military Corps would go to Bosnia-Herzegovina and the Maribor Corps to Serbia.

General Konrad Kolšek, the Yugoslav general who had been dismissed early in the attack on Slovenia because he was a Slovenian, remembers the effect of this epic moment in Yugoslavia's history: "The news about the withdrawal from Slovenia shocked the YA officers. Some of them cried, others took it as treason, while the rest of them accepted it as a harsh reality. Many of them had spent decades in Slovenia: there were a lot of mixed marriages, they accepted Slovenian customs, while mutual achievements of the YA and the local people left a deep imprint on every officer, regardless of his nationality, which in the past did not play a big role. Everything that was positive and unforgettable in the past was now destroyed …" he wrote.[418]

The European Community was surprised by the decision on the Yugoslav Army to withdraw from Slovenia. It was a sobering moment for those of their leaders who had defended Marković's Federal government to the bitter end. Now it was clear that Yugoslavia was only a dead tree which threatened to do much damage as it fell. Confused ambiguous European bureaucrats began to change their tune, and took a less hostile attitude towards Germany and other countries that were favourably disposed towards Slovenian independence.

The U.S. Ambassador to Belgrade, Warren Zimmerman, remained under the influence of the Yugoslav and Serbian lobbies. He accused Slovenia of being selfish and bearing great responsibility for the bloodshed that followed its independence.

Eight years after the dramatic events in former Yugoslavia, Lawrence Eagleburger, a career diplomat who for just over three months served as a Secretary of State in the Government of George H. W. Bush, explained how Americans had their minds elsewhere: "I remember that your politicians took advantage of every opportunity to ask us whether we prefer democracy or a unified Yugoslavia. We tried to explain to them that the two were not mutually exclusive because Marković's reforms looked promising. I think that the Slovenians were really angry ... You know, President Bush was very engaged in other events, like the Arab world, the Indo-Pakistan relations. The priority areas were also Russia, China, and Asia in general, so Yugoslavia was somewhat in the background. We had to explain to the President several times who the Slovenians were and where they came from, and what Croatians and Serbs wanted. Bosnia was even more difficult ... We left the initiative to Europe because the problem was in their corner and because Europe was deeply interested in solving the problems in the Balkans. In Washington, we followed the European lead."[419]

\* \* \*

The war in Slovenia generated myths, reproaches and despondency among the various protagonists. Slovenian Defence Minister Janez Janša could posture as Saint George, the mythical figure said to have slain a dragon with his lance. This image traditionally symbolises spring and resurrection – and George is the patron saint of Ljubljana and Piran on the Slovenian coast. Janša wanted to strike hard and soon against the Yugoslav Army, because he knew from his defence studies that the Yugoslavs would need 30 days or more to move from a peace-to-war-standing by updating war planes, preparing stored weapons for action and mobilising additional units.

Janša's view that the YA was a "paper tiger" prompted him, in the spirit of Saint George, to order the destruction of YA barracks and tank columns. This belligerence set him in conflict with the doves in the Slovenian leadership, who publicly complained that the Yugoslavs were the aggressors and they themselves were dedicated to peace. Slovenians are still divided in their opinions: some are grateful for the restraint, while others are proud that Slovenians developed a true fighting spirit for the first time in this struggle for independence.

---

419) Author's conversation with Lawrence Eagleburger at the Republican Party Convention,

Croatian leaders were appalled by the miserable retreat of the defeated YA from Slovenia, since they feared that the Yugoslav military would now turn its barrels on them. Perhaps concerned that the Croatian public would question their strategy, people around Croatian President Tuđman spread rumours that the war in Slovenia was a sham, and there had been a secret agreement between Serbian and Slovenian politicians. Even today, some in Croatia still believe this fantasy. A more credible reason for the Army's collapse however was faulty judgment by the Yugoslav Defence Minister, General Kadijević.[420]

The war in Slovenia strengthened the Serbs' idea of a Greater Serbia, but as events later showed, this was merely an illusion that helped Slobodan Milošević to seize power in Serbia. Like Tuđman, Milošević committed a series of strategic mistakes, and his main skill was in tricky tactics that obscured this failure from fellow-politicians, international mediators and diplomats. His only constant political objective was to keep power at any price, even at the expense of his closest political allies or at a cost of starting another war.

After the defeat in Slovenia, the Yugoslav Army found itself in a miserable position. Its officers no longer had a homeland to protect, or anybody to pay them for their services. Families all over Yugoslavia, except to a small extent in Bosnia-Herzegovina, started refusing to send their sons into the army as recruits.[421] Milošević's dream of a Greater Serbia offered a home to dispirited YA officers, who after the humiliating pullout from Slovenia were asking why they should defend a nation that did not want it. Milošević rallied these tattered remnants of the YA to join him in his "defence" of the Serbian minority in Croatia. This gave a new mission to generals, and attracted volunteers who brought new blood to the ranks of the Army. It also meant that Serbs in the Krajina region of Croatia received additional weapons from YA depots.

---

420) The historian Davor Marijan wrote that Kadijević believed that a demonstration of military force, as witnessed in Kosovo and Croatia, would be also successful in Slovenia. However, the intelligence-security system of the Yugoslav Army failed, because it was not able to anticipate the armed resistance in Slovenia. Davor Marijan, *Slom Titove armije* (The Breakdown of Tito's Army), Golden marketing – Tehnička knjiga, Zagreb, 2008.
421) Admiral Brovet reported that parents of military recruits from Serbia did not want to send their sons to Croatia. The Macedonian government decided not to heed calls for mobilisation, only

In August 1991, the weakened Yugoslav Army started to pull out of Macedonia, which declared its independence on September 8. The Army moved more than 90 per cent of the armoured vehicles stationed in Macedonia to crisis locations in Croatia. Slovenian intelligence intercepted a message that the YA received from Romania about 40 Soviet Luna rocket systems, together with trucks for their transportation. It is very likely that the YA also received MiG-25 strategic warplanes, together with Soviet pilots who came to Yugoslavia via China.[422]

After these massive reinforcements, bloody war engulfed the Croatian regions of eastern and western Slavonia, Banija, Kordun and Lika, the outskirts of the port city Zadar and the city of Knin, and other parts of Dalmatia on the Adriatic coast. Attacks followed a similar pattern. The local Serbs would first receive "help" from Serbian "patriots". Paramilitary units from Serbia, aided by armed local Serbs, would intimidate the Croatian population, steal their possessions and burn down their homes. After that, the Yugoslav Army would move in under the pretence of separating the two warring sides. In reality it protected only the Serbian population, it killed or expelled Croatians, stole the rest of their possessions and then fortified seized areas. A YA offensive usually started with artillery shelling, after which infantry advanced, backed by armoured units. When the territory was "cleansed", representatives of the "Serbian civilian authorities" would move into the "liberated" areas, which would be gradually filled with a Serbian population.[423]

European diplomacy needed much time to recognise the real intentions of the YA. Tuđman's central government also failed, leaving the main burden for defending Croatia on municipalities and individuals, who organised themselves as best they could. Weapons were scarce, and their importation from abroad was impossible because Serbs controlled all important roads and the main Croatian sea ports.

---

422) From a report by Defence Minister Janez Janša to the Slovenian Presidency, recorded transcription of the meeting, Ljubljana, November 29, 1991.
423) On a few occasions the Serb irregulars and the YA fell out with each other. In November 1991, Slovenian intelligence intercepted a message that an armed confrontation broke out in the small town of Slunj. Serbian police and Yugoslav Army exchanged fire in a dispute over dividing spoils of war.

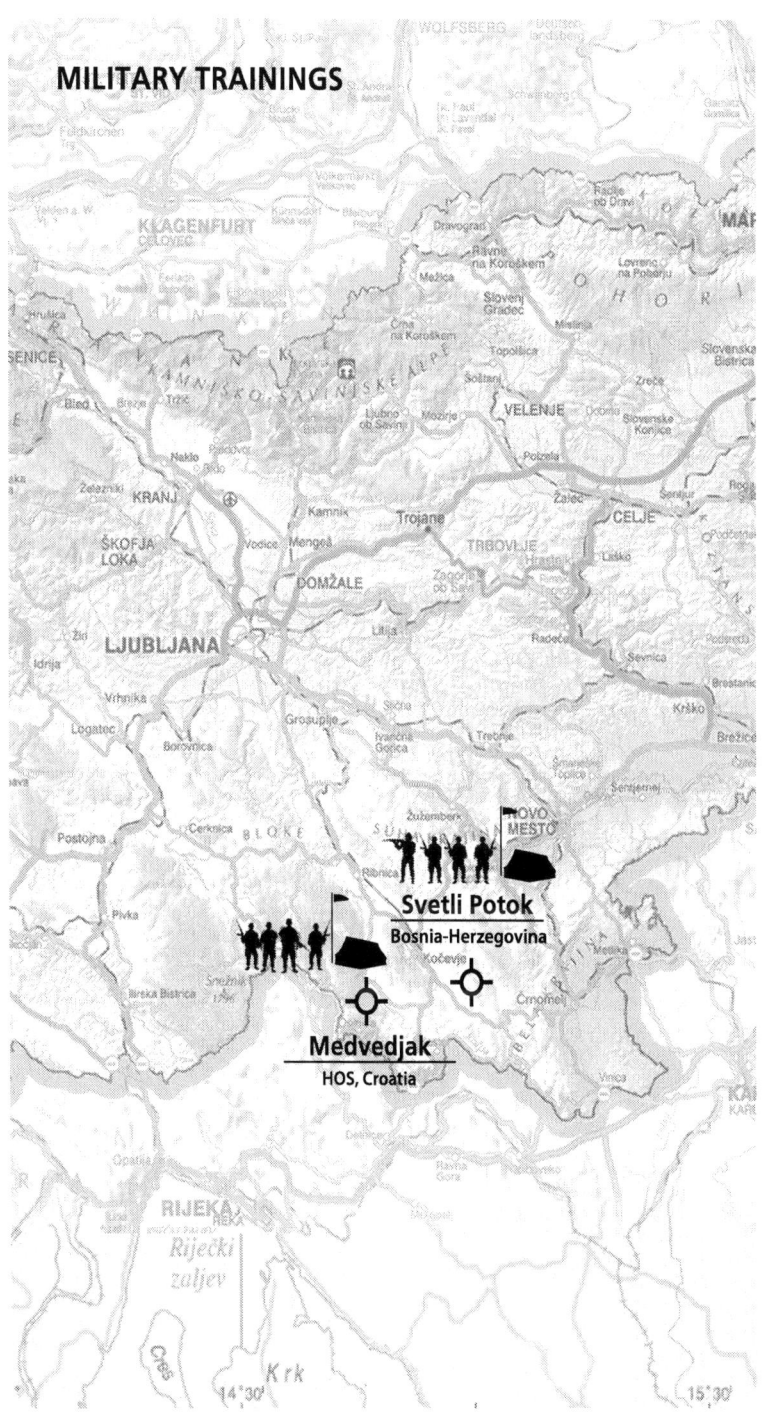

- *Chapter 11* -

# SLOVENIANS SWINDLE CROATIAN DEFENDERS AND LINE THEIR OWN POCKETS

In summer 1991, the medieval port city of Zadar in northern Dalmatia on the coast of Croatia came under fierce attack by Serbian rebels and Yugoslav Army led by Ratko Mladić, a combative Serbian Colonel in the ranks of the YA. After he failed in an attempt to capture another Dalmatian town, Šibenik, Mladić redirected his forces to destroy Zadar. The defenders of Zadar urgently needed weapons, especially anti-aircraft missiles. To get help, they contacted Orbis, the Slovenian company based in the town of Velenje.

In mid-September 1991, Ante Ivković, the President of the Zadar municipal government, received a letter from Orbis informing him that they could buy Strela 2M missiles for 12,000 U.S. dollars, and launchers for 6,000 dollars. On September 12, citizens of Zadar acting through their Jugotanker company, transferred 1,812,000 dollars to the account of a company named Scorpion – International Services, whose director was Konstantin Dafermos.[424] The company had a bank account in the Central-European International Bank (CEIB) in Budapest.[425] Orbis assured its counterparts in Zadar that the consignment would be delivered within 10 days. However the people of Zadar did not receive their weapons in September, and also waited in vain for the whole month of October, while Mladić's wild soldiery mercilessly destroyed Croatian homes in Zadar and its surroundings.

When the weapons from Slovenia did not show up in November either, the desperate defenders of Zadar had had enough. They decided to go to Slovenia and pick up their weapons themselves. Before departure, they were advised by Orbis Director Ivan Draušbaher to fetch their weapons at the Slovenian Defence Ministry.

---

424) Konstantin Dafermos founded numerous companies, one of which was Scorpion – International

In the Ministry's building in Ljubljana on Župančičeva Street 3, they were greeted by government advisor Ludvik Zvonar. He assigned to the Zadar's delegation twelve Strela 2M anti-air missiles and three launchers. They received their weapons from the Territorial Defence warehouse near Postojna. On the same day, November 22, 1991, Andrej Lovšin, head of the military Security organ (VOMO), sent a request to the Security and Information Service (VIS) director Miha Brejc and to a representative of the Operations-Communications Centre (OKC) at the Interior Ministry, to allow the Croatian delegation transporting the weapons to cross the border at Jelšane near Ilirska Bistrica.[426] From there, it was only a few kilometres to the Croatian port of Reka (Rijeka).

Thus, after waiting for 72 days, the defenders of northern Dalmatia received just a portion of the weapons they paid Orbis for, and then only after having to pick them up themselves.

That autumn of 1991 citizens of Zadar were also buying other weapons and ammunition at extremely high prices.[427] But, Zadar's municipal leaders were astounded at the invoice they received from Slovenia on December 18, 1991 for the weapons they purchased. For the Strela 2M rocket launcher, which in September Orbis was offering for 6,000 U.S. dollars, it was now charging four times as much, or 26,200 dollars, and the missiles initially priced at 20,000 dollars were now charged at 42,500 dollars.

The sellers of weapons were clearly overcome by greed, so they violated the agreement with the Croatian buyers. In a letter dated December 24, 1991 to the Orbis director Draušbaher, Šime Prtenjača of the Zadar municipal government complained at the outrageous prices and claimed return of 642,000 U.S. dollars. However, the Velenje company's managers would not budge. They attributed the astronomic price increase to the Slovenian Defence Ministry and advised the buyers to contact Janša directly.

---

426) Request of Andrej Lovšin to the VIS director and to the OKC representative at the Interior Ministry, documents no. 881/14-1329 and no. 881/14-1330, Ljubljana, November 22, 1991.

427) Automatic rifles ($450 apiece), machine guns ($9,500 apiece), SPG-9 anti-tank launchers and rockets ($15,000 and $530 apiece), 82 mm mortars and shells ($5,500 and $280 apiece), small-arms rifle grenades ($112 apiece), hand grenades ($35 apiece) and

The dispute dragged on into 1992. To settle the shortfall, Orbis offered Zadar defenders sniper rifles and hand-held RPG-7 rocket launchers, but the people of Zadar turned down the offer because the prices demanded were again irrationally high.[428]

In December 1992, a delegation from Zadar went to the Orbis headquarters in Velenje, and there Orbis agreed to return 262,000 dollars, and said the the rest would be returned by the Slovenian Defence Ministry. Draušbaher informed the Zadar delegation that the main reason for the delay in repayment was an unsettled debt of 766,300 dollars owed by the Grude municipality of Bosnia-Herzegovina to Orbis for another large consignment of weapons.[429] The Orbis director assured the defenders of Zadar that his company's debt to them would be settled by January 30, 1993 at the latest, which however did not happen.

In Yugoslavia, Slovenians were considered to be hard-working and honest people, even if they were a little stuck up and looked down on other Yugoslav nationalities. If somebody branded a Slovenian as a Balkan, he would be offended. In Slovenia, they would say that when you make a promise, you are bound by it. They scorn the mentality of the rest of the Balkans who abide by the pithy proverb: *obećanje, ludom radovanje,* which means "Only a fool can be made happy with a mere promise".

By their behaviour towards the weapons buyers from Zadar, the Orbis weapons merchants cast a big shadow over the reputation of Slovenians for honesty. They broke their promises many times and in 1994, the district of Zadar – Knin filed a civil lawsuit against Orbis and the Defence Ministry at the circuit court of of the Slovenian town of Celje. The lawsuit demanded repayment of the debt of 642,000 dollars, plus the interest for late payment.

The lawsuit dragged on for six years, and in the end Orbis avoided repaying its debt by declaring bankruptcy. At that point, the Celje

---

428) For a sniper rifle, Orbis demanded 35,000 DEM, for a RPG-7 rocket launcher at first $2,500 and later $5,500 apiece. Source: letter from Šime Prtenjača of the Zadar municipal government to the director of Orbis, Ivan Draušbauher, Zadar, December 27, 1991.

429) From a Puškarna company in Kranj, Orbis delivered to Grude municipality 15 handheld rocket launchers, 950 AK-47 automatic rifles with 114,000 rounds of ammunition, 31 RPG-7 anti-tank

circuit court judge had no option but to dismiss the lawsuit because Orbis ceased to exist.

The Defence Ministry was represented in court by the state attorney's office, which denied any connection between the Ministry and Orbis. It also challenged the jurisdiction of Celje's circuit court. In arrogant answers under interrogation by a Criminal Police officer, Defence Minister Janša denied any connection with Orbis and laughed off any suggested connection of his Ministry to the sale of weapons to Croatia.[430]

In December 1999, the plaintiffs from Zadar finally withdrew their lawsuit against the Slovenian Defence Ministry. Instead of trusted Slovenian justice, all they got were legal bills.

Several years later, when the statute of limitation expired for all legal proceedings, indisputable evidence about cooperation between the Defence Ministry and Orbis surfaced. A number of surviving incriminating documents showed that in the middle of October 1991, Orbis director Ivan Draušbaher asked Janša in writing for an "importation"[431] of a large quantity of infantry weapons and ammunition, which were to be sold to 13 municipalities in Croatia and one in Bosnia-Herzegovina.[432] A few days later, he requested an additional 2,000 AK-47 Kalashnikovs, accompanied by half a million rounds of ammunition. The Ministry granted his request. At first, Orbis received weapons from former Army depots captured by the TD during the fighting with the YA.[433] Zadar received first shipments from the Ložnica depot and later from Eastern European countries, when weapons were coming to the Slovenian

---
430) Interrogation of Defence Minister Janez Janša by Criminal Police officer Ljubo Jovanovič, no. 0221-21, Criminal Police Headquarters at the Interior Ministry, Ljubljana, November 20, 1995.
431) From June 1991 to the end of 1992, 40 arrivals of ships loaded with weapons were recorded in port of Koper. Ships were coming mostly from Bulgarian, Polish, Ukrainian and Romanian seaports.
432) On the Orbis list were mortars, grenades, machine guns, anti-tank and anti-aircraft weapons, rifles, handguns and ammunition, destined to be sold to Croatian municipalities: Zagreb, Zadar, Split, Osijek, Krapina, Garešnica, Virovitica, Valpovo, Ozalj, Klanjec, Vrbosko and Grude (Bosnia-Herzegovina). Source: Orbis' requests for import and sale of weapons, addressed to the Slovenian Defence Ministry, Velenje, October 18, 1991.
433) Among the evidence, there is a "material list", dated October 18, 1991 about the receipt of 1,650 AK-47 Kalashnikovs accompanied by 169,400 rounds of ammunition, and 202,940 rounds of ammunition for Makarov handguns. The recipient of the weapons was the head of the Orbis' sales department, Stane Vičar. The order for handover of weapons from the warehouse to Orbis was signed

port of Koper. Later Orbis also resold to Croatia several containers of infantry weapons and ammunition which arrived aboard the ship Scotia at Koper from the Bulgarian seaport of Varna.

The Defence Ministry did not issue invoices for the weapons handed over to Orbis, and Draušbaher was free to sell them at inflated, predatory prices to customers in the south of former Yugoslavia.

Experience from other world crisis areas confirms that war profiteering is one of the best ways to make a lot of money. In this case, the profits from the sale of weapons ended up in the pockets of influential people at the Slovenian Defence Ministry and Orbis. Investigators could not find any trace that the western currency brought into Slovenia by the Croatian buyers ended up in the Slovenian treasury, or that it ever benefited the common good of the Slovenian people. It is not clear how the profits were divided up.

At the time, Orbis did not have a licence for importing military weapons and was thus violating the law. The company received such a licence only on May 22, 1992 – the day Slovenia became a member of the United Nations – and by then, all the weapons depots in Slovenia had been more or less emptied.

Orbis was established in August 1990 as a subsidiary of the Gorenje corporate group in Velenje, a leading manufacturer of household appliances which now exports around Europe. The designated purpose of the subsidiary was manufacturing and selling hunting and sporting weapons. However the main reason for forming Orbis was to manufacture and market a renowned new machine gun, the MGV-176. The founder of the subsidiary was the first director of Gorenje, long-serving Ivan Atelšek, one of the most successful Slovenian entrepreneurs. He too was connected with the arms trade.

The Panama company registry shows that in 1989 Atelšek and the weapons magnate Konstantin Dafermos became heads of a company named Ark Ferralit-Guss.[434] It is to that company's account in the name of Dafermos in the CEIB bank in Budapest to which Croatian buyers of Orbis's weapons transferred their foreign exchange.

---

434) Ivan Atelšek was registered as treasurer of the company Ark Ferralit-Guss, the company's secretary was Konstantin Dafermos. The director, Austrian citizen Dieter Rabus, was son-in-law of the director of the Polish company that was manufacturing T-72 tanks. Source: an extract from the

The first director of Orbis was Ivan Draušbaher, a member of Gorenje's board and one of Atelšek's closest colleagues, who was also mayor of the municipality of the Muta[435] and a member of Janša's Slovenian Democratic Party (SDS). Despite enormous profits from the resale of weapons, under Draušbaher's direction the company drowned in debts and losses. In 1998, Orbis was deleted from the Slovenian registry of companies and with that, its debt disappeared also.

According to Jože Stanič, who headed Gorenje from 1990 to 2003, Draušbaher managed Orbis in his own way: "He always hid his business decisions under a cloak of military or state secrets. He was talking of his good connections with the Slovenian leadership. Many times, he was in the company of Janez Janša and the Slovenian Defence Ministry was among Orbis' most important business partners."[436]

In 1994, Draušbaher became a consultant for a company named Snežnik in Grosuplje headed by Ludvik Zvonar. This was supposed to be manufacturing rifle barrels, ammunition and hand-grenades, but in fact it was largely a front for reselling weapons. When Jelko Kacin replaced Janša as Defence Minister in March 1994, he disbanded the company.

Draušbaher was often the focus of Criminal Police investigations. In October 1994, during a search of his house, police confiscated a large quantity of weapons.[437] Even though several criminal complaints were filed against him and he was indicted, he was never found guilty. Indeed, despite the enormous dimensions of the illegal arms trade in 1990s, nobody in Slovenia was ever found guilty of it, and for that the responsibility lies with the office of the prosecutor.

Manufacturing and re-selling weapons was now centred on an informal but tightly connected group of leaders at the Defence Ministry and the Gorenje conglomerate. In summer 1992, Defence Minister Janša granted Draušbaher's request to use the Moris brigade's facilities in Kočevska Reka for manufacturing revolver

---

435) Muta is a small town alongside the Drava River, close to Austria.
436) A conversation of the Criminal Police officer, Ljubo Jovanovič with Jože Stanič, UKS MNZ, Ljubljana, September 7, 1995.
437) In Draušbaher's house, criminal police seized 60 different rifles, 38 handguns and revolvers, as

ammunition. In December 1992, three representatives of the Defence Ministry reached an agreement with Orbis's director to manufacture ammunition at a secret location X.[438] The Ministry was represented by the director of the logistics administration, Jože Zagožen, the head of the logistics administration's procurement, Ivan Crnkovič, and Moris commander Anton Krkovič.

\* \* \*

Several years later, the same people were actors in another story, the largest weapons transaction in independent Slovenia. In 2006, Gorenje was chosen to assemble Finnish Patria armed vehicles. The director general (CEO) of the Gorenje group was Draušbaher's son-in-law Franjo Bobinac, and its board of directors was headed by Jože Zagožen, who had become an important member of the Janša's Party (SDS). Ivan Črnkovič became director of a private Company, Rotis, which, with the help of SDS, was chosen to represent the Finnish company Patria in Slovenia. In November 2004, SDS president Janez Janša became Prime Minister of a new Slovenian Government.[439] Brigadier Krkovič, whose Moris brigade had been disbanded, allegedly acted as a lobbyist.

This group of merchants was joined by an infamous international businessman, Walter Wolf, who holds citizenships of Austria, Canada, Croatia and Slovenia. He became rich through oil trading and became owner of a Formula 1 racing team. He said his trademark was being the best-known man from the Balkans after Marshal Tito. To avoid risk of criminal prosecution, he subsequently escaped to his hideout called Ranch Wolf near Vancouver, Canada.

In a role of sidekicks were Jure Jurček Cekuta, a so-called "kind industrial spy",[440] who in his free time was also a painter, and Brigadier Peter Zupan, an individual well known from the weapons affairs of the 1990's. Ten years later, the latter was in charge of procurement of new weapons at the Defence Ministry ...

---

438) Brane Praznik, *Trgovci s smrtjo* (Merchants of Death), self-publishing, Ljubljana 2007, page 274-277.
439) The first time Janez Janša led the Slovenian Government was from November 9, 2004 to

In December 2006, the then Defence Minister, Karl Erjavec, signed a contract for delivery of 135 infantry armoured vehicles – eight-wheeler Patrias worth 278 million euro. Erjavec boasted in public that it was the most "transparent business deal so far", even though the bidding process was modified to favour the representative of the Finnish company Patria.[441] Slovenian "horse traders" made sure the leading people in Patria realised that there would be no business with Slovenia unless Rotis were chosen as a middleman, even though this small family business did not have any manufacturing capabilities. Gorenje was chosen for the manufacturing – that is to say, the assembly of the armoured vehicles.

Before announcing the call for proposals, they also agreed on the amount of bribes.

An important middleman role was played by Austrian weapons merchant Hans Wolfgang Riedl. The sellers from Finland and the buyers from Slovenia reached an agreement with him, giving him a 7.5 per cent commission on 162 million euro – a bit more than 12 million euro.[442] The intermediaries were supposed to divide this commission as follows: Wolf 6.7 million, Riedl 3.7 million and Cekuta 1.6 million euro.

However, this was not enough. The Slovenian group also wanted to line their pockets with gains made by Rotis, which would collect five per cent of the profit, with the rest going to Zagožen, Črnkovič, Krkovič and the SDS party.[443] They agreed on their shares before autumn 2005, oblivious to the fact that their secret deal would greatly harm Slovenian taxpayers.

The Slovenian participants were in a hurry to receive a 30 per cent advance payment from Finland. The most persistent was Zagožen, because Parliamentary elections were approaching fast and the "soccer team needed training", which is how one of the Patria's representatives described Zagožen's words during questioning in Helsinki.[444]

---

441) Karl Erjavec made this statement at the session of the Slovenian Parliament on September 9, 2008. Source: "Erjavec: the state budget was not adversely affected in any way," *STA*, Ljubljana, September 9, 2008.
442) 162 million euros is the net sum, after taxes and payment to Rotis are subtracted from the amount of 287 million euros.
443) Minutes from the preliminary investigation, no. 2400/R189/07, National Investigative

The intermediaries were under great pressure, and in their zeal they made a mistake. When on February 6, 2007, 3.6 million euro were transferred from Finland to the bank account of the Riedl's Company RHG, Wolf and Riedel rushed the following day to the office of the Steiermärkische Bank und Sparkasse in Lipnica (Leibnitz), Austria,[445] to spread the money around so they could hide the traces of corruption. However, a meticulous employee considered the transfer suspicious, refused to disburse the funds and reported the transaction to the Austrian Anti-Corruption Office.[446]

Soon afterwards, in February 2007, the Austrian Interpol Office started an international investigation into a network that it suspected was corrupt, informing police in Finland, Canada, Thailand, Liechtenstein and Slovenia.

Fortunately for the involved Slovenians, Slovenian police succeeded in "losing" the Interpol dispatch.[447] Instead of answering the inquiry within the statutory 24 hours, by February 22, 2007, the police confirmed its receipt only the following year, on May 19. In the meantime, telecommunication companies had erased all telephone contact data of the suspects in Slovenia and their business partners abroad.

On February 7, Riedl was able to withdraw from the bank in Lipnica only 300,000 euro in cash, 100,000 of which he handed over to Wolf. Subsequently, through fictitious contracts Riedl and Wolf were able to disperse commission from the Patria business deal on to several different bank accounts. 700,000 euro went to an account belonging to two citizens of Thailand, Apichat Sirithaporn and his wife, Chuangchan. A few days later, using a power of attorney from the Thai couple, Riedl withdrew that amount from the Vienna office of the Bank of Austria. Together with the net 200,000 euro that he took from the bank in Lipnica, Riedl was thus in possession of 900,000 euro.

During the subsequent investigation, Riedl told the Austrian police that he handed the cash he had withdrawn in Vienna to the Thai couple when they met at Vienna Airport. However it could be established that

---

445) Lipnica (Leibnitz) is a small town in Austrian Styria.
446) From the indictment against the accused, H.W. Riedl, Walter Wolf and three others, no. 601 St

he was not telling the truth. Under questioning in Bangkok on March 17, 2011, Sirithaporn admitted to Finnish and Slovenian Criminal Police investigators that he did not receive the money and that Riedl tried to bribe him to give false testimony.[448]

Indeed, on February 15, 2007, Riedl drove with the money to Ljubljana, parked his car in a public parking garage on Trdinova Street and spent the night in the Hotel Lev. Like the pedantic businessman that he was, he kept all the receipts, the most important of which showed he handed over 900,000 euro in cash to Jože Zagožen.[449]

In January 2012, Hans Wolfgang Riedl was put on trial at the Austrian Regional Court in Vienna. Walter Wolf was also on the list of the accused, but he did not participate because he was on his ranch in Canada. In April 2013, Riedl was found guilty of bribery and sentenced to one year in prison and two years of probation. He also had to pay a fine of 850,000 euro. Riedl appealed to the Austrian Supreme Court, but the Court confirmed the sentence, and so the Vienna weapons dealer had to go to prison.

Walter Wolf was also found guilty of bribery in Slovenia, among others counts, but he did not receive a sentence because he was not present at the trial.

The ruling of the Austrian Supreme Court stated: "In his quest for profit, Hans Riedl made a decision that by bribing a Slovenian politician in handing 900,000 euro to Jože Zagožen, an intermediary for Janez Janša, he would influence the bidding process in such a way that it would benefit Patria." The Supreme Court also confirmed the ruling of the regional court that "Janša, as Prime Minister of the Slovenian government, had an opportunity to interfere with the bidding process for purchasing Finnish armored vehicles."[450]

By coincidence or not: only one month after Riedl brought a large amount of cash to Ljubljana, Janez Janša bought a Volvo XC 70 for 46,000 euro. In a report published in 2013, the Slovenian Anti-Corruption Commission determined that Janša, in his disclosure of assets to the Commission (filed 12 months late), misstated the value

---

448) From the indictment against the accused, H.W. Riedl, Walter Wolf and three others, no. 601 St 5/09h, State Prosecutor's Office, Vienna, June 3, 2011.
449) Ibid.

of his Volvo as only 14,500 euro. The Commission also noted that Janša's assets had inexplicably increased by at least 210,000 euro.[451]

Because of the Anti-Corruption Commission's negative report regarding Janša, three coalition members left his government, and in February 2013 the government fell.[452]

Janša's 13-month long government (the second over which he presided since 2004) was one of the darkest periods in Slovenia's post-independence history. Severe austerity policies deepened social problems and people took to the streets in protests against the perceived arrogance of economic and political elites.

In Maribor citizens demonstrated in November 2012 against their corrupt mayor, and a violent response by the police ordered by Interior Minister Gorenak only led to fiercer protests, which then spread to the capital. To head off another protest rally on February 8, 2013, Prime Minister Janša branded the revolt against him as "left-wing fascism". However, by then his rule was melting away like the snowman effigy the protesters set up on Republic Square in Ljubljana.

\* \* \*

Riedl's conviction was a result of teamwork by investigators in Austria, Slovenia, and Finland. Further proceedings began on September 5, 2011, when Janša, Črnkovič, Krkovič and Zagožen were summoned to appear before the Ljubljana District Court. They were charged with giving and accepting gifts in return for illegal mediation.

Wolf avoided the trial by presenting a doctor's certificates that he was too ill to come to court. The District Court excluded him from the proceeding, but ordered his detention and filed an international warrant for his arrest.

In the pre-trial proceedings that were based on seized electronic messages, it was determined that Jože Zagožen had the most contacts regarding Patria with Janša, whom he called "the boss". As a director of

---

451) Report of the Anti-Corruption Commission, no. 06259-1/2013/1, Ljubljana, January 7, 2013.
452) The following parties withdrew from Janša's coalition government: Državljanska lista (Citizen's List), led by Gregor Virant; Demokratična stranka upokojencev (Democratic Party of Retirees), led by Karl Erjavec and Slovenska ljudska stranka (Slovenian People's Party), led by Radovan Žerjav. In the

Slovenian Power Plants, he did not have a "boss", but because he was a member of Janša's SDS party there is no doubt whom he meant.

During his long-time political and business career, Zagožen frequently attracted the attention of law enforcement agencies because of suspected violations of legal norms. "He acted like a typical representative of mafia structures. For them, a lie is a virtue," said Mitja Klavora, former director of the Slovenian Criminal Police.[453] Zagožen for his part said he does not lie but sometimes he has a poor memory.[454]

In Patria's case, the Ljubljana's District Court delivered a ruling, later confirmed by the Superior Court, that Janez Janša committed a criminal offence, together with Zagožen, who mostly acted as a middleman between Janša and Patria. In its later ruling, the Superior Court stated that Janša and Zagožen were influential individuals who, with the help of Riedl and Wolf, were able to create a belief that they could bring about a successful outcome of the Finnish company's bid. Without their substantive actions, coordinated in timing and among themselves, the criminal offence could not have been committed. Janša actions were critical to the success of the criminal offence, the ruling stated.[455]

Besides coming into conflict with legal authorities, Jože Zagožen also battled a serious illness. He died on September 13, 2013, at the age of 62, before the court announced its ruling. In April 2015, a Special State Prosecutor's Office in Ljubljana started legal action to seize illegally gained assets amounting to 837,000 euro from Zagožen's family – his wife, son, and daughter. The Defence Ministry terminated its contract with Rotis and the Finnish company Patria. Slovenia received only 30 armoured 8x8 vehicles, which was less than a quarter of the planned amount.

In December 2012, the Prosecutor's Office in Finland brought charges against senior Patria managers, whom it suspected of offering bribes and committing industrial espionage during the sale of the armoured vehicles to Slovenia. However a year later, a Finnish court

---

453) Author's conversation with Mitja Klavora, Borovnica, November 19, 2009.
454)

acquitted them, and this was confirmed by an appeal court on June 30, 2015. The defendants were reimbursed their legal charges and received compensation.[456]

When "judgment day" came to Slovenia on June 5, 2013, several hundred Janša sympathisers gathered in front of the District Court building in Ljubljana. At around 11 am, news of the outcome spread: angry shouts could be heard and some people had tears in their eyes. When the three individuals on trial came out of the building, people showered them with flowers.[457]

The county judge found Janez Janša, Ivan Črnkovič and Anton Krkovič guilty of the criminal act of receiving and offering gifts for illegal interference in the procedure of choosing and purchasing the Patria armoured vehicles.[458] Janša was sentenced to two years in prison, while Črnkovič and Krkovič received one year and 10 months. The judge ordered each of the three to pay a fine of 37,000 euro.

Ten months later, Ljubljana's Superior Court rejected appeals from the three convicts as unsubstantiated and upheld the ruling of the lower court.[459] The ruling of the county court thus became final and legally binding, and Janša, Črnkovič and Krkovič went to jail.

Almost the same group of people that gathered just over a year before in front of the Ljubljana court now assembled on June 20, 2014 to accompany their leader and idol to the prison of Dob pri Mirni in the Dolenjska region. The tearful farewell ceremony outside the prison was a mixture of a country fair, political rally and funeral. After Janša had disappeared behind bars, the same people protested for days on end in front of the Ljubljana court.

In the meantime, the two other convicts had also started to serve time in the Dob prison. In November 2014, the Slovenian Supreme

---

456) The acquitted managers were: Patria's Director – General Jorma Wiitakorpi; the former Executive Director, Heikki Hulkkonen; Patria's representative for Slovenia Reijo Niittynen and the Director of Sales, Tuomas Korpi.
457) The Ljubljana county court had jurisdiction for the trial. But because it did not have enough space, the proceeding took place in the District Court in Ljubljana.
458) All three were sentenced according to the first paragraph, Article 269, of the Penal Code (*Official Journal of the Republic of Slovenia*, no. 95/4, Ljubljana, January 1, 1995). Article 269 states: "A person who, whether for himself or for somebody else, asks for, or receives a reward, gift or some other benefit, or a promise of such a benefit, so that he would take advantage of his position or influence, and ensures

Court rejected the petition of the three convicts for so-called protection of legality. But early Parliamentary elections were set for July 13 and there was speculation that Janša would exchange his prison cell for the Prime Minister's office.

In the end, the SDS came second and its leader Janša was elected as a member of the Slovenian Parliament.[460] After two months, the lower house of Parliament removed Janša's mandate as a member, but the Slovenian Constitutional Court temporarily gave it back to him. The Constitutional Court then unanimously accepted Janša's appeal and suspended the County Court sentence. After 176 days in an open section of the Dob prison, Janša walked out as a free man. In their ruling, the Constitutional Court judges wrote: "The appellant is a member of Parliament and at the same time, President of the largest opposition party in Parliament. A well-functioning opposition to the leading party is one of the pillars of democracy."

Since they were ordinary citizens, Črnkovič and Krkovič had to stay behind bars. When on April 23, 2015 the Constitutional Court unanimously set aside all the courts' decisions in the Patria case and returned the case back to the County Court for a new trial, Črnkovič and Krkovič were also able to leave prison. Janša thereupon mercilessly attacked the prosecutors and the judicial system – branding it an "injudicial system".

In the meantime, a 22,000-page file on the Patria case was given to another judge of the Ljubljana County Court. However, at the beginning of September 2015, she declared that the statute of limitations for the case had expired. So the "trial of the decade" ended without an epilogue.

To many, this seemed a judicial farce such as had never happened before in Slovenia. Or had it? Not according to the Slovenian novelist, Josip Jurčič, writing 130 years ago.[461] In his satirical novel *Kozlovska sodba v Višnji Gori* (Judging the Goat in Višnja Gora), Jurčič described how ordinary people became

---

460) The Parliamentary elections on July 13, 2014 were won by the Miro Cerar Party (SMC). The Party's president, Miro Cerar, formed a coalition government with DeSUS (Democratic Party of Retirees), and SD (Social-Democrats). His government has the support of 52 of the 90 members of Parliament.
461) (Judging the Goat in Višnja Gora), by Josip Jurčič (1844-1881) a

entangled with schemes of individuals and groups. The main character is a goat named Lisec, who escapes from his owner and intends to eat vegetables in the garden of his owner's neighbour. The neighbour does not like the goat's owner, Lukež Drnulja, so he files a lawsuit against him saying that "in the Bible, an evil intention is forbidden, just the same as an evil deed". However, the judges, elders and wisemen cannot agree whether to hang the culprit or to show him mercy. The conundrum is solved by a Solomon's judgment delivered by Višnja Gora's fortune teller and beggar. Here is how the story ends:

> *Three times did the sun rise above Višnja Gora and when it rose for the fourth time, they brought Lisec the goat and Lukež Drnulja out of town to the Peščenjak Hill, where the famous Višnja Gora gallows stood, and there, in front of the crowd, they beat the shadow of the goat, and Lukež watched this beating with his eyes blindfolded, while they swung above his head with sticks.*[462]

\* \* \*

The Patria case split Slovenian politics, divided the legal community and provoked passionate arguments among the public. The seed of discord planted by the biggest Slovenian arms merchant had thus borne devil's fruit.

Walter Wolf, "after Marshal Tito, the most famous man from the Balkans", will most likely never see Slovenia again. Because of a warrant for his arrest issued by a Slovenian court, he was detained for a few days in Vancouver, Canada. However in June 2015, the Canadian authorities announced that they would not be able to carry through the extradition process before expiration of the statute of limitation.

---

462) Quoted from: Josip Jurčič, *Spomini na deda in druge zgodbe* (Memories About My Grandfather and Other Stories),

At first Wolf cited bad health as a reason for not attending the trial in Ljubljana and in a later interview he said he would not receive a fair trial in Slovenia. He claimed that nobody has any evidence that he offered or paid bribery to Janša. His only "sin" was his 35-year friendship with Riedl.[463]

So Walter Wolf, which is also a trademark for cigarettes and perfumes, got away without punishment.

---

463) "I swear that Stipe Mesić didn't take a single kuna as a bribe from the Finns," interview of Walter

- *Chapter 12* -

# SLOVENIA'S WEAPONS BAZAAR

*In war, the loudest patriots make the largest profits.*

August Bebel[464]

In the last days of July 1991, the situation in Croatia dramatically worsened. TV images of killed soldiers and civilians, burnt houses, columns of refugees, crying mothers, and fighters swearing they were going to destroy their enemy slowly awakened sleepy Europe. The Slovenian leadership decided to act, and that action soon had repercussions.

"We learned from reliable sources that the leadership of Slovenia until now sold more than 30 per cent of weapons and equipment seized from the Yugoslav People's Army to the Croatian National Guard (ZNG) and the Croatian Ministry of Interior. With this sale of weapons which did not belong to them, the leadership of Slovenia illegally gained material benefit," said an announcement by the Yugoslav Army's Ljubljana Corps published in the Slovenian newspaper *Delo* on August 27, 1991.

The same day, on the same page of the newspaper, the Slovenian Defence Ministry published a denial: "The leadership of Slovenia was not selling any YA weapons to anybody."[465] The denial was soon forgotten in the deluge of other, more dramatic news. However it was untrue.

Under questioning by a Parliamentary investigative commission in January 2000, Anton Peinkiher, head of VOMO intelligence department, affirmed that "in July 1991, a very secret decision was made by the Council of the Slovenian Presidency on Defence[466] to

---
464) August Bebel (1840-1913), a German Socialist, writer and co-founder of the Social-Democratic Party of Germany.
465) "The Territorial Defence is not selling YA weapons",    , Ljubljana, August 27, 1991.

offer Croatia help in weapons, so that the battle front would move outside Slovenia."⁴⁶⁷

That decision had a fateful impact on the events in Slovenia. Despite that, or because of that, there is no transcription of the audio recording of the meeting, while the minutes – if they exist at all – are still not available. Only a few people with a high degree of security clearance were able to see the document referring to the decision, and only for a few moments. An employee of the VOMO remembers the following words: "... that with the intention to deter the endangering of Slovenia, giving help to Croatia in weapons and military equipment ..." And also: "... that for the partial equivalent of sold weapons, we purchase new weapons and equipment ..."

The decision to help Croatia with weapons was apparently taken on August 26, 1991, when the Presidency of the Republic of Slovenia met in an enlarged session.⁴⁶⁸ Although there is no transcript of the audio recording of that session, the minutes show that a unanimous decision was reached confirming a decision that the Presidency's Council for Defence had adopted a month and a half earlier, which stipulated that Slovenia should help Croatia in the military area, but only if it would not endanger Slovenia's own capacity for defence.⁴⁶⁹

At the end of 1993, President Kučan stated: "When the war against Croatia started, the opinion of the Minister of Defence that helping Croatia is the best defence of Slovenia, was adopted. We all agreed with that. We also agreed with a viewpoint that we should help Croatia with weapons, but only to the extent which would not endanger defence capabilities of Slovenia."⁴⁷⁰

---

Slovenian Presidency for Defence monitors and evaluates defence preparations, advises the Presidency and helps it with carrying out duties regarding defence and protection. The Council is headed by the President of the Presidency, while its members are Presidents of the Parliament and the Government, Ministers of Defence, the Interior, as well as Commanders of respectively the TD and Civilian Protection, (*Official Journal of the RS*, no. 15, Ljubljana, April 6, 1991).
467) Minutes of testimony by Anton Peinkiher before the investigative commission of the Slovenian Parliament, no. 0610-12/99-00, Ljubljana, January 26, 2000.
468) Session (67th of a series) was chaired by President of the Slovenian Presidency, Milan Kučan, while the participants were members of the Presidency Matjaž Kmecl, Dušan Plut, Ciril Zlobec (Ivan Oman was absent), Prime Minister Lojze Peterle and Ministers of Foreign Affairs, Interior, Defence and Information, respectively Dimitrij Rupel, Igor Bavčar, Janez Janša and Jelko Kacin.
469) Minutes from the 67th enlarged session of the Slovenian Presidency, Ljubljana, August 26, 1991.

Thus, the sale of weapons to neighbouring Croatia was recommended by Defence Minister Janša, despite the fact that the Croatian authorities left Slovenia high and dry when the country was faced with an onslaught of Yugoslav tanks. Soon afterwards, war engulfed Croatia too and it needed Slovenian help with humanitarian aid, taking in refugees, reaching a cease-fire and diplomatic action. In that light, it is hard to understand the logic of sending weapons to Croatia and thereby becoming directly involved in a war in the neighbouring country.

Defence Minister Janez Janša, the Orbis Company and Interior Minister Igor Bavčar were selling weapons to Croatia already in 1990 – by agreement with the Ministers' Croatian counterparts, Špegelj and Boljkovac. And they were selling weapons again in July 1991, only a few days after the armed conflict in Slovenia ended. The decision of the Slovenian leadership thus accelerated the sales of weapons to Croatia which were already under way. Unbeknown to the citizens, the "arms tycoons" were taking over levers of power of the young, independent Slovenia.

Even though the Slovenian political leadership publicly emphasised that the provision of weapons was only in order to "aid a country that was attacked", the only real donation to Croatia was old commissary equipment not needed in Slovenia. All the rest – shells, ammunition, rifles and handguns of various calibres, anti-tank and anti-aircraft weapons – that is, everything found in the Yugoslav Army depots, was to be sold.

The Slovenian Presidency charged the Government with implementing the decision to give military help to Croatia, but Prime Minister Lojze Peterle had no authority to supervise the Ministries of Defence and the Interior. The secret state business of selling weapons was thus in the hands of those two Ministries, and their intelligence services led by Andrej Lovšin and Miha Brejc. In Parliament Peterle declared: "Slovenia is not sending weapons to Croatia because it doesn't even have enough of them for itself."[471]

In September 1991, Janša told the Parliamentary committee for defence that the sale of weapons was justifiable since it enhanced the defence capability of Croatia while the danger of a new attack on Slovenia had diminished.[472] It was not clear however that Slovenia's defence capacity was left intact. The sell-off of weapons, ordnance and ammunition from the former YA depots of Zgornja Ložnica, Borovnica, Šentviška Gora and Zaloška Gorica was done so thoroughly and without supervision that military experts started to warn of the risk of disarming the nascent Slovenian military.

While the Territorial Defence was standardising military equipment and getting rid of obsolete weapons, they received reports from the field that the Croatians were buying from Slovenia new types of weapons that Slovenia badly needed for its own defence.

In the Defence Ministry, the sale of weapons took place on a huge and chaotic scale. Most of the documents regarding quantity of weapons sold, prices and profit soon disappeared. To this day there is no indication where the money from the sale of weapons went.

The Slovenian political leadership made a big mistake in failing to set up civilian supervision over the Defence and Interior Ministries' implementation of the decision to sell weapons to Croatia. In July 1994, President Kučan said the leadership discussed the issue in detail and tried to learn more, but nobody ever made a detailed report to the Presidency, or to its Defence Council.[473] The Slovenian Presidency's good intentions thus degenerated and paved the way for the greed of a few and the shame of many.

The political burden of the ill-fated decision to "aid" the neighbouring country, regardless of circumstances, therefore also lies on the shoulders of Kučan and the other members of the collective Presidency; even though they did not enrich themselves from the sale of weapons, for a very long time they did not even know that the Defence Ministry was selling the weapons for cash.

"I admit I was naïve because I trusted Minister Janša that he would execute the task transparently and lawfully. We established this state together, and I thought that we could trust each other and that we would responsibly implement duties that the people

---

472) "Session of the defence committee of the Slovenian Parliament", *STA*, Ljubljana, September 24, 1991.
473) "What the President of Slovenia is discussing with his guests", interview with Milan Kučan,

entrusted us with after the referendum for independence. However, things turned out differently," said Kučan later.[474]

Kučan was the most important politician during the struggle for independence – a "crafty fox" with enormous experience. Despite his liberal inclination, he successfully navigated the rough seas of communist orthodoxy. Active politically since the late 1960's, when he became leader of the Slovenian Socialist Youth, he settled scores with the orthodox party elite – and survived politically.[475] To stay at the top, politicians in those gloomy times of the Yugoslav socialism used different skills, among them, the following, unwritten rule: Knew nothing, saw nothing, and heard nothing. Guilty of nothing! But was always there.

As the President of the Slovenian Presidency, Milan Kučan was the only professional politician in this body. Among other members, Matjaž Kmecl was a theoretician of literature, Ivan Oman a farmer, Ciril Zlobec a poet and Dušan Plut a geographer and ecologist. In February 2009, Plut agreed to be interviewed about the role of the Presidency and the way in made decisions.

"Do you remember the Presidency's decision on August 26, 1991, after a debate about offering help to Croatia with weapons?"

"Yes, we agreed that Croatia had to be helped in a military way. We decided that increasing Croatia's military strength also meant defending Slovenia."

"One month after that decision, the Security Council declared a weapons embargo on the whole territory of Yugoslavia. How did you …?"

Plut (interrupting): "Ahhh, I said that we had to help regardless of the embargo. But I was the first member of Presidency who asked the question about what was going on with the weapons. I think that Janša and Bavčar were not giving us enough information."

---

474) Author's conversation with Milan Kučan, Ljubljana, July 8, 2014.
475) Between 1968 and 1969, Milan Kučan was president of the Union of the Socialist Youth of Slovenia (ZSMS), until 1973 member of the Secretariat of the Slovenian League of Communists (ZKS) and until 1978, Secretary of the Socialist Union of the Working People (SZDL). Between 1978 and 1982, he was President of the Assembly of the Socialist Republic of Slovenia (SRS) and from 1986 on, a member of the Presidency of the Central Committee of the Yugoslav League of Communists. Before being elected President of the Slovenian Presidency in May 1990, he was President of the

"Did they give you any information at all?"

"They only said that everything was transparent and according to the agreement."

"Why did you …?"

Plut (interrupts angrily): "That's all I can tell you right now. At the Presidency, we didn't discuss this matter very much. Taking into account the international situation of Slovenia, this decision was very, very questionable."

"Questionable? From what point of view?"

"Mainly, from the perspective of international legitimacy."

"Are you aware that with this decision, the Presidency unintentionally also accelerated the sale of weapons to Croatia?"

Plut (very angrily): "I don't know anything about it … Ask somebody else …"

"In what way?"

"You're very unfair and misleading. You told me that we would be talking about the Presidency, but you are bothering me with weapons…I'm disappointed with you."

"I'm only asking about the consequences of the Presidency's decision."

Plut (a bit calmer): "I admit! We violated the legal norms, and our decision was outside the legal boundaries. However, helping Croatia didn't derive from the law but our morals. Slovenia too was arming itself illegally. Every nation has a right to self-defence."

"Nobody has a right to war profiteering."

Plut (red-faced): "Whaaat? Now I have enough of you. We're finished!"

"I didn't mean you. I only would like to know why you, as a member of the Presidency, didn't set up some supervision over the sale of weapons, when you already made a decision to sell it."

"I'm an anti-militarist. Not long ago, I signed a petition Abolish the Military. I only know that as a member of the Presidency, I was forced to act against my own value system … The discussion is over, please leave!"[476]

\* \* \*

After buyers of weapons started to come to Ljubljana in summer 1991, the upper floor of the monumental building on Župančičeva 3 for months looked more like a weapons bazaar than the headquarters of the Defence Ministry and the Security organ of the Ministry (VOMO).[477]

Among all the buyers from Croatia in the years 1991 and 1992, Josip Vukina – or Joža as his colleagues called him – was among the most frequent visitors. He was seen on Župančičeva so many times that some people thought he worked there. Vukina was a short, stocky man, from a town called Kačkovec in the Croatian region of Zagorje, not far from the birthplace of Marshal Tito, and very close to Slovenia.

Despite his apparent good nature and quaint Zagorje dialect, he was a merciless negotiator. At the beginning of the war, he bought weapons in Slovenia for his home region, and in autumn 1991 he became an advisor to the Croatian Defence Minister, with promotion to Brigadier.

"Almost all Croatian buyers of weapons in Slovenia were subject to intense interrogation by Croatian intelligence services on their return. They had to tell everything, and show all documents, so that security and intelligence officers could carefully record information about quantity, prices and sellers," explained Milan Zorko.[478] So Croatia holds incriminating documents. Common sense says that precious documents like this should not be destroyed. Most likely, they are stored somewhere deep in the Croatian secret service archives, which only a few chosen individuals may access. The *éminence grise* of the Croatian secret services in the 1990's was Miroslav Tuđman, son of President Franjo Tuđman.

In 1991, Croatia found itself completely unprepared for the war. President Tuđman rejected defence plans drawn up by General Špegelj and fired him as Defence Minister, but could not find an adequate replacement. When Serbian weapons started to be heard in Croatia, most of the people in towns and villages had to fend for themselves.

---

477) In the beginning of 1992, the Defence Ministry moved to the former YA hotel in Ljubljana's Bežigrad municipality. The Finance Ministry moved to the building on Župančičeva 3.

Therefore in front of Lovšin's and Janša's offices on Županičeva, a colorful group of weapons buyers from Croatia started to gather. Besides representatives of the Croatian military and Defence Ministry, people came from so-called "crisis centres" of Croatian municipalities and districts. All they needed to purchase weapons was an affidavit of some Croatian officer or an authorisation by a local Croatian leader that they needed weapons.

Money for purchasing the weapons was collected by priests in churches while Croatian immigrants also contributed sizeable funds to help their old homeland. Many people were ready to make great sacrifices to protect their homeland, but not all of them. Some Croatian "patriots" considered the war in their country mostly as a chance to enrich themselves. They were looking for ways to anchor themselves among the Croatian financial and political elite.

"I got the impression that the people Croatians were sending to Slovenia to obtain weapons were the greatest of scoundrels, who were then reselling them to different crisis centres in Croatia. To impress the Slovenian sellers, they showed business cards of important Croatian generals," said Criminal Police investigator Drago Kos, who investigated the illegal weapons trade.[479]

Among those "defenders" that Croatia cannot be proud was General Branimir Glavaš, commanding the defence of Osijek, a town in eastern Croatia, where he was also mayor. In September 1991 he boasted that Slovenian Defence Minister Janez Janša sent him a telefax in which he offered him his support and help, but only if it would go directly to Eastern Slavonia and there would be no involvement of the Croatian government.[480]

In autumn 1991, Glavaš secretly negotiated with Janša about purchasing weapons. His presence in Slovenia was confirmed by Colonel Peter Zupan[481] and by a Slovenian citizen living in Osijek, Jože Balent.[482] On December 9, soon after Glavaš and Janša met

---

479) Author's conversation with Drago Kos, Ljubljana, December 19, 2014.
480) "Branimir Glavaš - head of Osijek's defence," *Slobodni tjednik*, Osijek, September 21, 1991.
481) Official note of the VOMO investigators about the interrogation of Peter Zupan, MO, Ljubljana, September 23, 1994.
482) After this mediation, unknown individuals from Slovenia threatened Balent to "keep his mouth

in Ljubljana, four articulated trucks loaded with artillery grenades and ammunition, and accompanied by a member of VOMO, exited Slovenia at the Zavrč border crossing and continued towards Osijek.[483]

Subsequently however Drago Hedl, a journalist from Osijek, revealed that Glavaš had committed insidious crimes against Serbs, with the intention of inciting hatred.[484] Hedl also wrote of Glavaš's despotic leadership in Osijek and his brutal exploitation of the war for his own benefit. In May 2009, after a long court trial, during which he went on several hunger strikes, Glavaš was found guilty of war crimes and sentenced to 10 years in prison, but escaped to Bosnia-Herzegovina. In September 2010, the Croatian Supreme Court reduced his sentence to eight years and the authorities in Bosnia-Herzegovina arrested him and put him in jail in Mostar. Croatian President Ivo Josipović stripped Glavaš of his military rank as general and all the accompanying rights.

However, in January 2015, the Croatian Constitutional Court threw out the Supreme Court's verdict and Glavaš was a free man again. In March that same year, Glavaš was arrested anew, this time in Osijek, but he was soon set free again, and throughout this time he remained political active.

In elections of November 8, 2015, war criminal Branimir Glavaš was voted into the Croatian parliament, this time as representative of the Croatian Democratic Alliance of Slavonia and Barania (HDSSB), which he founded in 2005. This is the man with whom Janša did business in 1991.

Other seamy characters followed to Slovenia to pick up weapons and military equipment. Glavaš's colleague Drago Tadić was in 2014 found guilty by an Osijek court of accepting bribes and abusing his authority.[485] Zlatko Horvat, a former member of the crisis group from Varaždin, and later its deputy mayor, was indicted in the same year for abuse of authority and accepting bribes.[486] In September 1991, Horvat

---

483) Dispatch from the VOMO director Andrej Lovšin (no. 881/14-1181) containing a request to the VIS director Miha Brejc, to ensure that four articulated trucks with Osijek licence plates may enter Croatia at the Zavrč border crossing on November 9, 1991, MO Ljubljana, November 8, 1991.
484) See Drago Hedl's book: *Glavaš - kronika jedne destrukcije* (Glavaš - Chronical of Destruction), Novi Liber, Zagreb, 2010.
485) From a report of        /Croatian Press Agency

helped Franci Cimerman, a Slovenian ministry clerk, to procure an imported car at a very favourable price by acting as a go-between.[487]

Another such buyer was Josip Brajković, from Sisak. During the war, he systematically robbed, tortured and killed Serbian nationals in the Sisak-Moslavina district. Brajković prepared lists of people for private military "death squads" to kill.[488] According to incomplete information of the Croatian Interior Ministry and non-governmental organisations, several hundred Serbian nationals "disappeared" in the area of Sisak between 1991 and 1992.[489]

Such "horse traders" from Croatia saw the distress of their countrymen as an opportunity for personal benefit. They bought weapons, ammunition and ordnance in Slovenia and offered them to crisis headquarters in Croatian municipalities. Desperate people sold livestock, cars, furniture and other belongings to contribute to the defence of their homeland, but much of their donations ended in the pockets of arms traffickers.

\* \* \*

Slovenian Defence Minister Janez Janša, who at the time was 33 years old, had little experience, but he did things on his own – often without the approval or knowledge of the Slovenian Government, Presidency or Parliament.

In September 1994, during an internal investigation, Milan Zorko explained that the selling and purchasing weapons happened in different ways: "With the knowledge of the Slovenian leadership and the Minister of Defence, only with the knowledge of the Minister of Defence, or only with the involvement of the VOMO director or other key VOMO officials."[490]

According to Zorko, Janša was pulling all the strings: "The Minister was negotiating the biggest deals with Croatian buyers. He let Lovšin handle only those deals that were worth less than one million German marks, and he also allowed him to handle things

---

487) Import Declaration, Zagreb Customs Office, September 23, 2011.
488) In 2002, Josip Brajković committed suicide.
489) Taken from: "Đuro Brodarac died: our mini Srebrenica is his work",

in his own way. Janša realized that it was beneficial to him to be lenient with Lovšin, because this is how he bought his loyalty. In the VOMO, the only person who dealt with cash from weapons sales was Anton Peinkiher. Lovšin's clerk Franci Cimerman recorded transactions and carried away suitcases, sacks and wooden crates in which they were keeping foreign currency."[491]

Peinkiher, head of the intelligence department and one of the people ordered by Lovšin to count the money from selling weapons, stated: "I received detailed instructions as to how to do my job and also an order to operate without any documents and within the framework of helping Croatia and Bosnia-Herzegovina, in order to keep the war away from Slovenia."[492]

Lovšin and Peinkiher had been classmates at the Military Academy in Belgrade. Lovšin entrusted Peinkiher with a very delicate and risky business, and with time Peinkiher figured out that it was illegal. He fell out of favour with Lovšin and in autumn 1993, he was suspended, transferred and demoted. Janša accused him of leaking information about the weapons business to the public. Peinkiher fell victim to malicious gossip and internal disciplinary actions, but they could not prove any charges against him. Janša's removal from the position of the Defence Minister in 1994 brought Peinkiher some relief, but he could not return to his old job, and eventually he took employment at the Slovenian mass retailer Mercator.

During an internal investigation at VOMO in 1994 and questioning before the Parliamentary investigative commission in 2000, Peinkiher heavily incriminated Lovšin and Janša. He testified that he counted cash received from selling weapons and handed it to Lovšin, who kept it in a steel safe in his office. When Lovšin stepped down from the position of the director of VOMO, he passed the safe and documents to the commander of Moris Brigade, Anton Krkovič, who took everything to the Škrilj military hideout deep in the Kočevje forests.[493]

---

491) Author's conversation with Milan Zorko, Ljubljana, November 17, 2014.
492) Statement of Anton Peinkiher, no. 881-600, VOMO, September 21, 1994.
493) Testimony of Anton Peinkiher, records of the 10th session of the investigative commission,

Anton Peinkiher told all this to members of the Parliamentary investigative commission. However, the sound recording suddenly disappeared. The Secretary-General of the Parliament admitted the following day "that a mistake happened because of the carelessness of workers in the recording room".[494] Using the minutes from the interrogation, the Committee Chairman Rudolf Moge prepared a summary of the interrogation and forwarded it to the witness Peinkiher for his review.

Franci Cimerman from Grosuplje worked for VOMO as a courier, driver and clerk, and as such was Lovšin's "gofer". Amongst other things, he stacked cash into briefcases and took them over the Karavanke Mountain to Austria. His colleagues say that he deposited money in five different Austrian banks, among them the Kärntner Sparkasse and Bank für Kärnten und Steiermark. Using the account in this bank, he paid for the purchase of 21 cars for VOMO employees,[495] while in autumn 1992 he used the account to pay for handguns purchased in Italy.[496]

From Austrian banks, Cimerman paid bills for merchandise coming to Slovenia through the British Company Hillbrook, the co-owner of which was Lovšin's sister, Nada Ilc Mortin. Cimerman was also the owner of a British shell company, IPE.[497]

In November 1991, Cimerman asked a fellow-employee to help him take 17 million German marks out of the Lovšin's safe and count them. He then picked up all the cash, stacked it in a big bag and took it to an unknown location. Even after all that money was taken out, the safe in Lovšin's office still "contained a lot of cash in different currencies".[498] During the move of the Defence Ministry within Ljubljana at the beginning of 1992, Cimerman put 30 million German marks in cash in five weapon crates, which were then taken

---

494) Report of Jožica Velišček, Secretary General of Slovenian Parliament, Ljubljana, January 27, 2000.
495) Money transfers to car dealers in Austria from the Bank für Kärnten und Steiermark in Celovec (Klagenfurt), September 17, 1991. Total amount of transfers was 3,576,250 Austrian Shillings (ATS). At the time, one German mark was worth about seven ATS.
496) Money transfers from the Bank für Kärnten und Steiermark in Celovec to Cassa Rurale ed Artigiana in Ločnik (Lucinico) near Gorizia in Italy. The first transfer on December 7, 1992 amounted to 40 million Italian lire and the second on December 9, 1992 amounted to 29 million lire. At the time, one German mark was worth approximately 1.000 Italian lire.
497) The company Industrial Project Engineering (IPE) was founded by Lovšin and Cimerman. The company was used for importation of an advanced American audio surveillance device to be used by VOMO.

to a new location.[499] Dragutin Mate, a counter-intelligence official at VOMO, claims that more than 42 million German marks were actually in those crates at the time of the move.[500]

During questioning, Cimerman experienced severe "amnesia". In January 2000, testifying before Parliamentary investigative commission, he could not remember any cash. He said that he was "only following Lovšin's orders".[501] His red notebook, where he diligently recorded quantities, prices and amounts of weapons sold to Croatia, was nowhere to be found. At the time of writing, Franci Cimerman was still employed at the Defence Ministry. After promotion, his last job was as an under-secretary.[502]

Andrej Lovšin from Ribnica, who in summer 1990 could afford only a rickety car, soon became a part of military structures. The following May, he was a captain, and within two years a Brigadier. However, he was by then indifferent to his high military rank, since in the meantime he had become a wealthy businessman. In autumn 1991, he was driving an expensive Audi 100 Quattro, building a large house in Ljubljana and was buying apartments and land.[503] The director of the military intelligence service had more money than most of the people in Slovenia. However, he was sternly emphasising to VOMO employees: there is no selling of weapons for cash! All insinuations about it should be forcefully denied. In public, he denied any connections with people who came to his office and gave him money for weapons. During questioning in courts and before Parliamentary investigative commission, he too experienced serious amnesia.

In January 1994, Lovšin stepped down from his position as VOMO director and left the Defence Ministry. His sudden departure prompted much speculation. Former colleagues claim that Janša tried hard to keep him and offered him a promotion, but Lovšin would not listen. Mitja Klavora, former head of the Slovenian Criminal Police, thinks differently: "Janša forced him out because Lovšin took too much money

---

499) Testimony of Bojan Babič in front of the VOMO investigators, no. 881-600, Ljubljana, September 19, 1994.
500) Dragutin Mate, a report on the sale of weapons and the money, made by VOMO, Ljubljana, January 14, 2000.
501) Testimony of Franci Cimerman, transcription of the 8th session of Moge's investigative comission, Ljubljana, January 14, 2000.

from the sale of weapons, after which Lovšin felt offended and complained to people of Drnovšek's circle."[504] If so, whatever Prime Minister Janez Drnovšek heard does not seem to have disturbed him much.

Twenty years later, during the first government of Janez Janša, Lovšin became director of the logistics company Intereuropa, headquartered in Koper. His three years at the helm of the company were marked by bad business decisions and shady deals, and in 2009 he was fired. Five years later, a Koper prosecutor indicted Lovšin, accusing him of abusing his authority during an Intereuropa investment in Moscow's Chekhov logistics terminal. Intereuropa filed a lawsuit seeking compensation for damages of 37.5 million euro arising from bad investments in Russia and Ukraine.

In the meantime, tax inspectors discovered 1.88 million euro worth of unreported assets belonging to Lovšin, his mother, and daughter. In June, 2015, a special prosecutor filed a lawsuit against all three, demanding confiscation of allegedly unlawfully obtained assets: two hotel rooms and an apartment in the Slovenian seaside resort of Portorož, two apartments in Ljubljana, a vehicle belonging to Lovšin daughter, shares of Intereuropa and other property, altogether worth almost 900,000 euro.

\* \* \*

Let's go back to the military Security organ (VOMO), where at the beginning of the 1990s they figured out a cunning way of selling weapons to Croatia and erasing all traces.

Colonel Vladimir Milošević remembers that Peter Zupan, assistant commander of the Slovenian Territorial Defence, received from Lovšin lists of types and quantities of weapons and ammunition to be handed over. "On these papers, there were no prices for weapons. In our facility, the warehouse employees only checked and confirmed items that were available and then sent the lists back to Ljubljana. Based on the confirmed quantities of weapons, Lovšin started to bargain with the Croatians about prices. Only after Lovšin received the payment would he issue a permit to Croatians to take possession of weapons from our warehouses. On documents that were kept, we can only find Croatian signatures and signatures of Colonel Zupan and individual officers who were in charge of issuing merchandise from warehouses. They had to return one copy of the document to VOMO, while the second copy stayed with the regional Territorial Defence headquarters. Lovšin and his people from the intelligence service were very careful not to leave any fingerprints."[505]

In autumn 1999, Peter Zupan testified to the Parliamentary investigative commission that the Croatian military was receiving weapons from depots, but added that three conditions had to be met beforehand: the permit had to be issued by the Defence Ministry or VOMO; an arrangement had to be made to open the border with Croatia for their transit; and weapons had to have been paid for. "Lovšin himself demanded that," declared Zupan.[506] Peinkiher is convinced that Lovšin was following orders from Defence Minister Janša, who "mostly set weapons prices and only a few times would he allow Lovšin to do it".[507]

Miloševič remembers that "Lacković, my friend from Varaždin, who was coming to VOMO to buy weapons, told me in confidence that one particular time he threw a stack of foreign currency on Lovšin's desk and Lovšin only grabbed it and put it into his steel safe, after which the astonished Croatian asked him: – aren't you going to count it? – Lovšin answered that if only one German mark were missing, the next time he wouldn't get a single bullet. –"[508]

Milan Zorko of VOMO believes Lovšin was making extra money with blackmailing: "When everything was already agreed and the Croatian buyer brought the requested amount of money, the VOMO director would all of the sudden come up with a higher price: – If you don't pay more, you won't take the weapons with you –, he would say."

Zorko remembers Croatian Josip Vukina sweatily wringing his hands as if he was going to have a heart attack. He was shouting

---

505) Author's conversation with Vladimir Miloševič, Murska Sobota, May 20, 2014.
506) Record of the investigative session of the Slovenian Parliament, headed by Rudi Moge, Ljubljana, November 3, 1999.
507) Official note of the VOMO investigators about the conversation with Anton Peinkiher, no. 881-

how he was going to justify the new price to his superiors. However, Lovšin would not budge – and the Croatian officer had to add 20,000 more German marks, which Lovšin put into his pocket."[509]

Once Vukina exclaimed to a VOMO official: "You Slovenians are bastards. You're taking advantage of the fact that we cannot buy weapons abroad ourselves, and you're ripping us off. You're selling us Zolja missiles for 2,000 marks, even though on the market they sell for 300 marks."[510]

Another of those peddling weapons to Croatia was Ludvik Zvonar, Advisor to the Government and logistics director at the Defence Ministry. In his office, people often saw large quantities of cash. According to two Defence Ministry officials, in autumn 1991 one of Zvonar's aides secretly drove towards Croatia in a truck loaded with anti-tank mines and rifle grenades. Since nobody met him at the border crossing, he decided to leave his consignment in the hay barn of a nearby farm and call the Croatian buyers the following day to pick it up themselves. However, during the night there was a terrible thunderstorm, and the aide could not sleep a wink for fear that lightning would strike his dangerous cargo and destroy everything in sight.

The Slovenian Territorial Defence never presented any detailed inventories of weapons available to them, or sold to Croatia. When tons of weapons "disappeared", nobody was held responsible. Nobody requested documentation from the Defence Ministry or the Headquarters of the Territorial Defence, and nobody ever asked any questions.

Only a few chosen individuals knew at the time that all those weapons – that belonged to Slovenian citizens – were sold to Croatia and that most of the revenue was kept secret. Even those who knew kept quiet, because they were themselves implicated, or they simply looked the other way.

The weapons trade in Slovenia proved the old saying: if you steal, steal a lot. With the proceeds, you will be able to buy everything – even your innocence.

---

509) Author's conversation with Milan Zorko, Ljubljana, November 12, 2014.
510) Statement of a VOMO employee (without a signature), requested by Defence Minister Jelko

\* \* \*

At the beginning of November 1991, the governments of Slovenia and Croatia reached an agreement on offsetting operations called Weapons for Oil. Josip Vukina, who was the main executor of the plan on the Croatian side, later admitted that most of the weapons deals between the two countries continued to be done by money transactions from one account to another.[511] When Croatians were loading a large quantity of weapons from the Ložnica depot one particularly foul night, Vukina laid his hands on a secret contract between the two Defence Ministries, signed on the Slovenian side by Lovšin. Vukina put the document in a black folder, and when he shook Lovšin's hand, he had a broad smile on his face. Lovšin, however, was concerned that Vukina got possession of something that he should not have had, and sought help from his intelligence officers.

That evening Vukina drove to Ljubljana for dinner, and after returning to his motel on the outskirts noticed that the folder he left on the back seat of his Volkswagen Golf was gone. When he returned to VOMO headquarters, criminal police officers who showed up said that a professional had broken into the car. Vukina complained that publication of the secret documents could damage both Slovenia and Croatia.[512] However, the Slovenians were unconcerned, since the two people who took the folder and its document from Vukina's car were officials of the military Security organ (VOMO).

Vukina did not wait long to take revenge on his "business partners" in Slovenia. "He was coming to the Defence Ministry offices, until we received information that he had copied the lists of all employees of VOMO and forwarded them to Croatian intelligence," stated VOMO employee Žarko Henigman.[513] The theft of the internal phone book left carelessly on the duty officer's desk was probably the last action of Josip Vukina in the Slovenian Defence Ministry.

---

511) Official note of the Criminal Police officers on a conversation with Josip Vukina in Dobova, UKS MNZ, April 7, 1995.
512) Official note of the Criminal Police officer about his conversation with Josip Vukina, UKS, MNZ, Ljubljana, November 5, 1991.
513) Statement of Žarko Henigman, no. 881-600, VOMO, Ljubljana, September 8, 1994.

The weapons purchases in Slovenia brought Vukina promotions to General and great wealth. He built an opulent villa near Zagreb and bought several expensive cars, but Slovenian police had been on his tail since autumn 1991. In April 1994, he was arrested on an international warrant on suspicion of illegal weapons trading. However, he was released the same day, and his international arrest warrant was cancelled, probably at the request of the Slovenian Prime Minister of the time, Janez Drnovšek.

Vukina told Slovenian police that all the documentation regarding weapons trading with Slovenia that he had been systematically collecting all those years was safely stored in at least three separate locations. Among them he said was the copy of a receipt for 321,000 German marks with Lovšin's signature.[514] If what Vukina says is true, his cache of documents represents a valuable archive about the darker history of the Croatian-Slovenian relationship. According to the Croatian historian Davor Marijan, there are very few written records about the weapons trade from that era, and these are kept in the central military archives in Zagreb.

\* \* \*

Janša wanted to leave nothing to chance, so he ordered Lovšin to collect from the central and regional headquarters of the Territorial Defence all documents regarding transfers of weapons: registers, return receipts, material lists, removal orders, records of material asset transfers – basically everything that could show that weapons disappeared to an unknown place.

Colonel Milošević remembers: "General Slapar and I found ourselves out of favour with Janša. Slapar even received threats. Then he called me and said: – Vlado, I'll give you the documentation about weapons in Gorenjska, and you give me yours about weapons in Štajerska. – We agreed and made the exchange."[515] Before retiring in autumn 1994, Milošević handed over his documentation to the Military Intelligence and Security Service. (OVS) and the new Defence Minister, Jelko Kacin.

---

514) Official note of Criminal Police officers on their conversation with Josip Vukina in Dobova, UKS,

Officials of the Interior Ministry also remember that Lovšin's security people were looking at various police stations for dispatches about convoys of trucks loaded with weapons that crossed into Croatia. At police headquarters in Krško, near the motorway linking Ljubljana with Zagreb, they suspected that Janša was up to something, and figured that they had to keep those documents exactly because they were not allowed to keep them. They kept copies of numerous dispatches, which were a valuable source of information when they came into the hands of investigators of the weapons trade.[516]

Among documents are also notes referring to movements of Croatian military across Slovenian territory, which most likely resulted from an agreement that the Slovenian Defence and Interior Ministers reached on their own. By allowing foreign military to move through Slovenia, they violated basic tenets of international law and needlessly exposed Slovenia to danger from the Yugoslav Army and Serbian paramilitary forces.[517]

\* \* \*

"Between July and October 1991, the Republic's Coordination Group managed the border crossing of trucks with weapons and military equipment," wrote Janez Janša in his report on weapons in 1993.[518] At that time, most of the trucks were driving to the border over remote and badly maintained country roads, in order to avoid witnesses. There are no reports that European Community monitors in Slovenia ever noticed anything suspicious.

On October 7, 1991, the three-month Brioni moratorium expired, and Slovenians immediately placed mobile offices and Slovenian flags on border crossings. For the arms traders from the Defence

---

516) In April and May, 2008, the author of this book, and a journalist, Matjaž Frangež, (separately) requested from the Slovenian Parliament and the Defence Ministry documentation about weapons trading. After a long procedure, the Information Commissioner removed the classification designation from a large portion of requested documents. Therefore most of dispatches of the Police Headquarters in Krško, the Novo mesto's Office of the Security and Information Service (VIS) and some other documents about organised weapons transports to Croatia became publicly available.
517) It refers to a violation of the Hague Convention (V) about the rights and obligations of neutral states during war.

Ministry, the presence of possibly inquisitive customs officers at the makeshift border crossings was unwelcome.

At the Defence Ministry, they quickly found a way to overcome potential problems. Andrej Lovšin henceforth wrote a request for each border crossing, with one copy going to the Operations-Communications Centre (OKC) at the Interior Ministry. In his request, he stated the licence plate numbers of the trucks, the name of the border crossing and the time when it should be opened for certain trucks. Normally, that was late at night or early in the morning. Interior Ministry officials made sure that the weapons were paid for and that they crossed into Croatia without problems.

In only one month between October 28 and November 29, Lovšin issued 34 such requests. In that time, 87 trucks loaded with weapons, ammunition, ordnance and other military equipment crossed into Croatia.[519] Janša's Defence Ministry and Bavčar's Interior Ministry were operating in unison. However, then another problem appeared.

The import, transit and export of weapons into and out of the country was regulated by Article 45 of the Law on monitoring the state's border, which stipulated that weapons were allowed to be transported across the border only with a permit of the Interior Ministry, issued after positive opinions by the Ministries of Foreign Affairs and Defence. This was too cumbersome for Janša and Bavčar, so they decided to circumvent it with a new arrangement. They summoned a "coordination group" of officials of their two Ministries, and in meetings held on November 15 and 20, 1991, decided to "simplify" the law and eliminate the Ministry of Foreign Affairs from decisions about the arms trade.[520]

The two strongmen increased their power by means of that decree. It's not known why the Slovenian Government and Parliament kept quiet when they were faced with this "soft" *coup d'état*. Nobody mentioned that it constituted an illegal abuse of power.

---

519) Thirty-four requests of the VOMO Director Andrej Lovšin to the Security and Information Service (VIS) and to OKC MNZ, MO Ljubljana, from October 28 to November 29, 1991.
520) Minutes from the meeting of the Coordination Group of the Ministry of Defence and the Ministry of Interior, Ljubljana, November 15 and 20, 1991. Besides Ministers Janša and Bavčar, the following people were present: Deputy Defence Minister Miran Bogataj; Head of the Republic's Headquarters of the TD, Major-General, Janez Slapar; VOMO Director Andrej Lovšin; Director

- Chapter 13 -

# YUGOSLAV ARMY PULLS OUT: SLOVENIA RECOGNISED

On September 25, 1991, the United Nations was directly involved in the war in the Balkans for the first time. The Security Council unanimously adopted resolution number 713, which stipulated that "with the intention of establishing peace and stability in Yugoslavia, all countries shall immediately set in force a universal and comprehensive embargo on all shipments of weapons and military equipment to Yugoslavia".[521]

The resolution was contrary to the U.N. Charter, which recognises a natural right for individual and collective self-defence in case of armed attack.[522] It had a particularly adverse affect on Croatian defenders of their homeland, but benefited the Serbs, who received the majority of weapons from the Yugoslav military, including their artillery, air force and navy. Serbia also kept the majority of foreign exchange reserves in the National Bank of Yugoslavia, which it used to buy weapons from Russia.[523]

"The Slovenian Presidency decided to offer help to Croatia almost a month before the imposition of the U.N. embargo, and as Slovenia became a member of the United Nations on May 22, 1992, the U.N. Security Council Resolutions were not binding for Slovenia," stated Milan Kučan.[524]

It is true that in a formal sense, selling weapons to Croatia could not be considered a violation of the U.N. embargo, because at the time neither country had been internationally recognised. One way or another the Slovenian Defence Ministry was buying weapons abroad after September 25 and selling a large portion of them to the Croatians.

---

521) Resolution no. 713, U.N. Security Council, New York, September 25, 1991.
522) Article 51/Paragraph VII of the U.N. Charter.
523) Until September 1994, Russian delivered to the Serbs 83 122-mm Howitzers and several hundred railway cars loaded with other weapons. Source: *"Schweizer Käse"*, Der Spiegel Hamburg,

The U.N. ban on importation of weapons meant that the weapons merchants in the Slovenian Defence and Interior Ministries now had dubious legality, and they began acting as a state within a state, independently from other institutions in Slovenia. The embargo suited them well, since from then on all weapons channels had to be concealed, the number of intermediaries increased, prices went up, and the traders could look forward to higher profits.

Sensitive deals were supposed to be monitored by intelligence services of the Defence and Interior Ministries but, as it turned out, the intelligence services were the very entities most involved in weapons trading. In November 1991, the opposition Liberal Democratic Party warned that VOMO was acting without any Parliamentary oversight. Its members of Parliament requested that the Parliamentary commission for monitoring the Security and Intelligence Service (VIS) should perform its job. That this never happened had very negative consequences two years later.

Foreign intelligence services, among them the CIA, tracked the shipments of weapons coming into Slovenia, and moving from Slovenia to Croatia. In October 1991, Cyrus Vance, former U.S. Secretary of State and Special Representative of the U.N. for Yugoslavia, threatened Slovenia with economic sanctions because of its illegal importation and reselling of weapons, while at the same time warning Serbia that it could face sanctions for its violations of the ceasefire in Croatia.[525] The United States obtained information that Slovenia was violating the embargo by importing arms from Austria and Hungary. American intelligence officers found out that the main business centre for importation of weapons to the Balkans was Vienna. That was the operational base of Konstantin Dafermos, the principal arms supplier to the Balkans. His Scorpion International Services company had a bank account in Budapest, Hungary, where money came in to pay sellers of weapons from Eastern Europe.

On October 24, 1991, this issue was the main topic of a session of the Slovenian Presidency. Defence Minister Janša denied that "Slovenia imported anything from Austria or Hungary". In a sense, he was

---

525) Božo Repe,                (Tomorrow is Another Day),

telling the truth, because weapons were coming to the Slovenian port of Koper from Bulgaria, Poland, Romania and Ukraine. However, under questioning by members of Presidency, Janša admitted that Slovenia "had been importing weapons all along. No resolution forbade that, not even the Brioni Declaration. Things will change, when we'll be internationally recognised, and we'll become members of the U.N. Then, this question will be much more delicate, and Slovenia will not have any excuse anymore".[526]

The international mood after the Brioni agreement was not favourable towards Slovenia, since most foreign diplomats, analysts and observers were still watching the disintegration of Yugoslavia from Belgrade's point of view. Foreigners dealing with the situation in former Yugoslavia had a hard time figuring out complex Balkan relationships. Their findings were often dependent on their liking of certain individuals or nations, as well as on the hospitality or problems they experienced in various parts of the Balkans.

From July 1, 1991 until the end of the year, the Dutch Foreign Minister Hans van den Broek, who was leading the European Community's efforts to deal with the Yugoslav crisis, was bothered by the fact that he could not see Yugoslav flags at Slovenian border crossings, and openly showed affinity for Serbia and Milošević. Van den Broek was influenced by Britain, France and the United States, which at that time tried hard to keep Yugoslavia unified, and when their efforts failed, defended Serbia as self-proclaimed successor of Yugoslavia.

In autumn 1991, Serbian leader Slobodan Milošević was at the height of his power, and some people even wondered if he was acting in sync with American interests in the Balkans. At the end of 1970's, Milošević had been director of the New York office of the Yugoslav bank Beobanka, establishing contacts with influential American political and economic representatives.

Milošević's popularity rose, as paramilitary gangs and the remnants of the Yugoslav Army working now to further Serbian nationalistic goals, won one victory after another against the confused and poorly armed Croatian defenders. A delirium of conquest swept Serbian public

---

526) Recording transcription of the session of the Presidency of the Republic of Slovenia, October

opinion, silencing any voices of domestic dissent, and emboldening Milošević to scorn peace initiatives by international mediators.

Even after the Brioni agreement, the Yugoslav military kept the air space over Slovenia closed, but Kučan and Drnovšek insisted that they wanted to fly to the meeting of the top Yugoslav leadership in Ohrid, Macedonia. Their request was successful, and their flight broke the embargo on air movements: four days later the main Slovenian Airport of Brnik could reopen for all traffic. But Slovenia was not out of the wood yet. Hints came from Brussels, London and Paris that the three-month moratorium on Slovenian independence could be extended, and there was no talk of recognising Slovenia or Croatia.

In that spirit, an International Conference on Yugoslavia opened in The Hague on September 7, 1991, chaired by British diplomat Lord Peter Carrington.

The points of view of the representatives of former Yugoslav republics about the future of Yugoslavia were diametrically opposed. The Slovenian delegation – after an initial dilemma whether to participate in the conference at all – adamantly insisted that it was impossible to revoke the declaration of Slovenian independence. It expected its independence to be recognised, and that the conference should bring about a peaceful disintegration Yugoslavia.

The west Europeans were divided in their views too. While most sought to keep Yugoslavia together, Germany saw the main evil in Serbian aggression and let it be known it was considering recognising both Slovenia and Croatia. German Foreign Minister Hans-Dietrich Genscher knew that Slovenia was in a much better position than dismembered Croatia. At the same time, he promised that Germany would not act hastily and would be looking for a solution within the framework of the European Community.

Lord Carrington and van den Broek hoped to put the Yugoslav republics back into a loose customs, monetary or some other similar union. However Milošević turned down every proposal that did not accept changes to borders of the republics. When on November 5, 1991 Milošević also turned down a fourth peace plan, Lord Carrington had had enough and adjourned the conference.

\* \* \*

While foreign countries continued their futile attempts to tame the Balkan dragon, Slovenia continued implementing its plan for independence. Slovenian recruits still serving in the Yugoslav Army returned home in the days following the Brioni agreement signed on July 7, 1991. Slovenia skillfully used the three-month postponement of its independence that was required by Brioni for negotiating with the YA about how and when its soldiers and officers would leave the country, what it would leave behind, and what it would take away.

The political decision that the YA would withdraw from Slovenia was taken, but many Yugoslav soldiers reacted with provocations. In summer 1991, unknown perpetrators used military explosives to blow up power lines in the towns of Zlatoličje near Drava and Krško, and in Maribor they blew up a petrol station. Elsewhere YA soldiers destroyed a barn with incendiary ammunition, and residents in the area of the Cerklje military airfield complained that the YA was provoking them. When the last Yugoslav soldiers evacuated the airbase on October 22, somebody threw a smoke bomb. The wind carried the gas cloud to the nearby village and as somebody put it, caused Slovenians to shed their only tears for the Yugoslav Army's departure.

At the end of July 1991, a Slovenian commission headed by Miran Bogataj, the Assistant Defence Minister, and a commission of the 5$^{th}$ Army Region headed by Lt-General Andrija Rašeta started negotiations about details of the YA pullout from Slovenia. Observers from the European Community were also present.

However these were negotiations between people who had been enemies only a few days before. The Slovenian side treated the Yugoslav officers as losers who should be pushed into a corner, humiliated and expelled from the country. Miran Bogataj was an arrogant, pedantic and merciless negotiator who witnesses remember many times pushing General Rašeta into despair. Legally well-versed and sporting a bushy moustache, he was a chain smoker, and uncaringly blew smoke into the face of the General, who had only recently quit smoking.

Fortunately for the Slovenians, the negotiator on the other side of the table was a crude, bad-tempered man, who at certain moments completely lost his head. Meanwhile Bogataj received information from Slovenian agents in Belgrade indicating a strained relationship between the top YA leadership and the conceited Rašeta, a Serb born in Croatia.

During the first meeting of the negotiators on July 26, 1991, both commissions reached a tentative agreement that the Yugoslav Army would give back to the Territorial Defence weapons it confiscated in May 1990, while Slovenia would return to the YA seized weapons and other assets.[527] However, before long problems appeared, because the YA insisted that it would return weapons to the TD only when the Army left Slovenian territory. On August 3, Bogataj answered that the "proportion of forces" between the two militaries should be maintained and "Slovenia would be returning to the YA military assets and weapons at the same pace as the YA returned weapons to the TD".[528]

The negotiators thus got themselves in a Catch-22 situation, whereby each side demanded that the opposite side return confiscated assets, and stated that only then would it do the same. Slovenian negotiators believed that weapons which the YA confiscated from them would be theirs sooner or later, because there was no way that the YA could take these weapons with them to Serbia and Bosnia-Herzegovina.

They thus drew the Yugoslav officers into an ingenious trap. To delay Yugoslav realisation of the ploy, Bogataj's team proposed that the Slovenians would also begin to return non-lethal military equipment, commissary equipment and ordnance. So the YA commission could at the beginning of August report the handover of individual items to the Yugoslav Army command in Zagreb.

The lengthy lists specified long and short underwear, pyjamas, winter and summer pants, cooks' uniforms, various socks, blankets, backpacks, masking equipment and military belts, as well as canned food, liver paste, white beans, green beans, peppers, goulash, dry soup mixes, salt and spices.[529] In the middle of August, the YA handed

---

527) Findings of the Commission of the Republic of Slovenia and the Commission of the 5th Region of the Yugoslav Army, Ljubljana, August 9, 1991.
528) Recap from final decisions stemming from negotiation records between TD and YA commissions. The author has in his possession the record of these decisions.
529) From a report to the 5th Army Region, no. 1079-2/31, Ljubljana, August 1991. The author has in his possession voluminous YA documentation on negotiations, taking place between July 24 and

over additional demands for return of military technical assets. The bureaucratic precision in the drafting of these lists was amazing, as were the lists they also prepared of destroyed or damaged military assets during the armed conflict. To make the handover last as long as possible, the Slovenian side also included museum artifacts from so-called memorial rooms in various garrison headquarters.

Just as the Slovenians intended, the YA side ran into time problems as the days counted down to October 18 – the deadline for the retreat of the YA from Slovenia. On the Slovenian side, nobody was in a hurry and members of the negotiation commission sometimes felt that during meetings, the main negotiator Bogataj was even talking slower.

At the collection sites in the Yugoslav barracks in Ljubljana, Maribor and Celje, assets that were to be returned to the YA were slowly coming in. Handed over were several dozen artillery pieces and mortars, smaller quantities of rifles, radios, several tons of explosives and shells, an occasional old military truck – and an incredible amount of military uniforms.[530] To a Yugoslav armoured brigade were returned some old, stripped-down T-55 tanks and a damaged M-84 tank. The Slovenians did not return the Gazelle helicopter that pilot Jože Kalan and air-technician Bogomir Šuštar used to escape to the Slovenian side. The YA commission proposed in vain that they should "look for it together".[531]

Yugoslav Army barracks in those days looked like anthills, as the military prepared to depart. Under the watchful eyes of tripartite commissions consisting of the TD, the YA and the Slovenian police, they cleared out border posts and billets.

In railway stations, long trains of railway wagons could be seen, loaded with military equipment and ready to move away. Slovenian people were astonished at the amount of material the Yugoslav Army had accumulated over the decades, thought what they saw was only a small proportion of everything the YA really owned.

In those feverish days of August and the first days of September, 1991, the first convoys left Slovenia. On September 13, Janša informed

---

530) Colonel Nikola Stajčić's report to the Command in Zagreb, no. 1079-2/106, from October 19 shows that the YA took away from Slovenia about 200 non-combat vehicles.
531) Conclusions from the meeting of the Commission of the 5th Army Region and the Commission

the Slovenian public that remaining in Slovenia were less than 4,000 members of the YA, among them 500 Special Forces, only 70 tanks and 30 armoured vehicles. According to his estimate, that was "only one tenth of the firepower that was at the YA disposal during the attack on Slovenia, so it did not represent a real danger to the country".[532] During following weeks, another 1,000 Yugoslav Army soldiers left Slovenia, leaving behind weapons, military equipment and other belongings.

Then the question arose: what to do with the equipment belonging to the YA remaining in Slovenia? It was plain that anything further sent back would be used in the Serbian repression in Croatia. Dušan Plut, a member of the collective Slovenian Presidency, suggested that "armoured vehicles and other assets that the YA still had in Slovenia should be kept in the country until the cessation of hostilities in Croatia, or until a creation of some kind of a distribution system for dividing Yugoslav assets among former Yugoslav republics". He stated: "We are obliged to help Croatia despite the fact that it might cause us problems. The excuse that Croatians did not help us in June is not valid, because the circumstances in Croatia now are significantly different than they were in Slovenia."

Other Slovenian leaders pointed out that it was known all along that the arms and equipment the YA took out of Slovenia would be used by the Yugoslav military in Croatia. They were also concerned how the YA leaders in Belgrade might react, in particular since keeping hold of the military equipment violated a provision of the Brioni agreement.

The dilemma was solved by France Bučar, President of the Slovenian Parliament. He recommended that the military assets[533] would simply stay in Slovenia, without the Slovenian leadership telling anybody in advance that it was not going to hand them over to the Army: "I would play dumb and tell them: – You still haven't taken all this stuff away even though you promised you would. There aren't any good options available to you anymore to get what you want –. We should pretend to cry from all the pain we suffer because we couldn't give them all the cannons."[534]

---

532) "Janez Janša thinks that Slovenia is not faced with danger", *STA*, Ljubljana, September 13, 1991.
533) Military technical assets encompass all items that are used for military purposes (weapons, ammunition, vehicles …).

So the head of the Slovenian Commission, Miran Bogataj, began stone-walling and calmly rebuffed enraged protests by the officers left in Slovenia. Their only consolation was a promise by the YA Supreme Command that they would be rewarded with promotions for their loyalty. This enacted the age-old saying in Belgrade that "ever since the time of Duke Putnik, you only could buy a Serb by giving him a higher rank".[535]

Some frustrated Yugoslav officers destroyed everything they could not take with them. They pulled electric wires out of walls and smashed electric switches. In the Šentvid barracks on the outskirts of Ljubljana, an officer ordered his soldiers to shovel all the clay from the tennis courts into bags. When the Yugoslavs destroyed a bowling lane in a military club by drilling holes in it, a watching Slovenian quipped: "So we'll play mini golf here."

As the day of the Yugoslav Army departure from Slovenia approached, the responsible officer, Colonel Nikola Stajčić, mostly had to write in his reports to the Zagreb Command only one sentence: "There has been no hand-over today." In one of the last reports on October 9, 1991, another colonel added: "As for the further handover of assets, there has not been a concrete agreement yet, even though all the assets have not yet been returned."[536]

Three days later, a delegation from the Yugoslav Federal Defence Ministry headed by Admiral Ljubivoj Jokić visited Ljubljana to finalise agreement on the Yugoslav Army departure from Slovenia, and five days later a Slovenian delegation traveled to Belgrade to conclude negotiations. On October 18, Jokić and Bogataj reached an agreement which precisely set dates for the withdrawal of the last YA soldiers from Slovenia.

Under this agreement, the Slovenian side "generously" accepted the YA request for a seven-day extension of the Brioni deadline for withdrawal. Within 24 hours after signing of the new agreement, the YA had to de-mine all the facilities and return all weapons and

---

Ljubljana, October 7, 1991.
535) Serbian General and Duke, Radomir Putnik made a name for himself with a victory over the Turks in the battle at Kumanovo in October, 1912. He was a founder of the modern Serbian Army and became a model for many Serbian officers.

material equipment to the Slovenian TD. Military assets and material-technical assets had to be taken to the Vrhnika barracks and the air force base of Cerklje, and the ordnance and infantry weapons were to go to the military depot of Mačkovec near Postojna. All of these and other movable and unmovable military assets in Slovenia were supposed to remain the property of the YA until implementation of a future agreement on distribution of Yugoslav assets among former Yugoslav republics. Slovenia also took over maintenance of military apartments on Slovenia.[537]

A day before Jokić and Bogataj signed the agreement, Minister Janša already reported to the Slovenian Presidency about the substantial breakthrough regarding the YA withdrawal: "The Yugoslav Army more or less accepted all of the Slovenian demands and agreed to move out of Slovenia through the Port of Koper. They also agreed that their officers and soldiers may take with them only their personal weapons and equipment."[538]

Some officers tried to smuggle a gun or two, but the Slovenian Special Police officers took all these weapons away from the YA officers and soldiers before were able to board the ship.

On October 25, 1991, 15 minutes before midnight, the last Yugoslav Army soldier boarded the ferry Venus and headed towards the new Balkans' battlefields. In the Port of Koper, the Slovenian leadership, military and police gathered. When at four minutes to midnight, the ferry sailed off, people felt relaxed because their homeland was finally free from adversaries.

A Yugoslav officer on the ferry looked towards the Slovenian coast and exclaimed: "Thus far Janša was our problem, from now on he will be yours."

\* \* \*

Slovenia defeated "the fourth strongest army in Europe" on the battlefield and around the negotiating table. But the victory had a tragic and shameful sequel. The departure of the Yugoslav Army

---

537) The agreement between the Slovenian delegation and the delegation of the Federal Secretariat For

was a traumatic decision for the officers' wives and children who considered Slovenia their "homeland", and now had to choose to be separated from their husbands and fathers, or go with them into war, uncertainty and despair.

Numerous officers' relatives who did not return to Serbia or Bosnia-Herzegovina were erased from the register of Slovenian residents. This was a disgraceful and disturbing action by the democratically-elected Slovenian authorities. After February 26, 1992, which was the deadline for obtaining Slovenian citizenship, the Slovenian Interior Ministry deleted from the registry of permanent residents 25,671 people who did not submit petitions to obtain citizenship, or whose petitions were rejected. Clerks at Interior Ministry counters destroyed their documents in front of their eyes. Overnight, they became strangers in their own country, "erased" because they were not Slovenians by nationality, and left without personal property, rights to work, health insurance and education.

Behind this outrage were Prime Minister Lojze Peterle and Interior Minister Igor Bavčar, both known to be xenophobes, together with some other reactionary Slovenian politicians.

Seven years afterwards the Slovenian Constitutional Court ruled that the action was unconstitutional, and in the next few years, the judicial system took further steps to correct their status. In June 2012, the European Human Rights Court ruled that Slovenia violated the rights of the erased people and ordered Slovenia to pay them compensation. However, the battle of these victims for full correction of the injustice and restoration of their dignity is still going on.[539]

\* \* \*

With the Yugoslav Army soldiers gone, Slovenians turned their attention back to the material left behind. The YA had shown itself slow and bureaucratic in its response to events in Slovenia, but its rigidity also meant that it monitored its assets with great thoroughness.

---

539) *Društvo izbrisanih prebivalcev Slovenije* (Society of the Erased Residents of Slovenia) was established and operates in Slovenia and *Mirovni inštitut* (Peace Institute) is offering free legal help to people, who lost their citizens' rights. Their legal rights are also being defended by the former

Every Army military facility and weapons depot had to keep on the inside of the entrance door a plastic pouch containing a detailed inventory of everything stored inside, so that the General Inspection of the Yugoslav Army (GINA) could at any moment obtain detailed information about the equipment of a particular garrison. When there was an inspection, officers would find themselves in big trouble if only one weapon or piece of equipment was missing.

Vladimir Milošević said: "From the Zagreb military region, we received vast lists of equipment and requests that every rifle, every bullet ... from all border posts and barracks, located in our regional headquarters area, should be returned. Lists even contained bread and pork ribs they were supposed to send back."[540]

However, the colonels were counting military underwear and cans of beans, instead of using the three-month withdrawal period to retrieve the most valuable assets. When YA representatives and European Community monitors came to the Borovnica military depot on August 21, they saw a ransacked and almost empty place.

The YA commission reported that the Slovenian Territorial Defence returned only two per cent of weapons and ammunition that were taken away after the TD seized the depot on June 28. In the Šentviška Gora depot above Tolmin, the TD brought back less than 10 per cent of all assets that were taken away.[541]

The TD headquarters never prepared lists of weapons collected, combat technical assets and other military inventory on the territory of Slovenia. They could not have done so even if they wanted, because weapons in those months were quickly shipped off to Croatia, and by the end of 1991 most of those military assets had been sold off there.

If somebody ever added up all the YA military assets in Slovenia at the beginning of the war, and then subtracted what stayed in the TD warehouses afterwards, the result would be embarrassing for Slovenia.[542] Therefore, many people in Slovenia hope that the balance

---

540) Author's conversation with Vladimir Milošević, Murska Sobota, May 20, 2014.
541) From the YA list: Technical assets and weapons of the YA that the TD did not return, without a signature and a date.
542) Several VOMO employees confirmed official findings of the arms trafficking investigators,

sheet of the YA assets and liabilities to be distributed among former Yugoslav republics will not be available for many more years and that nobody will want to check it too closely.

* * *

At almost the same moment that the Brioni moratorium expired – at midnight on October 8, 1991 – Slovenia also opened 26 road and eight railway border crossings with Croatia and deployed 1,458 police officers in the area.

The same day, Slovenia introduced its new currency – the tolar – after extensive debates in the Parliament about its name. The first banknotes were provisional payment notes, with a picture of the Triglav, the highest peak in Slovenia and a traditional symbol of Slovenian identity. In the beginning, one tolar was worth one Yugoslav dinar.

With the introduction of the new currency, complications arose almost immediately. In a meeting on October 7, President Kučan warned that on the main farmers' market in Ljubljana individuals were illegally buying foreign exchange with Yugoslav dinars, and that the banks also joined in. Minister Janša added that the Yugoslav Army officers in Slovenia had at their disposal six billion freshly-printed Yugoslav dinars.

Two days later, the Defence Ministry warned Slovenian citizens that the YA soldiers were trying to exchange new dinar banknotes for foreign currency or the new provisional tolar notes, and there was a good reason to suspect that some of these dinars were counterfeit.[543]

The armed conflict in Slovenia was so brief that people responsible for purchasing new weapons for the emerging Slovenian military were caught short. The largest shipments of arms from abroad arrived only in autumn 1991 when the YA was already withdrawing. However those who wanted to justify megalomaniacal purchases of advanced weapons for the Slovenian TD – and at the same time make money for themselves by reselling them to Croatia – fueled fears of the war in Croatia spilling over into Slovenia and bringing back the Yugoslav soldiers.

On July 9, 1991, the Slovenian Finance Ministry allocated 400 million dinars to Ludvik Zvonar for purchasing weapons. The middleman was a notorious international arms trader, Nikolaj Oman, who, according to Zvonar, was brought into the business by foreign intelligence services.

Funds from the Slovenian budget were deposited into Oman's account at the Slovene Investment Bank (Slovenska investicijska banka – SIB). The reason for the deposit was for use in buying "medical material". The international embargo on the importation of weapons into Yugoslavia was still in force, and for foreign countries this still included Slovenia. The money was channeled through private companies set up by Zvonar and his assistant, Daniel Anžič.

Yugoslav dinars were worthless on the international arms market, so they had to be exchanged into convertible foreign currencies. Three years later, in a report to Defence Minister Jelko Kacin, Zvonar stated: "In Slovenia, there was a shortage of foreign exchange,[544] so I was tasked with exchanging dinars into German marks in neighbouring countries. With a lot of difficulties, we received an exchange rate of 26 Yugoslav dinars for one mark."[545]

On July 10, Nikolaj Oman transferred 36 million dinars into the account of a private Zagreb Bank Promdei owned by Ibrahim Dedić, a banker with a dubious reputation,[546] and Dedić was able to procure five million German marks in cash through the Social Accounting Service in Zagreb.[547]

The German marks arrived in Slovenia without any problems, but the Croatian Criminal Police officers got wind of the transaction and asked for help from their Slovenian counterparts. The latter had to decide whether to send to Croatia the requested documents, which would show that this was a transaction of the Slovenian Defence Ministry or, "in the name of the national interest", to be quiet about the whole thing. They chose the latter course and the investigation in Croatia stopped.

---

544) According to the Slovenian Central Bank Governor, France Arhar, Slovenia had only the equivalent of 170 million dollars of foreign exchange.
545) Ludvik Zvonar's report to Defence Minister Jelko Kacin: reference to the decree no. 2578-M, Ministry of Defence, Ljubljana, September 24, 1994.
546) In 1992, Ibrahim Dedić was detained in Slovenia on suspicion that he counterfeited tolars, but the Slovenian authorities could not prove it. He was the target of several assassination attempts and in July, 1999, he was murdered in front of his apartment in Zagreb. His killer has not yet been found.

Zvonar and his team were exposing themselves and their foreign currency to considerable danger, but exchange rate differentials enabled them to make additional profits. According to one of his colleagues, they made three to five dinars for every exchanged German mark.[548] To exchange the rest of the dinars received from the Slovenian budget, Daniel Anžič turned to Albanian foreign exchange dealers. Creation of a parallel, illegal state in Kosovo at that time opened the way for Kosovar Albanians to take control of most of the foreign exchange black market in Yugoslavia. With their help, Anžič bought substantial quantities of German marks and U.S. dollars for the Slovenian Defence Ministry.

With the help of his aides, Zvonar was eventually able to exchange all of the 400 million dinars he received from the Slovenian treasury. In exchange, he received just over 15 million German marks, equivalent to a little less than nine million dollars.[549]

Some money for purchasing weapons for Slovenia was contributed even by the Italian mafia. At least that was stated by one of its members going by the name of Aldo Fanelli, who at the beginning of 1990s was laundering money in casinos at the Slovenian seaside resort of Portorož.[550] Local police received information that in 1991 at the beginning of war, he contributed 400 million Italian lire to the Slovenian Defence Ministry.[551] Fanelli boasted that he gave the money to Janša to purchase weapons.[552]

When on October 8, Slovenia introduced the provisional tolar payment notes and abolished Yugoslav dinars, the old notes were officially said to have been taken to the Vevče paper mill near Ljubljana and ground into pulp. However, that was is not what happened to them. Soon after October 8, people in Croatia and Bosnia-Herzegovina noticed an unusually sharp drop in the value of the dinar. The German

---

548) Jernej Čepin's letter about the activities of Nikolaj Oman, Ljubljana, January 2, 1994.
549) A report from Ludvik Zvonar to Defence Minister Jelko Kacin: referencing the decree no. 2578-M, Ministry of Defence, Ljubljana, September 21, 1994. The exchange rate between the German mark and the US$ was calculated in the report as: 1 DEM = .572 US$; 1 US$ = 1.747 DEM.
550) He is also known to Slovenian and Italian police as Luigi Ciccarelli and Gino Chiessa. He was a member of the Venice criminal group known as Mala del Brenta.
551) At the time, 400 million lire were worth a just below 300,000$, or just over half a million DEM.
552) Information from the Koper police headquarters for the Ministry of Interior, Criminal Police

Consulate in Zagreb reported that one German mark there was worth between 70 and 90 dinars, and in Sarajevo even 150 dinars.[553]

This was at least in part due to the real fate of the Yugoslav dinars "written off" in Slovenia. According to Ludvik Zvonar, "we transported those dinar banknotes, about four tons of them, with four trucks to the south. We unloaded two trucks in Croatia and two in Bosnia-Herzegovina. To make sure that the authorities would lift the gates to our trucks, I showed the guards my two official IDs – one, with me as a Major of the Counter-intelligence Service of the YA (KOS) and the second with me as an inspector of the State Security Service (SDV). We were successful … But we didn't get very much money, maybe one million German marks."[554]

In the "brotherhood" of nations of once flourishing socialist Yugoslavia, now war mongers, criminals and other scoundrels were not only killing each other – they were also competing among each other who will rob whom in a more ingenious way. In this respect, Slovenia was perhaps ahead of the others. For the foreign exchange that Slovenian war profiteers received on the black market, they were buying weapons and reselling them to Croatia and Bosnia-Herzegovina. But when those two countries were buying weapons themselves, the foreign exchange they needed was much more expensive because of the sudden deluge of Yugoslav dinars smuggled in from Slovenia.

In revenge, a group of Bosnian Muslims led by shady businessman Fadil Đozo in spring 1992 counterfeited provisional Slovenian tolar notes. Đozo wanted to use them to buy large quantities of weapons from Andrej Lovšin, but the latter quickly realised the money was fake and the deal fell through. Đozo's group began disposing of the counterfeit tolars in Slovenia, but Slovenian law enforcement quickly detected them and prevented them from entering circulation. On August 2, 1992, Fadil Đozo was assassinated on the Igman Mountain above Sarajevo.

During the chaotic disintegration of Yugoslavia, Croatia also tried to convert as many Yugoslav dinars as possible into foreign currency, and

---

553) "In Sarajevo, one German mark was worth 150 dinars",

resorted to similar machinations. At the end of 1991, Yugoslav dinars were replaced by the Croatian dinars but before that, Tuđman's regime entrusted vast amounts of Yugoslav dinars to a few chosen individuals to convert them into foreign exchange on the black market. Among them was Croatian businessman and criminal Hrvoje Petrač.

He later admitted that some people in Croatia objected to the risky operation because they were concerned about consequences regarding the legal succession of the assets of the former Yugoslavia. However the operation was approved by the Croatian Finance Minister Jozo Martinović. So Croatian trucks transported Yugoslav dinars during the night to Belgrade, to Sandžak and Banja Luka, and were returning with German marks and U.S. dollars. Petrač stated that in a few months, Croatia obtained a substantial amount of foreign exchange, which it used for purchasing weapons.[555]

These perilous financial manipulations by the two republics paled in comparison with Milošević's break-in to the Yugoslav monetary system in January 1991, and subsequent manipulations which ruined countless Serbs.

In May 1992, the United Nations, the United States, and the European Community imposed a total embargo on the rump of Yugoslavia, made up of Serbia and Montenegro, because of interference in the war in Bosnia-Herzegovina.[556] So the new state was more or less excluded from the international community.

While Serbs were wagging their fingers at the international community, they did not notice that they were being robbed by unscrupulous new savings banks operating pyramid schemes depending for solvency on a constant flow of new funds.[557] The Serbian authorities wanted to lure the Serbian people into giving up the last savings they kept in socks or under mattresses.

---

555) From the book *Gotovina, stvarnost in mit* (Gotovina, Reality and Myth), by Ivica Đikić, Davor Krile and Boris Pavelić, Sanje, Ljubljana, 2014, pages 104 and 105.
556) Serbia and Montenegro formed a new federation, called The Federal Republic of Yugoslavia, but other former republics did not recognise it as a sole successor of the disintegrated Socialist Federal Republic of Yugoslavia.
557) Dafina Milanović ("Serbian mother") was the owner of the Dafiment banka, while Jezdimir Vasiljević, ("boss Jezda") was the owner of the savings bank Jugoskandik. The two banks were founded in 1991 and were part of Milošević's state "project" of taking savings away from the Serbs to enrich

Borka Vučić, who in Serbia was considered "Milošević's banker", made sure that the foreign exchange was moved to Cyprus.[558] Hapless Serbian citizens however had to cope with hyperinflation that reached one of the highest levels the world had ever known. The money printing house in Belgrade's Topčider borough had to operate in three shifts to keep up.

On one particular day in December 1993, the price for an egg on the Belgrade's farmers' market was 30 billion dinars, while on the same day a penalty for speeding was "only" half a billion. Prices of eggs and other food staples were being set by sellers at levels calculated to cope with the expected level of inflation. Traffic fines and other fees, set by the government, were being changed only periodically, so they lagged behind "market" values.

That same month, two bank employees carried Yugoslav dinars in a big plastic basket from one bank branch to another. They left the money basket unattended in front of the branch and stepped around the corner to have a cigarette. When they returned they were surprised to see that somebody stole the basket but left behind the money, which was being blown around by the cold *Košava* wind[559] on the street.

At the end of 1993, a banknote for 500 billion dinars was put into circulation, but then the Serbian regime realised that the devastating inflation must end and that it was time for an "economic miracle". The person set to perform this task was Serbian "Super Grandpa" Dragoslav Avramović. Already by January 1994 the monetary measures he took caused a downturn in inflation, which had peaked at 313,563,558 per cent. At the age of 78, Avramović had 50 years of experience – including at the World Bank – and was sympathetic towards the impoverished Serbian people. But he soon came up against Slobodan Milošević, and found he could do nothing to prevent the old adage: that a fish rots from the head down.

---

558) Borka Vučić was Milošević's colleague at the Beobanka. When he became President of Serbia, she held several important financial positions and many of her activities are still wrapped in secrecy. In August 2009, she died in unexplained circumstances on a highway between Belgrade and Niš.

559)         (Koshava) is a cold and powerful south-eastern wind, typical for Serbia and

As Yugoslavia breathed its last breath, anything went as a means for survival and enrichment. Muslims in western Bosnia were buying oil from the Croatian INA oil company and were reselling it to Bosnian Serbs, who were transporting it to Serbia. Then Serbian tanks filled up with this Croatian fuel and set about destroying Croatian towns and villages.

Defenders of the ill-fated Croatian town of Vukovar bought their first weapons from Yugoslav Army depots in Batajnica near Belgrade and paid with gold confiscated from a courthouse in Osijek. Vukovar finally fell to the YA on November 18, 1991. The Army then engaged in negotiations with the Croatian government, European observers and the International Red Cross, but at the same time Army soldiers were already digging a large hole at the nearby Ovčara farm. Out of sight of the other negotiators, the soldiers secretly dragged 250 injured Croatians from the Vukovar hospital through the back door and massacred them at the farm.

This violated an agreement all parties had signed ensuring the neutrality of the hospital and safe evacuation of 400 wounded and sick people inside. For the YA, that devious agreement was signed by General Andrija Rašeta, the Serb born in Croatia who negotiated the exchange of weapons with Slovenia after Brioni. Irony had it that one of the victims of the massacre was his nephew, who worked in the Vukovar hospital.

\* \* \*

In the middle of December 1991, the U.S. Administration, United Nations Secretary General Pérez de Cuéllar and the European Community peace mediator for Yugoslavia, Lord Peter Carrington, all became concerned that Germany apparently intended to recognise the independence of Slovenia and Croatia, without first obtaining a joint agreement on the future of the Yugoslav republics.

The United States, the United Nations and the European Community – more or less the whole international community – were neutral in the war between Serbians and Croatians. As long as Croatia was part of the Yugoslav federation, the war was an internal

However Germany had a different point of view: the government in Bonn came to the conclusion that recognition would put an end to turning a blind eye to the destructive politics of Slobodan Milošević, whose military trampled over all cease-fire agreements and peace initiatives. After a long and heated discussion in Brussels on December 17, the Council of Ministers of the European Community backed the German initiative to recognise Slovenia and Croatia, even though two-thirds of Croatia was occupied by the Serbian militaries and paramilitaries.

Even Dutch Foreign Minister Hans van den Broek, who presided over the 12-member European Community, seemed to realise that all options to preserve Yugoslavia were exhausted. Twenty years later, he explained: "Deep German involvement in the fate of Croatia at the time was probably connected with developments within Germany itself, namely the fall of the Berlin Wall in 1989, German unification at the end of 1990 and the right of self-determination for all East Germans ... Those emotions that came to the surface at that time were reflected by the Germans on to the Balkans. They developed a frame of mind that the Croatians should have a right to self-determination, because the Serbs, who wanted to achieve a dominant role in Yugoslavia, were oppressing and exploiting them."[560]

However, emotions were not the only factor that influenced Germans politics. Germany had for a long time been trying to bring its influence to bear in the Balkans, especially in the region of the Adriatic, which is the closest warm sea to Germany. Because of the bloody nazi occupation of that area during the World War II, German politicians had to wait for almost half a century for a chance.

The government of Chancellor Kohl saw the disintegration of Yugoslavia as an opportunity, and it has borne fruit. Today, the largest economic power in Europe is taking over strategic sectors of business in Slovenia and Croatia, buying key state-owned companies and reaching deep into their infrastructures.

At the beginning of December 1991, Slovenian diplomats were saying that German recognition of Slovenia was more or less a done

---

560) Hans van den Broek's letter to the Slovenian leadership upon the 20th

deal. While waiting for that recognition, they were amicably trying to persuade other states to do the same, in particular the Vatican, since they reckoned that other Catholic countries would then follow suit. A high Church official, who wants to stay anonymous, remembers that the Pope John Paul II. was favourably inclined towards recognition, but with the cautious proviso that the Vatican should not be the first to do so, though it could be the second.

Resourceful Slovenian diplomats searching for a country which would break the logjam hit upon Iceland far out into the Atlantic, where people reputedly knew how to think with their own heads. They were thought to favour independence for the Baltic states, which were just then trying to liberate themselves from the failing Soviet Union.

When the officials in the Icelandic capital of Reykjavik received a letter from Ljubljana, they studied maps and documents regarding Slovenia. After that, they sent a delegation to Slovenia, which established that not only that Slovenia really existed, but it was a newly established state and was really serious about its independence. On December 19, the Foreign Ministry of Iceland duly announced recognition of the independence of Slovenia.[561] By accident or intent, Germany and Sweden followed suit on the same day. The Vatican did likewise on January 13, 1992, and then two days later the European Community as a whole. Russian recognition came on February 14, 1992, allegedly after being granted a sneak viewing of American technology built into Slovenia's Krško nuclear plant.

On December 23, 1991, as the German Consul in Ljubljana handed over a document formalising German recognition of Slovenia to representatives of the Slovenian government, the Slovenian Parliament passed a new Constitution. France Bučar, President of the Parliament, at the last moment prevented efforts by conservative deputies to strike out an article about freedom of choice regarding birth of children. Bučar thus averted the possibility that abortion would be outlawed.[562]

---

561) "Iceland, as the first western country, officially recognised Slovenia", *STA*, Ljubljana, December 20, 1991.
562) It is Article 54 of the Slovenian Constitution. President of the Parliament Bučar convinced members

The Demos coalition which dominated the government tried to limit the authority of a future President of the Republic by transferring his supreme command of the armed forces to the Defence Ministry. The attempt failed, not least because of the great popularity of President Milan Kučan.[563] In independent Slovenia's first Presidential elections in December 1992, Kučan swept aside his opponents, and did so again in the next elections in 1997.[564] A week after the passing of the new Constitution, the Demos coalition which had overseen Slovenia's passage to democracy disintegrated.

As Slovenia's first recognition came through in December 19, 1991, another fateful event was in the offing. On Christmas Day, December 25, Soviet leader Mikhail Gorbachev resigned and on December 26 the Soviet Union – the world's greatest socialist imperial force – ceased to exist and moved into the history books.

---

delay in its passing. A two-thirds Parliamentary majority was required to adopt the Constitution.
563) Article 102 of the Constitution states that the President of the Republic is supreme commander of its armed forces. Available on http://www.us-rs.si/media/ustava.republike.slovenije.pdf (last access: June 1, 2015).
564) Eight candidates participated in the Presidential elections on December 6, 1992. Milan Kučan, running as an independent, received 63 per cent of votes already in the first round. In the second

- Chapter 14 -

# THE VANISHING PROCEEDS FROM ILLEGAL ARMS SALES TO CROATIA

*In Notranjska land there is a village called Vrh. In olden times, there lived in this village a big strong man named Krpan. Bigger than anybody far and wide. He didn't care much for work, but he carried English salt on his mare from the sea, which was strictly forbidden at the time ...*

That is the beginning of the great Slovenian novel *Martin Krpan* written in 1858 by Fran Levstik.[565] The story goes that Krpan is smuggling English salt from the sea when he encounters a carriage with the Austrian Emperor himself. To move away from the path, he grabs his mare together with its load, and moves them both to the side. The Emperor is impressed by such a strong man. Later, when fearsome Brdavs threatens the people of Vienna, the Emperor's courtiers think of Krpan, who soon arrives and defeats the dreaded Brdavs. Very soon Krpan exasperates the Emperor's snooty courtiers headed by Minister Gregor with his simple wit. In the end, Krpan convinces the Emperor to write him a permit for transporting English salt, and he returns satisfied and content to his village, Vrh near Sveta Trojica. The novel concludes with the words: *"Good luck!" says the Emperor, but Minister Gregor says not a word.*

Slovenians have mixed feelings towards this fictional personage who has come to embody the national character. Martin Krpan is a peasant standing up against noblemen, a provincial muscleman and a hero, but also a slacker, a smuggler and a crude xenophobe. The novel was written when Slovenians were ruled by Austria, and it depicts the

---

565) Fran Levstik (1831-1887) was a Slovenian storyteller, playwright, essayist, literary critic and

typical drive of people under foreign hegemony for self-preservation through cunning resourcefulness and petty crime. Just like Martin Krpan, who got his concession for transporting salt, they are often satisfied with crumbs from the master's table.

After independence, Slovenia had an opportunity to break with this mediocre past. But it did not. The beginnings of the new Slovenian state were marked by systematic fraud and the shady peddling of weapons purchased "for Slovenia". Those involved continued with these unsavoury activities for some time afterwards. And what is worse, they created a new "rule of law" according to their seamy standards.

\* \* \*

Ludvik Zvonar was a resourceful man. He started his career under Communism as a police officer and later taught military tactics to cadets, police colleagues and members of the State Security Service (SDV). Gradually he moved up the hierarchical ladder of the state and security structures. He became head of Protocol at the Castle of Brdo pri Kranju, a state residence frequently visited by Josip Broz Tito, who felt comfortable there thanks in good part to Zvonar's organisational skills. Ten years after Tito's death, Zvonar was still performing protocol functions in Slovenia and Belgrade.

When Yugoslavia still existed, he was considered a reliable cadre and was procuring weapons and equipment for the uniformed police and organs of state security in Slovenia. He became well versed in that role and knew influential people throughout Yugoslavia and abroad, so in May 1990 after free elections he was named as an Advisor to the Slovenian government. He was in charge of weapons purchases abroad and headed the logistics administration at the Defence Ministry.

Zvonar was brought into the government by Deputy Prime Minister Jože Pučnik, who had joined the League of Communists but only because, as he put it, he wanted to destroy the system from the inside. Pučnik was a fierce critic of Yugoslav socialism, and was sentenced to seven years in prison for his political dissidence.

After he was released from prison, he moved to Germany and returned home in 1989 when the League of Communists began liberalising. He presided over the first new opposition party in Slovenia, Socialdemokratska zveza (Social-Democratic Union), but in the the Presidential elections of 1991 he was defeated by Milan Kučan. A few years later he left politics, and leadership of the party went to Janez Janša, who moved it away from the social-democracy based on enlightenment envisioned by Pučnik. Janša presents himself as a political student of Pučnik, but has moved the party (now known as Slovenska demokratska stranka – SDS) well to the right.[566]

According to Professor Tine Hribar, Janša pushed Pučnik out of the party's leadership. "Jože Pučnik was a philosopher and a theoretician, who on the operational level wasn't very effective. He simply didn't know how to preside over meetings, and that's probably also how he operated in the party. Janša took advantage of Pučnik's shortcomings and quietly took over the leadership of the party. When Pučnik noticed it, it was unfortunately too late. He withdrew from the party in disappointment."[567]

After Pučnik's death in Germany in 2003, Janša named Ljubljana's airport after him.[568]

\* \* \*

The merchant vessel carrying the first "official" shipment of weapons "for Slovenia" arrived in Koper on June 20, 1991.[569] Automatic rifles, ammunition and hand-held anti-tank rocket launchers from that ship were the only weapons that Slovenian defenders received before armed conflict broke out with the Yugoslav Army.

---

566) In 1996, SDSS changed its name to Socialdemokrati Slovenije (Social-Democrats of Slovenia) and in 2003, to Slovenska demokratska stranka or SDS (Social Democratic Party). The President of the party remains Janez Janša.
567) Author's conversation with Tine Hribar, Tomišelj, May 6, 2014. Tine Hribar (born in 1941) is a philosopher and professor of phenomenology. He was the first editor of the famous *Nova revija* (The New Magazine) and is one of the authors of the Slovenian Constitution.
568) In September 2014, the Jože Pučnik Airport was sold to the German company Fraport, which operates Frankfurt Airport in Germany.
569) The name of the ship was Herman C. Boye, operating under a Danish flag and loaded at the

Another four "official" shipments arrived some considerable time after weapons in Slovenia fell silent. The second ship arrived two months after the agreement for withdrawal of the Yugoslav Army from Slovenia, and the last one docked the following year in April.

For all these shipments, Slovenia paid over 82 million German marks.[570] The June 1991 shipment worth eight million marks was paid for from the budget several weeks before arrival. The Defence Ministry obtained about 15 to 16 million marks by converting Yugoslav dinars from the Slovenian budget into foreign exchange on the black market. Probably a smaller amount was a "donation" from the Italian mafia.

According to Anton Peinkiher, Zvonar received from Security organ of the Defence Ministry (VOMO) between 10 and 15 million German marks to purchase new armament. He remembers that for that money, Zvonar bought Igla anti-aircraft missiles and weapons samples from Singapore, and he also settled port fees.[571] Fifteen million marks is approximately a tenth of what VOMO "earned" with selling weapons seized from the YA to Croatia.

To obtain additional foreign exchange that was still needed to pay for weapons, the Defence Ministry took out trade loans in the name of the Republic of Slovenia.[572]

The procedure was complex, mostly because they wanted to cover tracks that would lead to the lender. The business deal was arranged through the Munich foreign trade company Unimercat founded in the mid-1980s. At the time, Slovenia was trying to move away economically from Belgrade, and Unimercat, a trading company with mixed ownership, was established to facilitate entrance of important Slovenian companies to foreign markets. Among those companies was a construction company, Slovenija ceste Tehnika (SCT), headed by Ivan Zidar.

The loan which financed the first purchase of weapons was actually assigned by the Slovenian Parliament for construction of motorway

---

570) A report on activities of the Parliamentary Commission in charge of supervising intelligence and security services (KNOVS) for 2006, no. 020-02/93-36/26, Ljubljana, June 28, 2007. The total value of purchased weapons was 82,586,833 DEM or 47,273,545 US$.
571) Official note of the VOMO investigators about the conversation with Anton Peinkiher, no. 881-600, Ljubljana, September 21, 1994.
572) In the case of a trade loan, the lender supplies the goods and the borrower later repays the debt

sections in the regions of Dolenjska and Gorenjska.[573] The relevant law enabled Zidar's SCT to obtain a loan of 20 million German marks from a German company ISC – Integrated Systems GmbH in Munich, which as it happened belonged to a group of companies also controlled by Zidar.

Only 15 years later did the Slovenian public learn that the money meant for roads was used to purchase arms. The weekly magazine *Mladina* published an article stating that in October 1991 ISC transferred the money into a bank account of another Munich company, Unimercat, rather than to Slovenia. With that, the loan metamorphosed into a trade credit.[574] On October 18, Ljubljanska banka provided a guarantee for repayment of a loan amounting to 28 million German marks to Unimercat.[575]

That day Nikolaj Oman, the middleman for arms purchases, and Zvonar's assistant Daniel Anžič arrived in Munich from Vienna. Their trip was tracked by Security and Intelligence Service (VIS) agents, who tapped their phone calls. The buyers of weapons were in a hurry to get the money as soon as possible and they then handed it over to Konstantin Dafermos in Vienna.[576] The account of his company, Scorpion International Services, in the Central-European International Bank (CEIB) in Budapest was almost empty, and Slovenian arms buyers had to replenish it before they could receive new shipments of weapons.

Today people at the Finance Ministry and the Company for Managing Highways of the Republic of Slovenia (DARS) cannot explain the financing of the two motorway sections. When DARS was established in December 1993, it took possession of all archives about building highways, but people at the company insist that they know nothing about the financing because at that time they took over only the maintenance of these sections.[577]

---

573) A law to finance construction of the two highway sections, *Official Journal of the RS*, no. 2/91-I, Ljubljana, June 25, 1991.
574) "Whatever I manage does not fail", *Mladina*, Ljubljana, May 28, 2006.
575) Loan guarantee no. III/3743-91, Ljubljanska banka, October 18, 1991. The document was signed in the name of the bank by Alojz Jamnik, who later became a member of the executive board. At the time, his brother Janez Jamnik, was financial director of the SCT and its co-owner.
576) Source: transcription of the phone tap, VIS, Ljubljana, October 18, 1991.

\* \* \*

In October 1991, Slovenia took out a new trade loan in accordance with a decision by Prime Minister Peterle's government. Just over 15 million U.S. dollars went for health care and 26.6 million for weapons. For the arms, the story repeated itself. Through the mediation of Unimercat, money from Munich went to the arms supplier's bank account in Budapest, where on November 22, a deposit of 10 million German marks was recorded.[578]

Ludvik Zvonar stated: "Don't look for shenanigans at Unimercat because there were none. Weapons for Slovenia were mostly paid with loans. We arranged them in different ways because of the embargo. No company was buying weapons for us. However, some companies helped with obtaining loans."[579] Despite his assurances about Unimercat being "clean", it found itself under investigation by Slovenian Criminal Police, and after failing to repay loans to banks, it went bankrupt in 2001. The construction empire SCT experienced a similar fate and went bankrupt in 2011.

Ivan Zidar was a businessman for all regimes. In the era of Socialist Yugoslavia, he was "only" a director of a company, but in independent Slovenia, which adopted capitalism, he became the company's majority owner, a practice followed by many other former "red" directors.[580]

Those managers took better care of their companies and their workers when the companies were still state-owned than later when they were privatised. Many directors ran the companies down during the privatisation process, so they could buy them as cheaply as possible while the workers were laid off.

In the 1980's, when Zidar's SCT had 11,000 employees, the company was building for Saddam Hussein in Iraq and Gaddafi in Libya. Zidar often met with Marshal Tito, and he also met Pope John Paul II. As a successful businessman, he received several Yugoslav and foreign recognitions, and he cooperated prolifically with Germany. During construction of the Slovenian highway system, his company was

---

Highways of the Republic of Slovenia (DARS), December 19, 2014.
578) A bank account statement of the company Scorpion I. S., no. 032616 500, CEIB, Ltd., Budapest.
579) Author's conversation with Ludvik Zvonar, Radovljica, December 22, 2010.

awarded the most public contracts. In Germany however, he was found guilty of accepting bribes and had to pay a fine of one million euro.

His SCT company was making huge profits by building apartments, commercial buildings and religious facilities. Under the code name *Čista lopata* (The Clean Shovel), Slovenian authorities launched criminal proceedings in 2008 against Zidar and other executives of the largest construction companies in Slovenia. They established that their cartel price fixing scheme during the construction of Slovenia's highways cost the Slovenian taxpayers two billion euro. Zidar allegedly caused SCT damage worth eight million euro. During one of the trials he said: "I don't know what I'm accused of because I've been working in the same manner all of my life."

On March 2013, the court found seven construction company executives guilty of bribery.[581] All the convicted managers had to go to prison except Zidar, who was able to avoid it because of alleged "permanent damage to his brain". As stated by his lawyer, "Zidar still thinks that he's the director of SCT" and worse, "he's convinced that his company is operating well".[582]

Thus was the fate of the businessman and the company which in autumn 1991 enabled the Defence Ministry to obtain two trade loans amounting to 46.5 million German marks. At the end of all these paths was Germany, the great ally of Slovenia and its people.

In 1991 and the year after, the Defence and Interior Ministries enjoyed special privileges, amongst them separate budgets. Financial inspectors were not supposed to enter their buildings. However in September 1992, the Social Accounting Service (SDK) did violate the "inviolability" of both Ministries, and found out that foreign loans amounting to 36.5 million German marks were not posted in books anywhere.

At the time, Industry Minister Dušan Šešok[583] justified huge malversations with taxpayers' money with the assertion: "If we did

---

581) Besides Zidar, the most important executives found guilty were Hilda Tovšak, director and part-owner of the construction company Vegrad, and Dušan Črnigoj, director and majority-owner of the construction company Primorje. Both companies ended in bankruptcy.
582) Statements of Zidar's lawyer, Boštjan Podgoršek in a news programme 24 hours of *POP TV* Ljubljana, July 3, 2014.
583) Dušan Šešok was Finance Minister from May 1991 to May 1992 and then until January 1993,

things the way SDK wanted us to do, maybe today we would not be an independent state. Slovenia would not be recognised, and they would have introduced another embargo against us."[584]

Mitja Gaspari, who at the time of discovery of these illegal loans was Finance Minister, took a steadfast legal stance on the issue: "Budget items are decided by Parliament, while the laws and resolutions set the legality of spending of the budget money. If anybody possesses sources outside this context, it means that these sources are illegal."[585] Nobody however was indicted for illegally obtaining and using loans, nor was any hearing held.

Ludvik Zvonar, in his report to the Defence Minister at the time, Jelko Kacin, mentioned the two loans that the Finance Ministry made available to the Defence Ministry.[586] However, he mentioned an amount of a little less than 36 million German marks, and not 46.5 million. There was no explanation what happened to the missing 10 million or so.

Three years later, when he started to quarrel with other weapons traders, Nikolaj Oman said individuals in Slovenia made an enormous amount of money in two ways: "For weapons they were buying for Slovenia, they quoted a higher price than they actually paid. At the same time, they re-sold a lot of weapons for high prices, which was in fact done by the Ministry of Defence."[587]

On September 21, 1991, the Danish ship Ardal brought modern Soviet-made anti-aircraft missiles from the Polish port of Gdynia to Koper. The bill of lading for this second "official" shipment stated that the three containers contained "technical goods" destined for Budapest.[588] In fact, they contained 240 Igla SA-16E missiles and 24 rocket launchers. In Koper, the containers were moved on to trucks, which were accompanied by Moris brigade personnel to Kočevska Reka, where they were distributed to Territorial Defence units.

---

584) "I don't know what the second part of the credit was used for", interview with Dušan Šešok, *Delo*, Ljubljana, September 29, 1992.
585) "A state without a timetable", interview with Mitja Gaspari, *Mladina*, Ljubljana, January 12, 1993.
586) Ludvik Zvonar's report to the Defence Minister, Jelko Kacin, Ljubljana, September 21, 1994.
587) Nikolaj Oman, in his conversation with the Criminal Police officer, Drago Kos, Ljubljana,

Zvonar stated that the shipment was worth 12,192,000 U.S. dollars[589] and added that he did not handle "transit documents." From this, it can be inferred that the ship also brought weapons for Croatia.

Igla anti-aircraft missiles, together with rocket launchers and Soviet anti-tank weapons, were also brought to the Port of Koper by the ship Hel on December 7, 1991. This was the fourth "official" shipment. There is no available documentation about it, but from interceptions of phone calls to the intermediary Oman and the supplier Dafermos, it can be deduced that enormous quantities of infantry weapons were unloaded.

Zvonar wrote down detailed quantities of Igla anti-aircraft missiles and AT-4 Fagot and AT-7 Metis anti-tank weapons, which were on that day taken with trucks to the barracks of Šentivid near Ljubljana.[590] According to Zvonar's notes, the shipment was worth 8,838,000 dollars.

The fifth and largest "official" shipment of weapons to Slovenia was delayed for several months and arrived only on April 1, 1992. The delay was caused by quarrels between the Defence and Interior Ministries over other weapons that were stored at Ljubljana's Brnik airport.

The ship which brought this fifth consignment was named Sabine and sailed under an Antiguan flag. The cargo of weapons was loaded at the Ukrainian port of Mykolaiv into four containers, and was declared as medical equipment meant for Austria. In fact, the cargo consisted of Igla anti-air missiles with launchers and Fagot and Metis anti-tanks weapons.[591] For the whole shipment, Slovenia paid US 12,100,500 dollars.

The fourth and the fifth "official" shipments were paid for with the trade loans already mentioned. Zvonar, in his report to the Defence Minister, quoted the same prices for items in all three shipments.[592] He added: "I do not name the sellers, banks and countries that were

---

589) US$ 12,192,000 equals 21,299,424 DEM (For his calculation, Zvonar used the exchange rate 1 US$ = 1.747 DEM)
590) There were 88 Igla SA-16E rockets with 21 launchers; 58 Fagot anti-tank rockets with 5 launchers, and 200 Metis anti-tank rockets, with 10 launchers.
591) In his report, Ludvik Zvonar wrote that in the shipment on April 1, 1991, there were 72 Igla rockets, together with seven launchers; 442 Fagot rockets, together with 45 launchers and 10 Metis launchers.
592) Igla rocket – US$ 48,000; launcher – US$ 28,000; Fagot rocket –US$ 15,500; launcher – US$

willing to deliver weapons to us, and thus helped us to gain our independence and enabled us to achieve a minimal defence readiness. This information would not in any way help the assessment whether the purchases were made correctly and in an honest way."[593]

In 1992, modern weapons for anti-aircraft and anti-tank warfare made in Russian plants in 1989 and 1990 strengthened the capabilities of the Slovenian Territorial Defence. Defence strategists in Slovenia were able to breathe a sigh of relief, at least for a bit, since they possessed weapons enabling them to shoot down Yugoslav fighter jets at an altitude of 3,000 metres.

Meanwhile Janez Janša was hailing a turnaround in the United States's attitude to arms shipments to Slovenia: "Assets were also acquired with the quiet consent of the United States, which after the first visit by the U.S. Military Attaché in Vienna to the Slovenian Ministry of Defence, changed its policy towards Slovenia and its defence. The U.S. Government in spring 1992, before official recognition of Slovenia, approved the Slovenian purchase of 3D LR radar for controlling Slovenian airspace."[594]

In his report, Janša said that Slovenia bought 350 Igla missiles and 50 launchers. These amounts were also mentioned in a report by a Parliamentary Commission in charge of supervising intelligence and security services (KNOVS), which stated: "Most of the countries that Slovenia asked for help refused it. With great difficulties, we were able to secure some offers, which were not realised early enough because of inability to pay, closure of airspace and the blockade of the port."[595]

However Zvonar, who had operational control over weapons procurement, said they purchased 460 Igla missiles and 52 launchers. 110 anti-aircraft missiles and two launchers were missing and presumably re-sold. According to Zvonar's price list, the difference in price is 5.3 million dollars, and it is not known at what price the missing anti-aircraft weapons were re-sold. Zvonar expressed surprise and said he did not know.[596]

---

593) Ludvik Zvonar's report to the Defence Minister, Jelko Kacin, Ljubljana, September 21, 1994.
594) A weapons report from May 15, 1990 to September 1, 1992, no. DT/90/91/92/93-DK, MORS Ljubljana, September 19, 1993, page 12.
595) Report about the activities of KNOVS in 2006, June 28, 2007 and report KNOVS for 2004,

Was this the price of patriotism that the "fighters for independence and defenders of the homeland" paid to themselves? More than 100 missiles that can be used to shoot down aircraft "disappeared", as did five million dollars. The mystery will remain unsolved. A quarter of a century later, who will count the money for some anti-air missiles?

\* \* \*

Slovenia's dark saga brings to mind another funny story by the aforementioned writer Fran Milčinski – about the *Butale* mayor and his shepherd, who takes 19 sheep to the pasture but comes home with only 18. The mayor scolds him, but the shepherd insists that they should count the sheep again. However, the shepherd still counts one more than the mayor, so they come to an agreement: they will call 16 town councillors plus a policeman. Altogether there will be 19 people, and each will grab his own sheep. This is how the story continues:

> *The mayor called "Now!" and everybody rushed and grabbed the closest sheep. The men were reliable and honest, and nobody grabbed two. And then it turned out that everybody had one sheep, and the policeman had none. "So you see, you stubborn man," said the mayor to the shepherd. "The policeman is without a sheep, so there are 18 of them!"*
> *And the shepherd gruffly retorted: "I don't care about the policeman! There are 19 of them! The policeman is a fool. Why didn't he grab a sheep in time, when there were still some there."*

The writer knew very well why the policeman was left without a sheep. The characters of the *Butalci* stories entered Slovenian culture as archetypes of simple-minded people. However those who were "counting" anti-aircraft systems can be accused of anything but simple-mindedness.

\* \* \*

Out of five shipments "for Slovenia", at least three also contained weapons for Croatia, which was not mentioned in Zvonar's report. Several years later, after he checked his notes, he stated that in September, in the second shipment brought by ship Ardal, there were also "14 or 16 containers with Kalashnikov automatic weapons and ammunition".[597] These were exactly the weapons that Draušbaher's company Orbis handed over to the defenders of Zadar.[598]

The fourth shipment of weapons, brought by the ship Hel on December 7, would probably stay a secret, had the Security and Information Service (VIS) not intercepted a phone discussion between Oman and Dafermos. Oman was not sure what was in the containers, and the supplier from the Scorpion Company explained to him that there were 10,000 automatic weapons and 10 million rounds of ammunition. Unaware that Slovenian intelligence officers were listening in, Oman mentioned to Dafermos that he was accompanied by Franci Kosi, the advisor to Interior Minister Igor Bavčar and the main procurer of weapons at the Interior Ministry. The Greek arms dealer said his friends in Vienna knew the "department and its leader" very well.[599] This "leader" could not have been anybody else but Igor Bavčar, who a few months later would play an important role in re-selling a large quantity of weapons in the so-called "Brnik affair".

In reply to the question how many weapons "for Slovenia" stayed in Slovenia and how many were re-sold, Zvonar's deputy at the Defence Ministry, Elo Rijavec, gave different figures: "Ordinarily, every ship bringing weapons that were ordered by the Slovenian military brought only 10 per cent of weapons of interest to the Slovenian military, while 90 per cent were weapons and equipment that the Slovenian military was not interested in, such as 5.45 mm-calibre Kalashnikov automatic rifles, Makarov handguns, uniforms, helmets and similar stuff. They called all these weapons and equipment the 'ballast' which was later resold to buyers in Croatia."[600]

---

597) Author's conversation with Ludvik Zvonar, Radovljica, December 29, 2010.
598) Source: a letter from the company Jugotanker to the Orbis Company with a request to supply the invoice for US$ 249,600 and a statement of the Scorpion I.S. bank account no. 032616 500 in CEIB, Ltd., Budapest.

Zvonar became upset when challenged over this and said that Rijavec knew nothing, because he was only his deputy in charge of transportation of the weapons. He vehemently denied Rijavec's assertion that 90 per cent was "ballast". Zvonar and his deputy did however agree about something. In separate reports made at about the same time, they both confirmed that Janez Janša was present at all deliveries, and that he directed which weapons would stay in Slovenia and which would be sold to Croatians.

The facts about the third "official" shipment for Slovenia were closer to the assessment given by Rijavec. Unloaded from the ship Scotia, sailing under an Antiguan flag, was the largest cargo of rifles and ammunition brought so far to Koper, and the large majority was sold to Croatia. Just at the time the Yugoslav Army was returning weapons to the Slovenian Territorial Defence, its soldiers on board a ferry taking them out of Slovenia were probably able to see the Scotia navigating towards them. They did not know that it was loaded with weapons which in a few days would be in the hands of their new enemies – the Croatian Army.

The bill of lading stated that this cargo brought from Varna in Bulgaria consisted of technical items owned by a Polish company Cenrex, heading for Morocco. However, in Koper, they unloaded 79 containers, containing more than 29,000 Kalashnikov automatic rifles accompanied by 25 million rounds of ammunition. The price of the shipment was 9,654,456 U.S. dollars.[601]

Zvonar wrote in his report that they took the weapons from Koper to the Šentvid barracks. When he was asked whether the ship also carried a shipment meant for somebody else, he opened his notebook and started reading: "Three containers of clothes, flour and cans of food, and one container of weapons. That was meant to be sent directly to Croatia."[602]

But the documents tell a different story: the Kalashnikovs and ammunition went to Croatia too. Heavy trucks accompanied by police transported 67 containers to the "arms bazaar" in Kočevska Reka.[603]

---

September 27, 1994.
601) The ship Scotia brought to Koper: 29,217 AK-47 Kalashnikov automatic rifles; 24,327,100 pieces of caliber 7.67 mm ammunition and 767,760 pieces of ammunition for Makarov handguns. Ibid.

There is no information about the remaining 12 containers. The fact that the shipment was meant to be re-sold was confirmed by a telephone conversation between the intermediary Oman and Zvonar.

Zvonar was very upset and yelled at Oman about the delay in the shipment of weapons: "Last night, they threatened that they were going to burn down my house because the shipment is not here yet." Oman calmed him down saying: "The mail is sailing on the ship." They agreed to meet at Brdo pri Kranju, and go from there to see the Defence Minister, who, according to Zvonar, "wanted to know about his money".[604]

In the meantime, the Velenje company Orbis sent a request to the Defence Ministry for the importation and selling of weapons,[605] and on October 23, Defence Minister Janša gave his approval concerning that items stated in the annexe.[606] The request and Janša's approval confirm that Draušbaher's company was not only selling weapons from former YA depots, but was also re-selling imported weapons meant "for Slovenia".

Even more unusual is the fact that the annexe, with proposed weapons for export, also contained several Fagot anti-aircraft missiles together with launchers, for which the Slovenian government had to give assurances to suppliers that it would not sell them. Also, at the time, Orbis did not have a licence for re-selling military weapons. However, the demands for arms generated by the savage war further south in ex-Yugoslavia proved too tempting for the Slovenian merchants. The arms were purchased at low prices in Eastern Europe, and there were huge profits to be made by re-selling them.

In his report, Zvonar wrote that AK-47 Kalashnikov automatic rifles from the Scotia were purchased at the price of 185 U.S. dollar or 320 German marks apiece. If the Defence Ministry or Orbis were re-selling them for 450 marks, as stated by Janša in his report, the profits exceeded 20 million marks. However, the Orbis merchants were charging up to 1,300 marks for Kalashnikovs, as can be seen

---

604) Parts of the transcription of the phone tap, VIS, Ljubljana, October 8, 1991.
605) Approval of import and sale: correspondence of Orbis to Minister Janez Janša, Velenje, October 18, 1991.

in their correspondence with Croatian buyers.[607] The actual prices that the buyers paid for these weapons are not known.

A few weeks later, the Croatian military came to Kočevska Reka to pick up their Kalashnikovs and ammunition. Brane Praznik from VOMO remembers: "Each time, Croatians also brought with them 30 workers to help with loading trucks. The Slovenian side would not hand over weapons to Croatians until they received a fax message that they had been paid for, and only then they would allow the Croatians to load their trucks. I don't know who sent those payment receipts, I only remember that they were transporting weapons on Croatian trucks, as well as on trucks that belonged to a Slovenian company, Avto Kočevje."[608]

Among buyers of weapons from the Scotia was Croatian Army Lt-Colonel Ivan Bećir, who in the summer had been buying ordnance from the Ložnica former YA depot. Another buyer was Josip Vukina, who on October 24, two days after the arrival of the guns and ammunition at Kočevska Reka, showed up at the Obrežje border crossing between Croatia and Slovenia, accompanied by a Croatian Interior Ministry official. He told Slovenian border officials that he and his colleague were on their way to visit the Slovenian Defence Ministry. When he also mentioned that he was carrying a briefcase containing three million German marks, the Slovenian official became suspicious and would not let them enter Slovenia. They were only allowed to enter after the border officials received a phone call from Rijavec to admit them.

Vukina and his aide drove away towards Ljubljana. However, they were being monitored by Slovenian police, who video-recorded that the Croatians met with people in front of the Defence Ministry building and then at the nearby Opera bar, and that they made phone calls from the post office. No action was taken.[609]

On that day, the Security and Information Service (VIS) tapped into several discussions among the arms dealers, revealing that

---

607) Approval of import and sale (correspondence of Orbis to Minister Janez Janša), Velenje, October 18, 1991.
608) Official note of the VOMO investigators about the discussion with Brane Praznik, no.881-600, VOMO, September 2, 1994.

Vukina was caught in a complex network of "tied selling" created by arms supplier Dafermos and intermediary Oman.

The same day, Zvonar and Oman were recorded in Vienna, which indicates that Slovenian intelligence was also intercepting international conversations. Vukina complained to Zvonar that he brought money, for which he had had to wake up the Croatian Finance Minister in the middle of the night, and now they were "playing games" with him.

Zvonar knew that the purchase of weapons from the Scotia was linked to Oman's offer to sell the British anti-aircraft system Blowpipe to the Croatian Armed Forces[610] because in an earlier call Defence Minister Janša allowed the sale of automatic weapons and ammunition to Vukina on condition that the latter also bought Blowpipe anti-aircraft missiles and launchers. Everything seemed taken care of so that the deal could go through. Well, almost everything … Soon after, an angry Vukina called again: "I received a telefax and the prices for ammunition are not the ones we agreed upon. We said 180 and not 190!"

"Well, that's for a particular commission," calmly answered Oman.[611]

The Slovenian Defence Ministry thus was not only selling and re-selling weapons, but it was also involved in tied trading – for which it was collecting commission. In 1991 and 1992, the Finance Ministry did not register any increase of Slovenia's budget arising from selling or re-selling weapons to the Balkan battlefields.

Although the temperature in Ljubljana on October 24, 1991 was only a few degrees above freezing, Vukina stepped out of the phone booth with a warm feeling – he had only to turn over the cash and then he could go home. However when he stepped on to the street outside the Defence Ministry he realised to his consternation that his Honda Civic car was no longer there. Eventually a Defence Ministry official to whom he ran for help discovered that the car was in a pound for illegally-parked vehicles. After 20 minutes he had retrieved his car, which fortunately for him still contained a bag of his with three million German marks.

---

610) Offer of Oman's company Orbal M.S., addressed to Vukina at the Croatian Ministry of Defense, invoice no. 91716, October 28, 1991.
611) Source: transcription of the wiretap. Phone conversations include Rijavec and Vukina, who were

The following morning, on October 25, a heavy truck of the Slovenian Viator company able to carry 24 metric tons drove away from the arms bazaar in Kočevska Reka towards the Croatian border and on to Zagreb, carrying the guns and ammunition that Josip Vukina had purchased for the Croatian military.

From autumn 1991 until the following spring, Slovenian transport companies made a lot of money carrying weapons.[612] Viator invoices submitted to the Defence Ministry show that every transport of weapons to Croatia was worth at least 1,500 German marks, that is, more than three times the average monthly salary in Slovenia at the time. The condition for making this money was keeping quiet. If a truck driver wanted to know what he was transporting, he was told by his superiors that it was better for him not to ask questions.

---

612) Preserved invoices show that Viator, hired by the Defence Ministry, carried out 87 transports of weapons from Slovenia to Croatia between August 28, 1991 and April 29, 1992. The total value of

A day in the war in Bosnia-Herzegovina, June 1992.

Photo: Igor Mali

Andrej Lovšin, Director of the Security organ of the Defence Ministry (VOMO), entering the Parliament building. During questioning under oath before an investigative commission in January, 2000, Lovšin experienced "amnesia" regarding his arms dealings.

Photo: Tomaž Skale, Dnevnik

Ludvik Zvonar, adviser to the Slovenian Government and primary purchaser of military weapons abroad. He claimed that Janša and his aides were dealing with arms behind his back.

Ivan Zidar, the CEO of construction company SCT, was heavily involved in various financial machinations connected with the arms trade. Known as the "businessman for all regimes", he also did business with the Slovenian Catholic Church. Picture shows Zidar meeting with Slovenian Cardinal Franc Rode in 2006.

Photo: BOBO

Ivan Draušbaher, head of the Orbis Company. He charged extremely high prices for weapons sold to Croatia. However, his company went bankrupt due to his mismanagement.

Photo: Tomaž Skale, Dnevnik

Smiling faces after the Defence Ministry decided that the Gorenje Company was to assemble armed vehicles of the Finnish company Patria. From left to right: Ivan Črnkovič, Director of Rotis, a small private company which represented Patria in Slovenia; Heikki Hulkkonen, Patria's Executive Vice-President; Jorma Wiitakorpi, Patria's CEO and Franjo Bobinac, Gorenje's CEO, June 21, 2006.

Photo: Tomaž Skale, Dnevnik

Koper, Slovenia's main seaport. From mid 1991 to 1992, at least 40 ships, loaded with weapons, sailed into the port. They arrived mostly from Bulgaria, Poland, Romania and Ukraine.

Croatian fighters resting in Posavina, northern Bosnia-Herzegovina, June 1992.

Photo: Igor Mali

Destruction and demolition in Mostar, Herzegovina, in June, 1992. Muslims and Croats clashed in the city.

Left: Prime Minister and leader of the Liberal Democrats of Slovenia (LDS) Janez Drnovšek, with his colleague from the LDS, Rudi Moge, head of the Parliamentary investigative commission on arms trading. Drnovšek was opposed to anyone looking into this case.

Photo: Tomaž Skale, Dnevnik

Nikolaj Oman, a Slovenian-Australian businessman and arms dealer, lived in opulence and luxury. The picture was taken in his small chateau of Grimšče near Bled in 1996, shortly before he was deported from Slovenia.

Photo: Tomaž Skale, Dnevnik

From left to right: Miha Brejc, Janez Janša and Lojze Peterle in Kočevska Reka, December 17, 2005. Brejc was the first director of the Security and Information Service (VIS) and later became a member of the European Parliament. After he fell out with Janša in 2003, he

- *Chapter 15* -

# THE FOX GUARDS THE HENHOUSE

In autumn 1991 the arms trade in Slovenia gathered pace as soldiers, officials, businessmen and truck drivers emptied captured Yugoslav Army depots, re-sold weapons, purchased more abroad, and moved armaments in transit across Slovenian territory towards Croatia and later Bosnia-Herzegovina. Between the middle of that year, when the first ship loaded with arms tied up in Koper until the end of 1992, at least 40 ships carrying weapons, ammunition, ordnance and other military equipment sailed into Slovenia's main seaport.

The main Croatian harbours of Dubrovnik, Split, Šibenik, Zadar and Reka were still controlled by the Yugoslav Navy, so the only practicable seaport for Croatia was Slovenia's Koper. These clandestine movements and the re-selling of military material arms through Koper are even today considered a deep state secret and would probably have escaped notice if it were not for a few individuals who gathered courage to speak out.

One of them is Mirko Slukan, commander of the border police station in Koper. Among the documents that drew the attention of investigators was a letter in which Slukan explained the circumstances of weapons transports through the Port of Koper. He stated that based only on the port's official documentation, it would be impossible to prove that weapons were involved. When documents of his police unit or another body contained something even slightly suspicious, they were immediately destroyed. Silence was in the interest of all involved, since according to Slukan, the Port of Koper, the customs office and companies involved in shipping, logistics and transportation increased their income by up to 100 per cent through dealing with weapons.[613]

---

613) Information about weapons transportation by ships through the Port of Koper, Mirko Slukan,

The commander of the border police station brought his knowledge to the attention of his superior, who was surprised that his subordinate knew as much.[614] Several containers were left behind in the port after being emptied of their weapons. Dockers knew when arms for Croatia arrived, because by the entrance gate to the port they could see numerous heavy trucks with Croatian licence plates or with Croatian markings on their tarpaulins.

Slukan told police that once a partition collapsed in one of the ships, so workers had to physically carry weapons; then some shipping crates broke apart and bombs rolled on the pier; also around 120 handguns allegedly went missing during reloading. No investigation was ever launched into these incidents, because officially there was no cargo. "At that time, only three people were making decisions about all that: Defence and Interior Ministers Janez Janša and Igor Bavčar, respectively, and the Director of the Security and Information Service (VIS), Miha Brejc," explained Slukan.[615]

As for the customs service, Slukan said that in dealing with the weapons, shamefully "it did not do its business". From June 1991 and for another 16 years, the Customs Administration of Slovenia was headed by Franc Košir, who at the time was closely connected with Ministries of Defence and the Interior and their two intelligence services.[616]

Mirko Slukan wrote down dates of arrival and names of certain ships which docked in Koper between October 1991 and January 1992. He was convinced that they were transporting weapons. Criminal Police gathered information about other suspicious ships and compiled a dossier.

So for example, the ship Ali-B, supposedly from Lebanon, whose bill of lading stated that it was carrying "zinc coated tubes", brought about 400 tons of weapons for Croatia. Would Slovenian intelligence officers and Croatian Defence officials have been present for the unloading, if they were only metal tubes?

---

614) His superior was Dušan Moljk, head of the Koper police headquarters.
615) Conversation of the Criminal Police officer, Ljubo Pirkovič, with Mirko Slukan, UKS, MNZ, Koper, September 23, 1994.
616) Franc Košir was a director of the Customs Administration of the Republic of Slovenia until November

On the third arrival of the Danish ship Herman C. Boye on October 19, 1991, police ascertained that it brought "an unknown quantity of weapons".[617] The vessel Opatija arrived on October 13 from Buenos Aires, Argentina, carrying 16 containers with 256,980 pieces of "general cargo". The original destination was the Croatian seaport of Reka, but because of the Yugoslav Navy blockade, it had been changed to Koper.

On November 20, 1991, the Hel brought to Koper 50 tons of weapons for Croatia, and five days later, Edith landed a further 90 tons. Both ships loaded their cargoes at the Polish seaport of Gdynia.

These quantities in October and November were nothing compared with the shipments that arrived in Koper in December 1991 and January 1992 from the Romanian seaport of Constanza on three huge ships: Dina, Ani and Rima. In more than 200 containers were almost 3,000 tons of infantry weapons, ammunition and several Romanian trucks with mounted multi-barrel rocket launchers.[618] Croatian Defence Minister Gojko Šušak informed the Slovenian Interior and Defence Ministries about the types and quantity of weapons loaded in Constanza on to the Dina and Ani.[619] There was no information about the Rima.

Dina's bill of lading stated that the ship was carrying "machines for fruit processing",[620] which was true only in that they later tore up the fruit orchards of war-stricken Dalmatia and Slavonia. More cynical was Rima's bill of lading, which stated that it carried "hospital equipment".[621] All three ships from Romania were met in Koper by Croatian officers accompanied by Slovenian VIS intelligence agents, and on two occasions also by Croatian Interior Ministry officials.

---

617) Report about weapons transports over Slovenian territory, Headquarters of the Criminal Police Office of the Ministry of Interior, no. 221/12, Ljubljana, June 7, 1994. The report was signed by the Interior Minister, Andrej Šter.
618) Dina arrived in Koper on December 12, 1991. It unloaded 15,000 automatic rifles, 10 million rounds of ammunition and 3,000 rockets. On January 2, 1992, Ani brought 20,000 artillery grenades, 10,000 shells, 10,000 RPG-7 anti-tank rockets, 250 rocket launchers, 3,000 rockets and five trucks with mounted multi-barrel 122 mm-calibre rocket launchers. Rima arrived in Koper on January 1992 bringing 986 tons of weapons.
619) Two notifications from Gojko Šušak to the Slovenian Ministries of Interior and Defence, December 13, 1991 (without a number) and January 2, 1992 (no. 512-16/92).
620) Dina's bill of lading, voyage no. 15/91, La Valletta, Malta, December 6, 1991.

The peaceful days of Christmas and the New Year of January 1992 were disturbed by the rumble of heavy trucks driving from Koper to the Obrežje border crossing and from there to Zagreb. Only five trucks were allowed to leave the port at the same time, so as not to attract unwanted attention.

Why did the information about mishaps and coverups in the Port of Koper provided by Mirko Slukan not prompt any action? Maybe because his superior officer forwarded them to the VIS, which in monitoring and organising the weapons transport was itself acting illegally. The fox was guarding the henhouse.

Only after Janša was removed from his position of Defence Minister did the investigator from Ministry's VOMO respond to appeals of individuals who drew attention to weapons containers still stored at the Port of Koper. In June 1994, the military intelligence service found in the duty-free part of the Port of Koper about 20 containers, six of which were still sealed and most likely filled with weapons. "We found them in a part of the port which was considered a secure area. Dockers called it Janša's village."[622]

\* \* \*

In a 1994 investigation into illegal arms trafficking, Interior Minister Andrej Šter requested an explanation from the civilian secret service, which in the meantime changed its name from Security and Information Service (VIS) to Slovenian Intelligence and Security Agency (SOVA),[623] about the role of a certain VIS official regarding weapons transports. Drago Ferš, who at the time was the deputy director, stated that a VIS agent, whose identity was classified, was informed about some of the weapons transports.[624] However, he

---

622) Special notation, Ministry of Defence containers in the duty free area of the Port of Koper, no. 90/6, Ljubljana, June 27, 1994.
623) On June 17, 1993, a government decree changed the name of the Security and Information Service (VIS) to Slovenska obveščevalno-varnostna agencija, (the Slovenian Security and Information Agency - SOVA).
624) From January 1992, Drago Ferš was in charge of the arrivals of ships into the Port of Koper. Before him, the position was occupied by Miklavž Lukman, head of the operational area of VIS,

emphasised that he knew only of those that received prior written approval of the Interior and Defence Ministries, and that his duty was solely safeguarding by counter-intelligence.[625] The weapons transits were indeed at risk after the international weapons embargo came in force on September 25, 1991.

In November 1991, SOVA received information that a group of employees of two Koper shipping companies organised the transport of containers with weapons from Varna to Koper on a ship named Aghia Varvara. They loaded the containers on to two Croatian trucks, which set off for Croatia accompanied by the Croatian National Guard and Croatian Interior Ministry officials. However, a Special Police Unit stopped the first truck on the way up from the coast and the second on Ljubljana's southern ring road.

In its report on the incident, SOVA stated that the organisers of the transport did not have the right permits for transit over Slovenia, and the cargo was falsely declared as "technical merchandise". The Criminal Police took possession of the cargo but the following day the Defence Ministry issued a permit to take it to Croatia.[626] Before it got there however, the Slovenians played a dishonourable trick, replacing all 30 Strela 2Ms and two launchers with other infantry weapons. The exchange was confirmed by Nikolaj Oman.[627]

"Without permission of the VIS, nobody in the port was allowed to touch the containers with weapons, while the VIS could not send a cargo anywhere, especially not across the border, until Janša signed the permit," stated Zvonar. Today, he is amazed that so many ships with weapons came to Koper. He also believes that he was occasionally "taken for a ride".[628] He informed Defence Minister Janša that on October 20 two ships arrived in Koper with merchandise for Croatia, without him knowing about it. Zvonar also mentioned problems regarding the ship Edith, which failed to arrive at the expected time. Zvonar was surprised to learn that ship already arrived several days beforehand and that Croatian representatives had taken possession of

---

625) SOVA's note to Interior Minister Andrej Šter, no. Z-27/246-94, Ljubljana, October 18, 1994.
626) Information about closed deals in the arms trade, SOVA, no. KI, Ljubljana, July 2, 1993.
627) Nikolaj Oman, talking to the Criminal Police officer, Drago Kos, UKS MNZ, Ljubljana,

four containers on board.[629] Zvonar was shocked. Later he learned that intelligence officers were issuing permits themselves, and he is convinced that VIS stole four containers.

In August 1993, a Slovenian nationalist politician Zmago Jelinčič was asked by a journalist whether Janša, VIS Director Brejc and Bavčar were selling weapons for the benefit of their ministries or their own pockets. He answered: "If they were selling it for the benefit of their ministries, it would have been recorded somewhere." Jelinčič asserted that all three were engaged in arms sales and they divided the profits afterwards."[630]

Because of this statement, he was sued for slander by Janša and Brejc. In June 1999, a court established that the plaintiffs suffered "mental distress"[631] and ordered Jelinčič to pay damages of 800,000 tolars[632] to Brejc and 700,000 tolars to Janša. Brejc and Janša lodged appeals to be paid at least two million tolars in damages each, but this was rejected by the Superior Court in Ljubljana in March 2001.[633]

\* \* \*

Miha Brejc has a doctorate in self-management – the economic ideology of former Yugoslavia – as did many others of those who at the end of the 1980s sensed a new political wind blowing, and started to attack so-called "continuity forces", meaning hardliners of the communist regime.

Brejc was a reliable and proven cadre – his father was a communist before World War II, a Partisan during the war and a political activist afterwards, while the son was a military officer in reserve. From the middle of the 1970's, he worked as a security officer in the Territorial Defence at the closed area of Kočevska Reka.

In November 1990, he became head of the State Security Service (SDV) – which a month and a half before the outbreak of the

---

629) Ibid.
630) "I will not be a political dead man for a very long time", interview with Zmago Jelinčič, 7D Weekly, Maribor, August 4, 1993.
631) A ruling of the Circuit Court in Ljubljana, no. IP 175/98, Ljubljana, June 10, 1999.
632) Official exchange rate 1 EUR = 239,64 SIT (tolars).

military conflict with the Yugoslav Army changed its name as the Security and Information Service (VIS). Even though he did not know much about the intelligence business, he "cleansed" about 200 operatives in a thorough reorganisation. He wanted to transfer a repressive "iron fist" service tasked with defending the communist regime into a modern intelligence service.

The most important question of any intelligence activity is whether it truly operates in the interest of the defence of a country – or only in the name of the country for some other interests. As far as Brejc was concerned, he took the view that intelligence agencies should act according to the principle that "the end justifies the means". The legal framework and formal supervision was of secondary importance, or just a front to deceive naïve people.

The counter-intelligence activities of VIS aimed at preventing leakage of classified information gave rise to particular abuse. Under the guise of monitoring foreign spies, VIS also followed individuals of interest because of their politics or for completely private reasons. Courts permitted VIS to eavesdrop phones only of people deemed "interesting from the point of security", but they also apprehended other people after tapping telephones. In his role as Director, Brejc was thus constantly walking on the edge of legality and fighting allegations of illicit eavesdropping.

Just before Brejc took over, many of the files from the old regime went missing from the SDV archive. In 1992, VIS determined that at the beginning of 1990 a full fourth of the archive containing documentation regarding anarcho-liberalism, clericalism, the bourgeoisie, right-wing emigrants and political criminals had disappeared.[634] According to estimates by Interior Minister Bavčar, 17,000 – 18,000 personal files were gone.[635]

\* \* \*

---

634) "The SDV archives will be taken to the archives of the Ministry of Interior", *STA*, Ljubljana, May 19, 1992. The Director of VIS reported about this to the Parliamentary commission for monitoring the legality of VIS, headed by Peter Bekeš.
635) Igor Bavčar, "Some new elements in the Hit affair",

Among the two dozen parties that competed in Parliamentary elections held in December 1992, the Liberal Democratic Party (LDS) headed by Janez Drnovšek emerged as the clear winner, and in January 1993, he formed a strong coalition of four parties. One of the first measures of the new government was to move the Security and Information Service from the Interior Ministry and place it under direct government control.[636] A new law turned VIS into a government service.

The following month, Prime Minister Drnovšek replaced Brejc with a close aide, Janez Sirše, former Secretary of Tourism and Hospitality Industry. This was after Brejc's deputy, Roman Jeglič, alleged improprieties in VIS operations and accused Brejc of misusing VIS for political aims.

What followed was a surprising turnaround. On April 23, 1993, Miha Brejc brought two suitcases containing 1,500 pages of VIS documents to the Parliamentary commission for monitoring the legality of VIS. They alleged improprieties in Casino HIT in Nova Gorica and Casino Portorož, for which his VIS successor Sirše had been responsible.

In his letter accusing Brejc, Jeglič referred to a report prepared in autumn 1992 by the head the anti-terrorism sector at VIS, Rafael Mohorko. The 100-page report entitled "Overview and evaluation of activities of certain individuals in weapons trading" mentions 130 people said to be connected with the illegal trade, from small middlemen to important representatives of political, military and security intelligence structures of Slovenia, Croatia and Bosnia-Herzegovina.[637]

Mohorko's report alleges that the main protagonists – among them Bavčar, Brejc and Janša – tried to keep their weapons deals secret, because they feared the cover could be blown at any time and they would find themselves under investigation.

---

636) Officials, monitoring VIS were: Interior Minister Ivan Bizjak, Minister for Science and Technology Rado Bohinc and the Justice Minister Miha Kozinc.
637) Mohorko's report mentions many individuals known from arms affairs, among them: Danijel Anžič, Hasan Čengić, Ivan Čermak, Konstantin Dafermos, Ivan Draušbaher, Janez Janša, Franc Kosi, Anton Koščak, Andrej Lovšin, Nikolaj Oman, Josip Perković, Miro Predanič, Vladimir Zagorec,

Controversy focused on the alleged casino scandals, with right-wing politicians leveling wild accusations that the old communist Party nomenclature was using the casinos to launder money and sending it abroad. Eventually, two managers of Casino HIT went to jail and the government curbed further development of casinos on the coast.[638] In the meantime, Jeglič's warnings to Drnovšek about the weapons trade were forgotten.

Brejc then told the investigating Parliamentary commission that several leading members of the government were incriminated in the secret service dossiers. The committee decided to check the contentious dossiers, and also suggested that the dossier of Ljubljana Prosecutor Tomaž Miklavčič should also be looked at to see whether he used to be a member of the old communist secret service, UDBA. However Miklavčič vehemently opposed the idea.

In May 1990, Tomaž Miklavčič became a member of Pučnik's SDSS party, a predecessor of Janša's SDS, and in March 1991 he was appointed to head the Ljubljana Prosecutor's office. He was a son of Franc Miklavčič, a reputable Ljubljana's judge and a Christian Socialist, who in 1976 was detained because of an article in which he demanded the investigation of the liquidation of Slovenian Home Guards collaborators after the WWII.[639] He was only released after several protests from abroad.

The career of Tomaž Miklavčič was controversial and early in 1993 five of his deputies quit the Ljubljana Office of the Prosecutor. The reason: disagreements over how Miklavčič managed the office and unprofessional decisions. Miklavčič had disbanded the economic section of the office saying that in Slovenia it is not clear what an economic criminal really is.[640]

---

638) The Circuit Court announced its ruling on June 12, 2000. Former director of HIT Danilo Kovačič was sentenced to four years in prison for abusing his authority, while the former head of the loan service Danilo Kodrič was sentenced to three and a half years in prison and had to return illegally gained assets amounting to 325 million tolars. After their appeal, the Supreme Court lowered both sentences by four months. In February 2006, Slovenian President Janez Drnovšek pardoned Kovačič. In November of the same year, Kodrič died in a traffic accident. The former director of Finance in HIT, Darko Makuc, committed suicide while he was in detention.
639) Franc Miklavčič (1922-2008) was an activist of the Liberation Front and a Partisan during the war. He was never a member of the Communist Party, but he was very active in the Christian-

In 1994, a report on his activities by his superior, senior Public Prosecutor Pavle Dolenc mentioned serious irregularities, from unacceptable efficiency, slow legal proceedings to hindrance of the court's activities and many baselessly dismissed criminal complaints. The most egregious was his dismissal of a criminal investigation into Miha Brejc and some of his colleagues regarding alleged improprieties in directing the VIS. Miklavčič did not explain his decision, even though he was supposed to according to the law.

Miklavčič also dealt with a complex case of weapons trafficking. On July 20, 1993, the Defence Minister Janša "discovered" at Maribor Airport a large quantity of weapons sent by Muslim friends in Sudan to the government of Bosnia-Herzegovina, with logistical help by the United States. With this move, Janša wished to divert the attention from his own weapons dealings, and he also wanted to settle scores with political enemies. Janša was convinced that Slovenian President Kučan knew about this shipment of weapons because Kučan was a friend and ally of Alija Izetbegović, President of Bosnia-Herzegovina, who several times asked the Slovenian leadership for help.

At a court in Maribor, Janša was a witness for the prosecution against nine individuals charged with illegal weapons trading and abuse of authority. He tried to prove that weapons from Maribor Airport did not have a Defence Ministry permit for transit through Slovenia and that the whole shipment was thus illegal. However, when Janša was asked whether the United Nations weapons embargo was complied with during other weapons transits through Slovenia, he was caught in a trap.

This caught the attention of Dušan Požar, a principled senior State Prosecutor from Maribor, who drew up a voluminous file of evidence against Janša, concentrating on two particular cases. He showed that after May 22, 1992, Janša signed at least nine permits for transfer of weapons and ammunition across Slovenian territory.[641] On that day, Slovenia became a full member of the United Nations and was obliged to respect the weapons embargo decided by U.N. Security Council on September 25, 1991.

---

641) There were seven transit permits for Croatia and two for Bosnia-Herzegovina. The author has

In the second case, Požar presented evidence which showed that Janša sold 5,560 Kalashnikov automatic rifles for cash to the plenipotentiary representative of the Government and Army of Bosnia-Herzegovina, Hasan Čengić. Even though Kalashnikovs were of inferior quality, Čengić had to pay 1,345,520 U.S. dollars for the shipment.[642] Prosecutor Požar sent the evidence material against Janša to the Ljubljana Prosecutor's Office, which had jurisdiction over the matter. However, Tomaž Miklavčič dropped the proposal for a criminal prosecution. In the justification for his decision he wrote that the U.N. Security Council Resolution was not legally binding for Slovenia, because Slovenia had not prepared an implementation act which it could use to prosecute violators. As for the Janša's sale of Kalashnikovs to Čengić – he simply ignored it.[643]

Senior Public Prosecutor Pavle Dolenc determined that Miklavčič's decision was incorrect, because it was based on wrong legal reasoning. He ruled that the Ljubljana Prosecutor unjustifiably overlooked Požar's evidence about Janša's delivering weapons to Čengić. However, Dolenc's report was pointless, because it got stuck in a drawer of the Slovenian Chief State Prosecutor – General Anton Drobnič.

In spring 1995, Miklavčič left the Prosecutor's office and became Slovenian Ambassador to Argentina. In his previous post, he was replaced by his former deputy, Barbara Brezigar, who was later instrumental in ensuring that none of the charges connected with illegal weapons trading between 1991 and 1993 reached court.

What motivated Miklavčič's behaviour remains a mystery. In summer 1993, when the VIS – HIT affair was at its peak, Peter Bekeš, a former president of the Parliamentary Commission for monitoring the legality of VIS, remarked that the consequence of the furore is that today nobody apart from a very few individuals knows whether the intelligence services are trustworthy and operate according to the Constitution and the law.[644]

---

642) Evidence about selling weapons to Čengić was found during the house search on July 30, 1993 of the home of VIS employee Miro Predanič. The author has in his possession relevant documents and official notations of the conversations with Predanič, as well as his court testimony.
643) Prosecutor's decision to drop charges against Janez Janša, no. Kt 76/94-2-TM/SR, TJT, Ljubljana, February 1, 1992.
644) "Who is the righteous man who can pass judgements", interview with Peter Bekeš,

When in March 1994, Prime Minister Drnovšek removed Janša from his position as Defence Minister, Miha Brejc joined Janša's SDS party. A year later, he became the party's vice-president and in 2004, he was elected to the European Parliament from the SDS's list of candidates.

However in 2011, Brejc helped his son-in-law Gregor Virant to form his own party. Janša was furious and treated his colleague as if he did not exist. "Occasionally, the human side of Janša fails," said Brejc laconically.[645] In 2012, he left the SDS, and on the party's web site appeared the words "His exit from the party did not touch anybody ... Miha Brejc remains a pale memory."[646]

---

645) "In the SDS, Janša is the first by far, then nobody else for a long way back", interview with Miha

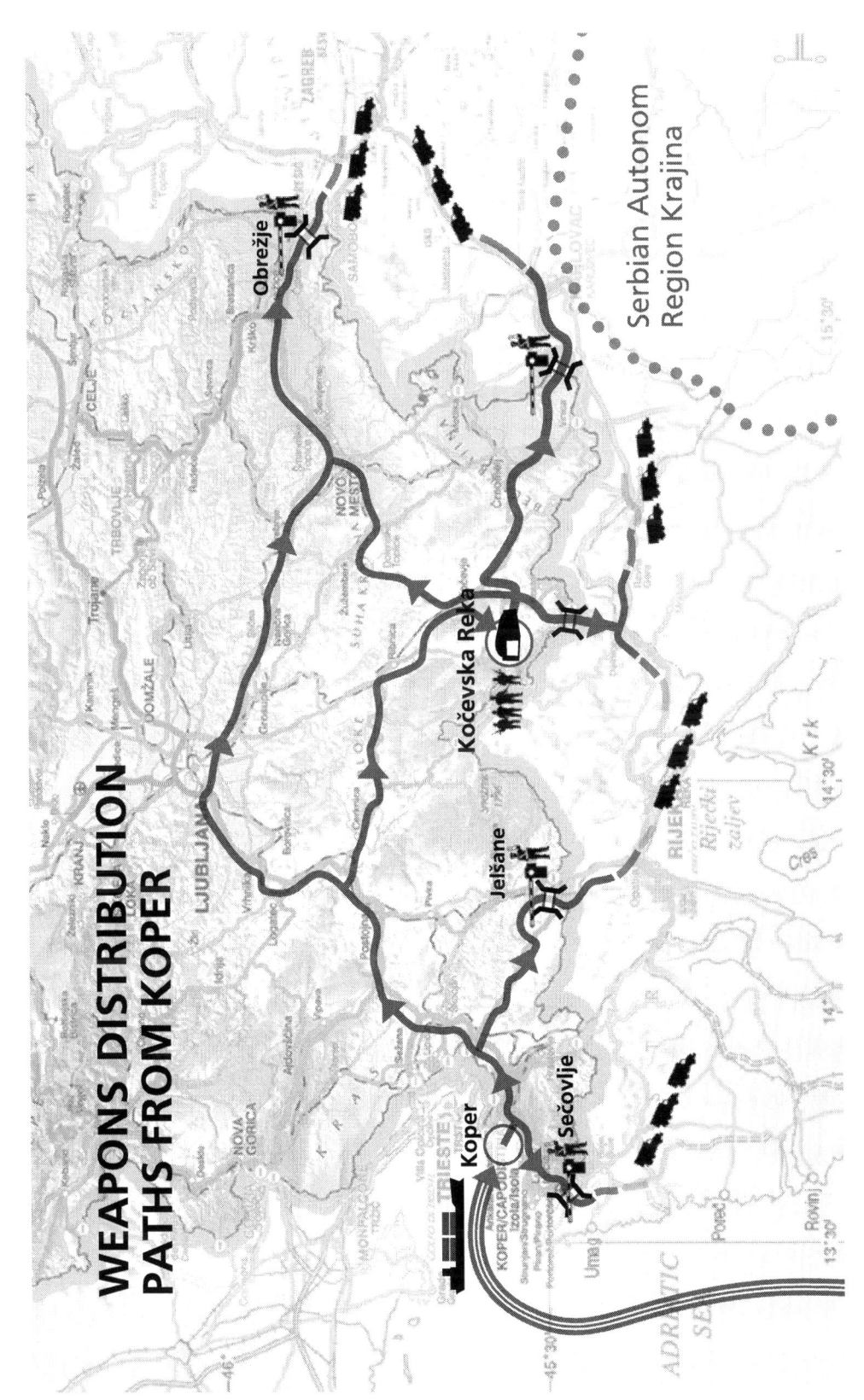

- Chapter 16 -

# SIX MEN CHARGED AND A PROSECUTOR INFATUATED

After the destruction of Vukovar, about a quarter of a million inhabitants escaped in just a few days from Osijek, located 30 kilometres away, leaving only about 10,000 defenders and the road apparently open for Serbian irregulars to move on towards Zagreb. However in the ensuing battles alongside the Danube River, Serbs suffered severe losses and the tide of battle then turned.[647]

As pictures of devastation on the battlefields came in, Serbs started to lose their enthusiasm for their war of conquest. In Belgrade, Novi Sad and other Serbian cities and towns, avoidance of military conscription by young men became almost a national sport. Volunteer soldiers paid more attention to drinking plum brandy and robbing homes of Croatian refugees than to "liberating Serbian lands".

Remarkably, some parts of Croatia remained completely untouched by war. It was if the people there did not live in the same country. A newspaper in beleaguered Osijek, *Glas Slavonije*, took people living on Istria peninsula in the northwest of Croatia severely to task: "Istrians experience the war only on TV screens. It is true that police officers there wear the Croatian coat-of-arms on their hats and fly Croatian flags, but for them, the biggest problem is a failed tourist season."[648]

However the war was by no means over. In the east of Croatia, Slavonian towns were suffering and Dubrovnik, a historic city on the Adriatic Sea treasured by tourists, came under attack by the Serbs and Montenegrins from November 1991 to June 1992. As black smoke rose over the ancient walls, one of the besiegers with a poetic inclination stood on a hill opposite and proudly declaimed: "If we have to, we'll build an even more beautiful, and if necessary, older Dubrovnik."

---

647) In Vukovar, Serbian forces lost about 8,000 soldiers, and 600 tanks and other armoured vehicles. Croatia lost 1,800 soldiers, while 2,000 people went missing. Source: *Jane's Intelligence Review*, 5/1993,

People in Slovenia were very concerned when heavy weapons started to be heard in the Croatian city Karlovec, only 10 kilometres from the Slovenian border. Serbian attackers tried everything they could to cut the road through the city connecting Zagreb with the Adriatic coast. From the middle of October 1991 to the beginning of January next year, Karlovec was the target of daily shelling by Serbs. That autumn, Croatian defenders in the north of the country fought heavy battles to capture barracks of the Yugoslav Army and seize weapons, ammunition and ordnance. As a result, the YA was forced to withdraw from Čakovec, Varaždin, Križevci, Samobor, Zagreb, Jastrebarsko and Karlovec.

In November 1991, Croatian forces captured the largest YA weapons depot in Croatia, located in Delnice in the Gorski kotar region, and seized more than 6,000 tons of military hardware. Slovenians breathed a sigh of relief, because that Croatian success more or less eliminated the danger that the war would spread across the border into Slovenia. By the end of 1991, the YA withdrew from their barracks in two-thirds of Croatian territory. However, in the self-proclaimed Republika srpska Krajina on Croatian territory, Serbs tried to reinforce their domination.

A 16th ceasefire agreement between the Serbian and Croatian militaries was just holding, when Croatian Defence Minister Gojko Šušak and YA General Andrija Rašeta signed an agreement on January 2, 1992, in the city of Sarajevo. The agreement implemented a peace plan drawn up by Cyrus Vance, Special Representative of the U.N. Secretary-General, and endorsed by Tuđman and Milošević. United Nations International Peacekeeping Forces, known as UNPROFOR, were deployed in the occupied areas of Krajina.[649]

What followed was a period of relative peace between Croatians and Serbs in occupied Croatian territory. There were only sporadic military skirmishes until the Croatians launched their *Oluja* (Storm) military operation in 1995 to clear all Serbs from occupied Croatian territory. That period of 1992, with weapons in Croatia silent and the war in Bosnia-Herzegovina not yet started, was a bad time for arms traffickers and dealers in Slovenia.

---

649) UNPROFOR (United Nations Protection Forces) were U.N. units deployed in Croatia and created in February 1992 by U.N. Security Council Resolution no. 743. According to Secretary Vance's plan, they divided a protected area (UNPA - United Nations Protected Area) into four sectors: north (Bania

In this "period of drought", police officer Franci was given the difficult task of selling a large arsenal of weapons from the ship Hel. And so began the infamous story of the so-called "Brnik weapons". For several years, it kept "heroes" of Slovenian independence, police and prosecutors busy, and sowed the seeds of new political quarrels.

At the beginning of 1991, at a meeting in the closed area of Kočevska Reka, leading officials of the Slovenian Defence and Interior Ministries expressed interest in obtaining German-made Heckler & Koch short-barrel automatic rifles. "Police officer Franci" was appointed to purchase the weapons. Franci, whose last name was Kosi, was an advisor to Interior Minister Bavčar.

Early in October 1991, Kosi placed an order with the intermediary Oman for about 6,000 H&K automatic weapons, with ammunition and several handguns.[650] He paid from the Ministry's account opened at the Celovec bank, Die Kärntner Sparkasse, using the password *Dragica*. In November, he transferred 1,200,000 U.S. dollars in two instalments into the Dafermos company's account at the Central European International Bank (CEIB) in Budapest. Both times, he falsely used the name of the Clinical Centre in Ljubljana as the sender.[651]

Kosi waited for a long time for his order to appear. At the end of 1991, he received a message from the Dafermos company in Vienna that together with the merchandise he ordered, he would also receive seven or eight containers of other stuff as a "camouflage". He was not surprised, because he knew that arms dealers often do this. However, the arrival of 46 containers in Koper on January 18, 1992, shocked even those familiar with the tricks of the merchants of death.

Weapons listed as "farming equipment in transit" were brought from Varna by the German ship Hel flying the flag of Antigua. Only two containers were loaded with the German automatic weapons that were ordered, and the question then arose where to put the other 46 containers loaded with the "ballast".

It was decided to take the containers to the military part of Ljubljana's Brnik airport, which is why the merchandise was nicknamed "Brnik

---

650) Interior Ministry order: 1,540 H&K MP A3, 60 H&K MP5K, five H&K MP5SD, 500 ammunition clips and 20 American Smith & Wesson handguns, Ljubljana, October 25, 1991.
651) A receipt for the transfer of US$ 1,2 million to CEIB, Ltd., Budapest, letter of credit no. 65019,

weapons" and the resulting complications the "Brnik Affair". Quarrels persisted for a long time afterwards about who was responsible for storing all these weapons at the main Slovenian airport.

Members of a Special Police Unit under the command of Vinko Beznik accompanied the convoys of trucks transporting the containers of arms to Brnik, and Special Police set to work recording the contents. In the containers numbered 30 and 34, they found 1,540 H&K MP5 automatic weapons, but not the handguns that were also ordered. Shot-barrel automatic rifles were distributed to the Defence and Interior Ministries, and before long the two ministries started to quarrel how many each should get. The rest of the containers were loaded with enormous quantities of automatic weapons, handguns and ammunition. For all these weapons, Oman submitted an invoice amounting to almost nine million dollars.[652]

Kosi, an athlete in his young years, honed his skills as a police officer during a state of emergency in Kosovo several years earlier. He believed in discipline and punctuality, and was convinced that extreme physical preparedness could solve many of the problems facing police at that time.

"Franci was not deceitful, and I don't believe that he would be capable of cheating anybody. In Kosovo, he was faced with very hard challenges. Around him, people were dying, even those he was responsible for. As far as I know, something broke inside him because of what he went through in Kosovo. I'm not sure that since then he was able to make demanding independent decisions, like ordering and selling weapons. When he was forced into doing it, he was afraid that he was going to get killed. He was afraid to go home, and he asked me if he could spend a night with us at the police station."[653] So said his friend, Vinko Beznik, commander of the Special Police Unit.

Franci felt helpless before the mountain of weapons that he had to sell to fighters in the south of former Yugoslavia. When Kosi

---

652) The shipment contained 13,000 AK 47 Kalashnikovs (US$ 195 apiece); 13 million rounds of ammunition (US$ 0.185); 10,000 Makarov handguns (US$ 225); five million rounds of ammunition (US$ 0.295); 2,000 uniforms (US$ 85) and 14 120 mm-calibre mortars (US$ 9,200). The total price was US$ 8,983,800. The invoice was submitted by Oman's company Orbal Marketing Services, headquartered at 113 Swanston Street, Melbourne 3000, Australia.

tried to reach Oman on the phone, his call was always answered by Oman's assistant, Jernej Čepin. Kosi complained that the sale of weapons was also made more difficult because of the "gentlemen from the neighbouring building, who were selling these things in large quantities through that gentleman from Velenje".[654]

He was talking about the Defence Ministry and Draušbaher's Orbis Company, which were selling weapons to Croatians with help from Zvonar. They were using established channels and all of the sudden, Franci was providing competition from the "neighbouring building" – the Interior Ministry.[655]

To make things more difficult, Croatia captured many classical infantry weapons by seizing Yugoslav Army barracks on its territory, and the Sarajevo Agreement had silenced most of the Croatian battlefields. As if all of this was not a big enough challenge, Nikolaj Oman decided to condition the delivery of Igla, Metis and Fagot missiles ordered and paid for with loans from the Defence Ministry with the selling of the "Brnik weapons". The missiles were the fifth "official" delivery of anti-aircraft and anti-tank weapons, which, because of the Oman's procrastination, only arrived in Koper on April 1, 1992 – April Fools Day.

In these circumstances, the "competing" Ministries of Defence and the Interior had to cooperate. So Zvonar offered his help to Kosi.

\* \* \*

During the sale of the "Brnik weapons", members of the Criminal Police and the Special Police Unit – both belonging to the Interior Ministry – almost shot at each other by mistake on the evening of March 4, 1992. Franci Kosi then convinced Interior Minister Bavčar to grant him immunity against actions by the Criminal Police while he was performing his duty – that is, smuggling weapons for the benefit of the Interior and Defence Ministries.

---

654) Transcription of the telephone conversation between Franc Kosi and Jernej Čepin, VIS, Ljubljana, January 31, 1992.
655) The Interior Ministry is headquartered on Štefanova 2 in Ljubljana and was only a few metres from the Defence Ministry, which until the beginning of 1992 was located on Župančičeva 3. After

That evening, Kosi's group was expecting Hasan Čengić, who would soon became the main buyer of weapons for the Army of Bosnia-Herzegovina, and Ivan Horvat, who was procuring weapons for the so-called Croatian Defence Council (HVO), the military formation of the Herceg-Bosnia parastate. At the time, Čengić and Horvat were still helping each other, but in autumn 1992 the two militaries became enemies and started to shell each other on battlefields of Bosnia-Herzegovina. The Slovenian sellers and re-sellers of weapons were not interested however where the weapons ended up, as long as they received payment for them. Nor did they care what happened with their sold merchandise on the Croatian side of the border.

Once Croatian authorities seized all the weapons that Horvat was transporting and detained him. He was a Croat ethnically however, which brought him some leniency. The Croatians treated Bosnians[656] transporting weapons from Slovenia through Croatia into Bosnia even worse. Čengić said that at several control points in Croatia, the authorities deliberately procrastinated with their procedures, harassed drivers, took away all or parts of shipments, and charged extortionate fees.[657]

On March 31, 1992, Kosi's employment ended. He could have left everything and retired, but he continued re-selling the weapons, and by early May he had got rid of most of the "Brnik" consignment, for which he was able to collect 5,737,169.60 U.S. dollars.[658] He transferred just over two million dollars into the Dafermos' bank account in Hungary, and he handed 300,000 German marks to Oman's assistant Čepin. The rest of the money he personally delivered to Ludvik Zvonar in three instalments.[659] Also present during the handover of the money was Daniel Anžič, who as an authorised courier, took cash over to Austria and handed it to Dafermos.

---

656) Besides Serbs and Croats, Bosniaks are one of the three constituent nations in Bosnia-Herzegovina. Most of them are Muslims, speak Bosnian and represent about 50 per cent of the population. In the former Yugoslavia, they were called Muslims but during the war in Bosnia-Herzegovina, they changed their name to Bosniaks, because they also included agnostics and atheists.
657) Author's conversation with Hasan Čengić, Sarajevo, May 20, 2008.
658) The complete overview of the financial situation on May 10, 1992, Kosi's report to the Interior Minister.
659) Notes about handing over the money, April 4, 13 and 16. The money was handed over by Kosi

After each handover of money, heavy trucks loaded with weapons crossed the Slovenian-Croatian border. Miha Brejc, Director of VIS, made sure that the transit was unobstructed.[660]

Because of his punctiliousness in dutifully recording everything he did, including accounting for the "Brnik weapons", Kosi was not a typical arms merchant, such as those from the Defence Ministry. When investigators saw notes with all the detailed information prepared by Kosi, they were surprised that he had written everything down and saved all the documents.

Was there not a practice among arms dealers in Slovenia that they talked about their business only one-to-one, that they avoided telephone conversations, did not leave tracks, and only used cash? In other words, that they did not talk much, and their business was conducted without papers – like drug dealers or criminals of the mafia? And that *omerta* reigned when some individual asked undesired questions?

Kosi himself ruefully remarked: "I don't know anybody who talked about his weapons dealings and survived."[661]

With this "achievement", Kosi aligned himself among state criminals. However, he did his dirty job very well, reselling weapons at a very inopportune time. Besides that, the Kalashnikovs imported from Pakistan were of bad quality. He sold them according to Oman's price list, except for the Makarov handguns, for which he had to lower the price, as almost nobody wanted to buy them. About two-thirds of the handguns and most of the ammunition for them stayed at the airport.

\* \* \*

At the time of the "Brnik affair", Bavčar and Janša were still considered heroes of Slovenian independence and the architects of the victory in the "war for Slovenia". They behaved as if if they were untouchable, and considered any summons to be interrogated by the Criminal Police or to appear in court as insults, and they rarely obeyed them.

---

660) VIS dispatches to the Obrežje police station about the exit of trucks, April 4, 9 and 16, 1992.

Before independence, when Janša was arrested in 1988 by the Yugoslav military and put on trial in a military court accused of treason, Bavčar headed a committee for defending Janša's rights.[662] Only four years after that trial, which triggered the Slovenians' revolt against Federal Yugoslavia, Janša returned the favour to his former protector by betraying him. He accused him of illegal importation of a large quantity of AK-47 Kalashnikovs, Makarov handguns, mortars, ammunition and uniforms, and of storing them in the military section of Brnik Airport without the knowledge of the Defence Ministry. Janša also accused Bavčar of illegal sale of weapons without the permission of the Defence Ministry.[663]

So Janša's destructive character surfaced. He abandoned his "friend" Bavčar, if indeed they ever were friends.

In his August 1994 letter to Bavčar, Franci Kosi assured him that Janša knew for sure that "Brnik weapons" did not arrive in Slovenia by arrangement of the Interior Ministry. Bavčar sent Kosi's letter to the new Defence Minister, Jelko Kacin, who five months earlier had replaced Janša. In a note accompanying Kosi's letter, Bavčar wrote: "Your colleagues know this situation well. To achieve his political aims, Janez Janša is willing to use every possible mean, lies, false accusations and insinuations, even if it means walking over his former friends. He obviously needs all of that, if he wants to make himself the only high priest of Slovenian independence. The rest of us who were there obstruct him so much, that he is willing to sacrifice not only his former colleagues but also his friends who at a certain time literally risked everything for him."[664]

Mitja Klavora, then head of the Criminal Police, said: "During the investigation, Janša could have said that weapons were legally imported in Slovenia. Then, nobody would have been prosecuted. But he preferred to demolish Bavčar publicly, and with that he put himself too into an uncomfortable situation."[665]

---

662) After the arrests of three other people involved in the JBTZ affair, the committee changed its name to the Committee for Protection of Human Rights.
663) A letter of Defence Minister Janez Janša to the Interior Ministry, no. 234-7-sz, Ljubljana, May 12, 1992.

\* \* \*

In order to show who was the boss, Janša secretly took away the remaining handguns and sold them to Croatia together with accompanying ammunition. He ordered the transportation of the five containers left at the airport to the Šentvid barracks. On October 26, 1992, a warehouse employee issued a receipt for 6,500 Makarov handguns accompanied by just over three million rounds of ammunition.[666]

Everything was arranged for transportation of the weapons across the border. On October 23, Janša signed permission for their transit across the border to Croatia, explicitly stating that it involved a shipment belonging to the Interior Ministry.[667] That day, Miha Brejc sent a dispatch to the Obrežje border crossing requesting opening of the border for two articulated trucks.[668]

The Makarov handguns and ammunition were purchased by Josip Vukina, the loyal buyer of Slovenian weapons. When four years later he was interrogated by the Slovenian Criminal Police, he remembered this import of the Makarov handguns from Slovenia.[669]

It is not known how much money Defence Ministry advisor Ludvik Zvonar made from selling these weapons to Vukina. Taking into account the lower prices quoted by Kosi, the weapons were worth approximately 1.5 million dollars. However, Janša did not mention these handguns and ammunition in any of his reports, and the Defence Ministry did not record any revenue from that sale. The money simply disappeared.

Franci Kosi openly stated that the Interior Ministry made one million German marks in profit from the sale of the "Brnik weapons", and with that they bought eavesdropping equipment for Special Police officers. Some of the money also went for "bribing customs officers at home and abroad, and also for buying various certificates, and for paying the truck drivers".[670]

---

666) Invoice of the Republic's Headquarters of TD. The weapons were handed over by the head of the Šentvid barracks Anton Perko.
667) Permission for transit: 6,000 Makarov handguns and 3,180,000 rounds of ammunition. MO, Ljubljana, October 23, 1992.
668) A dispatch from the VIS permanent operation centre regarding two articulated trucks, no. 911023, October 23, 1992.

Kosi's cooperation with the police was very costly for him. His colleagues and acquaintances branded him a traitor and when he was questioned by a Parliamentary investigative commission on arms trafficking, he said: "Somebody is threatening me that he will pour acid over my 21-month-old daughter unless I listen to him ... The man who is threatening me knows very well where he can hurt me."[671]

\* \* \*

In February 1995, criminal complaints were filed against people implicated in selling the "Brnik weapons": supplier Konstantin Dafermos, middleman Nikolaj Oman, Interior Minister Igor Bavčar, his advisor Franci Kosi, Defence Minister Janez Janša and advisor Ludvik Zvonar. They were suspected of illegal arms trafficking.

The criminal complaints were written by Criminal Police officer Drago Kos because nobody else would do it. There were complications from the very beginning. Which Prosecutor's office had jurisdiction? "First, we filed the criminal complaint at the Kranj Prosecutor's office, because it was the closest to the Brnik Airport," said Drago Kos. "But none of Kranj's Prosecutors wanted to take the case. They said they did not have geographic jurisdiction. Then we tried in Koper, where the original shipment arrived. There they accepted my criminal complaint against the accused and even commended me. However, they did not start any legal proceedings, and in May 1995 handed the case over to the Ljubljana Prosecutor's office."[672]

Finally, the criminal complaint against the six suspects was taken over by District State Prosecutor Damjan Gantar from Ljubljana, who started working on the case seriously. He requested another interrogation of Janša, but it was blocked by Barbara Brezigar, who at the time was the acting head of the Ljubljana Prosecutor's office. In the meantime, the Criminal Police officers were trying to obtain

---

Ljubljana, March 6, 1995.
671) Kosi's testimony in front of the Parliamentary investigative commission, headed by a member of Parliament from the Slovenian People's Party (SLS), Zoran Madon, Ljubljana, April 5, 1995.

Janša's report on weapons from the Defence Ministry,[673] because they were trying to figure out whether he recorded the sale of the Makarov handguns and ammunition. However, they were unsuccessful in trying to persuade the Ljubljana Prosecutor's office to forward the document. Andrej Šter, Interior Minister at the time, urged Prime Minister Drnovšek to declassify the document but to no avail.[674]

In February 1996, following the recommendation of Chief State Prosecutor Anton Drobnič, Barbara Brezigar took over the group of prosecutors for special investigations. In April, Drobnič took away the "Brnik case" from the Ljubljana Prosecutor's office and handed it over to Brezigar's group. Three months later, the Prosecutors dropped the cases against five of the six suspects, leaving Franci Kosi as the only one still being prosecuted. All the guilt was assigned to the "smallest fish" in the treacherous games played around the "Brnik weapons".

"With this kind of actions by prosecutors, the suspects don't need defence lawyers," said Drago Kos, and added that the greatest hindrance was that Barbara Brezigar was infatuated with Janez Janša."

"Infatuated? Really?"

"Yes, head over heels."[675]

When Prosecutor Brezigar dropped the case, everything was over. The charges never came before a judge. So what happened with the "Brnik case" in the end?

Franci Kosi was subject to criminal prosecution for another three and the half years after the prosecution of the other five was dropped. He decided that he would no longer be quiet and would tell the truth about everything he knew. In December 1999, the Slovenian Chief State Prosecutor dropped the charges against him, and the following March the investigative magistrate did the same.

\* \* \*

---

673) The Report on weapons from May 15, 1990 to September 1, 1992, no. DT/90/91/92/93-DK, MORS, Ljubljana, September 19, 1993.
674) A request from Interior Minister Andrej Šter to Prime Minister Janez Drnovšek, Ljubljana,

Bavčar's Interior Ministry was also selling weapons but – unlike the Defence Ministry – they kept a detailed inventory of quantities of weapons and money received.

At police stations, old weapons meant for reservists were meanwhile accumulating: rifles, machine-guns and ammunition. These surplus weapons that the police no longer needed were in October 1991 sold to Croatia[676] and six months later also to Bosnia-Herzegovina.[677]

For the weapons sold to Croatia, Interior Ministry official Miro Predanič received 400,000 U.S. dollars and 120,000 German marks. He handed the cash over to VIS Director Miha Brejc, who entrusted the money to Deputy Interior Minister Bogomil Brvar. Most of that money was spent on purchasing Puch SUVs and equipment for Special Police Units.

Selling weapons to Bosnia brought just over half a million marks into the coffers of the Interior Ministry.[678] For half this amount they bought automatic rifles and bullet-proof vests. The rest of the money was put into a secret VIS fund, which in part was later used to purchase police uniforms.

The sale of weapons by Bavčar's Interior Ministry appeared to be in line with the decision of the Slovenian Presidency to help Croatia and Bosnia-Herzegovina under attack, and it was in cooperation with the Finance Ministry. Bavčar's deputy Brvar kept detailed records. He later recounted to the Parliamentary investigative commission all the details about the operations, cash received and expenses. When he was faced with accusations of being a arms dealer, he was deeply hurt.[679]

\* \* \*

---

676) In October 1991 they sold to Croatia 1,849 automatic rifles, 851 rifles of different calibres, 72 machine guns and about one million rounds of ammunition. Source: a letter to Prime Minister Janez Drnovšek, MNZ, Ljubljana, September 30, 1994.

677) In March 1992, they sold to Bosnia-Herzegovina 924 rifles, 806 automatic rifles, 149 machine-guns, a quarter of a million rounds of ammunition and material for uniforms. Source: a letter to Prime Minister Janez Drnovšek, MNZ, Ljubljana, September 30, 1994.

678) It was equivalent to 575,000 DEM in different currencies (US$, CAN$, Swiss francs, French francs, Austrian schillings). Source: an overview of the foreign currency spending, received from the sale of weapons by the Ministry of Interior from October 15, 1991 to March 4, 1992, Interior Ministry Ljubljana, without a date.

679) Testimony of Bogomil Brvar during the session of the Parliamentary investigative commission

Like the Defence Ministry, the Interior Ministry also held a secret Austrian bank account at the Die Kärntner Sparkasse in Klagenfurt, with the password *Dragica*. To manage the account, they put in charge Franci Kosi, who on May 17, 1993 walked into the Ministry and threw on the table 338,900 U.S. dollars in cash. He left a note for the Deputy Interior Minister saying that he was handing over the rest of the money from a secret foreign currency account which was closed. He added: "All the funds from this account derived from the budget."[680]

However, this was not entirely true. Hrvoje Šarinić, an influential Croatian politician, deposited a little less than 300,000 German marks and just over 600,000 dollars in not one, but two accounts, under the same password. The money was sent from the Bank für Kärnten und Steiermark in Beljak (Villach), where deposits from Croatian immigrants were coming in to help Croatia in the war.[681] Nobody ever explained where the money came from and how it was spent.

Regarding the account under the password *Dragica*, former Criminal Police officer Drago Kos stated that it seemed relatively clean, which means that the money was used for purchasing weapons and equipment for the Interior Ministry: "However, I do think that there must have been another account, where money had been collected that was meant for certain individuals and political parties. However, we were never able to prove the existence of such an account."[682]

Kos is correct, and from all this it can be inferred that the Interior Minister – and maybe somebody else from his inner circle of confidants – was living a double life: an orderly one, according to the society rules, with an "official secret account", and another one, with a "secretive secret account"[683] for additional needs. It is similar to marital life when a married man affords himself a mistress. In this case, the mistress was the account *Dragica*.

---

680) Kosi's message to Brvar, Ljubljana, May 17, 1993. In the second Slovenian government, the Interior Minister was Ivan Bizjak.
681) Šarinić's telefax to the Bank Die Kärntner Sparkasse, directing it to deposit 295,600 DEM in the account no. 9981-582365 *Dragica* and US$ 609,226 in the account no. 9981-410526 *Dragica*, Zagreb, September 11, 1991.
682) Drago Kos' e-mail, February 17, 2015.
683) The number of this account is 9981-410365      . Another number appeared in the Ministry

Bavčar's Ministry handled the arms business in a more discreet and perhaps more cunning way than Janša's Ministry. However its share in the Slovenian arms trade was 10 per cent or less. The money made by Bavčar's men was peanuts compared to what was going on at the Defence Ministry.

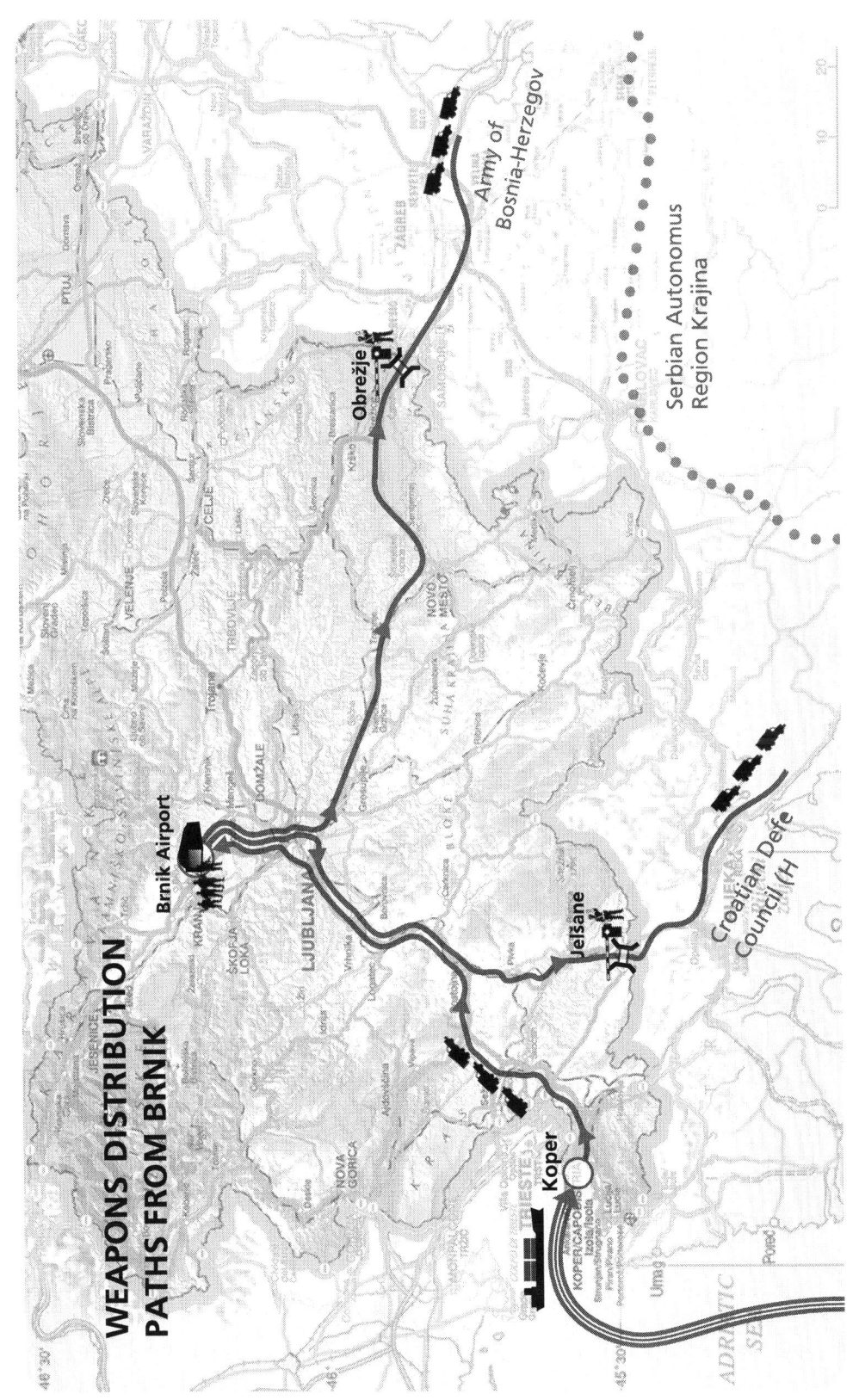

- *Chapter 17* -

# THE DUBIOUS MIDDLEMEN OF SLOVENIA'S ARMS TRADE

Konstantin Dafermos, Vienna businessman and a Greek from Spyridon, was in a bad mood when he walked into the offices of the Austrian police on August 29, 1995. He had been summoned to answer questions on behalf of the Slovenian Criminal Police acting through Interpol. They were investigating the illegal sale of weapons.

He answered the questions of the Austrian police reluctantly and informed them that the sale of weapons was not some kind of a private undertaking. "Franci Kosi always acted as an official plenipotentiary of the Slovenian Interior Ministry," he stated.

Dafermos became irritated when weapons were mentioned: "I never procured any weapons or technical-military assets, but exclusively bulletproof vests, military footwear and other merchandise arriving at the Port of Koper from Russia." He said part of the cargo that Slovenians referred to as the "Brnik weapons" was paid for with a letter of credit into the account of his company Scorpion, another part was settled with cash, and the last part had yet to be paid for by the Slovenian buyers.[684] The Slovenian Security and Information Service (VIS) estimated that at the time, Slovenia owed Dafermos a million and a half dollars, and there is no indication that the debt was ever settled.

Slovenian law enforcement did not find Dafermos' answers to the Austrian police very useful. So VIS started collecting information about him. Dafermos had considerable influence in arms trading circles and did not take on business deals worth less than a million U.S. dollars. He was also familiar with financial tricks – he reported his business deals in one country, and received money transfers in another. He was very favourably inclined towards the Slovenian buyers, offering them weapons at lower prices than to Croatians. Allegedly he also had Slovenian citizenship.[685]

At the beginning of the 1990s, Dafermos was an important contact for the resale of weapons to the Balkans. In the two years from September 1991 to September 1993, his account at the Central European International Bank (CEIB) in Budapest, which he opened on August 28, 1991, received about 86 million dollars or 150 million German marks in deposits from Slovenia, Croatia and Bosnia-Herzegovina.[686]

He used this money to settle bills of arms suppliers. On September 3, 1993, he transferred 13 million dollars to a mysterious company Confin.[687] A company with a similar name, Confin & Oiltrade – established in October 1991 – was one of the eight companies Dafermos founded in Panama.[688] Among those Panama companies was Scorpion International Company, registered in January 1986,[689] and Scorpion Navigation founded in November 1990.[690]

Before Dafermos started to sell arms to the Balkan battlefields, he had already fallen foul of Austrian law enforcement over alleged embezzlement, carrying weapons and other illegal acts. In September 1987, he was sentenced to six and a half years in prison. In December of the same year, the Vienna police detained and then released him after a few days. In April 1990, the Austrian Justice Ministry requested his extradition from Greece, because he had been further sentenced to 16 years in prison in absentia.[691] However, three years later, Interpol announced that his conviction had been overturned.

How did the Slovenian investigators first get on Dafermos' trail? On September 12, 1991, police stopped a Croatian articulated truck near the Slovenian city of Celje, carrying large quantities of mines and ammunition, and accompanied by a passenger car. Mladen Peša, a member of the Croatian National Guard (ZNG) from Nin near Zadar, told the officers that he obtained the merchandise through the Orbis

---

686) A certificate opening account no. 032616, CEIB, Ltd., Budapest, August 28, 1991.
687) A bank statement of the Scorpion Company I. S., account no. 032616 500, CEIB, Ltd., Budapest.
688) Listing in the public register of Panama's companies, no. 252080.
689) A certificate establishing the Scorpion Company I. S., General Directorate of public registries, Panama, July 6, 1990.
690) Listing in the public register of Panama companies, no. 240868.
691) Document from the Federal Justice Ministry of the Republic of Austria, no. 0.12634/19-IV/90,

Company and paid 911,000 German marks for it.[692] Franc Laubič, Ivan Draušbaher's deputy at Orbis, had taken anti-tank and mortar shells from ordnance depots at Pohorje and in Kranj. The defenders of Zadar had had to reach deep into their pockets: they paid an astronomical 1,071 marks each for 120-millimetre calibre shells, even though the Yugoslav Army priced them at only 254 marks.[693]

Such predatory prices were by now not unusual. More significant was the content of a document that the ZNG officer showed to the Slovenian police. It showed that the Slovenian Defence Ministry sent a fax to Orbis containing the name of the Scorpion International Services company and the address and account number of the CEIB in Budapest, together with an instruction that as payment for deliveries a deposit in foreign currency should be made "at this bank".[694]

Slovenian Criminal Police found out that the owner of the bank account was the weapons trader, Konstantin Dafermos. At least two million German marks were transferred to his bank account from Slovenia. However, that is where the investigation stopped. Today, they say that the reason was the lack of money and manpower, but it seems safe to assume that they encountered pressures "from above".

\* \* \*

The Security and Information Service's transcripts of taped telephone conversations included names of people who were selling weapons to Dafermos, mostly imported from Russia. Among them was a Polish national, Jerzy Dembowski, a.k.a *Wirakocza*,[695] former Lt-Colonel in the Polish Military Information Services WSI (Wojskowe służby informacyjne). At the end of the 1980s, with the blessing of the Polish authorities, he set up a company named Cenrex for selling weapons. He collaborated with weapons merchant Monzer Al Kassar, a.k.a. the "Prince of Marbella", and also established

---

692) IS Zadar order for Orbis, Velenje, without a date. 100 anti-tanks mines for 180 DEM apiece, 400 82 mm-calibre mines for 659 DEM, 400 120 mm-calibre shells for 1,071 DEM and 201,000 rounds of ammunition for one DEM apiece.
693) Pricelist of technical-material assets, no. VP3553, Major Petar Stojanović, Ljubljana, August 18, 1991.
694) A report on weapons transports over Slovenian territory, MNZ, no. 0221/12, Ljubljana, June 7, 1994.

contacts with Scorpion International Services. Dembowski had supplied weapons to Al Kassar, which the latter shipped to a war zone in Somalia, using forged end-user certificates (EUC).⁶⁹⁶

Dembowski and the WSI secret service were subject to a domestic investigation, in which a Polish Defence Ministry Commission discovered violations of the U.N. embargo, bribery of Polish officials, issue of fraudulent export licences and concealment of profits. The Commission found that the Polish Military Secret Service tried to cover up the participation of Polish military in illegal arms trading and thus contributed to the growth of organised crime. As a result, the Polish Parliament disbanded the Military Secret Service in 2006. In the same year, the Prosecutor's office in Gdansk brought charges against Dembowski, but he was acquitted by a military court in July 2011.

Slovenia was not mentioned either in the Polish investigation or U.N. reports on arms trading, despite undeniable evidence that in autumn 1991 at least four ships loaded with weapons arrived in Koper from Gdansk.⁶⁹⁷ Part of the cargo consisted of advanced Russian anti-aircraft systems. Documents that have been preserved concerning the three other ships do not show where they were laden, but it can be assumed that for them too it was Gdansk.⁶⁹⁸

Dembowski never came to Slovenia for meetings with the Defence or Interior Ministers. Dafermos would not allow it, because he did not want to reveal his suppliers. In a telephone conversation with his middleman, Nikolaj Oman, he explained the monopolistic workings of the arms trade market: "If Slovenians needs merchandise again, they will have to get it through us, because only Russians and Dembowski have it, and nobody else. They can only order it through the Polish Cenrex, but without me, Cenrex cannot buy the merchandise from the Russians because it possesses the import licence with my name on it."⁶⁹⁹

---

696) The End-User Certificate (EUC) attests that the buyer is also the final user. With this certificate, the seller makes sure that the weapons will not be sold to a hostile country, a terrorist organisation or a country that is under embargo.
697) The ships were: Lyn (September 1, 1991), Ardal (September 21, 1991), Hel (November 20, 1991) and Edith (November 25, 1991).
698) It is assumed that the ships Madget-H (August 9, 1991), Herman Boye (October 19, 1991) and Hel (December 7, 1991) also came from Gdansk.

The Polish company headed by Dembowski was also ordering weapons through other countries, so it would be harder to trace them. The company was involved in the shipment to Koper on the ship Scotia of a massive amount of merchandise, which was later resold to Croatia with the help of Orbis. Cenrex was also connected with the shipment of "Brnik weapons" brought to Koper by the ship Hel. The contracts concerning both vessels, with a start number 96633, prove the presence of Jerzy Dembowski's Cenrex.[700]

In September and October 1991, 9.4 million U.S. dollars was transferred from Dafermos' account in the Hungarian bank CEIB, Ltd, into the account of the Cenrex.[701] This was probably only a small portion, because the arms traders were also using other banks around the world. And of course, they also dealt in cash.

\* \* \*

In autumn 1992, Dafermos' company developed significant cooperation with a company named Global Technologies International (GTI), with offices in Panama, Moscow and Vienna. The company was headed by the Ukrainian businessman Dmitry Streshinsky, reputed to be a member of the "Odessa mafia". From CEIB, Ltd, bank statements, Slovenian Criminal Police found out that in September 1992 and May and September 1993, Scorpion I.S. transferred just over seven million dollars to the account of arms supplier GTI.

On May 1, 1994, Slovenian Criminal Police responded to an urgent dispatch from the Ukrainian Interpol Office, inquiring about Dmitry Streshinsky and the ship Jadran Express, which had been sailing under the Panamanian flag heading to the Croatian port of Reka.[702] Two months earlier, the Italian authorities had intercepted that ship in the southern Adriatic and taken it to the Italian port of Taranto. On board, they found 133 containers loaded with weapons,

---

700) The full contract number for the ship Scotia is: 96633/1/0021/i; and for Hel: 96633/1/0035/E; on the EUC certificate for PDR Yemen, the number is: 96633/2/0014; and on the EUC obtained in Latvia, the number is: 96633/2/0020/E.
701) Company's bank statement, account no. 032616 500, CEIB, Ltd., Budapest, from September 4, 1991 to August 13, 1993.
702) The Slovenian Criminal Police Office response to the Ukrainian Interpol office, name of the case:

of a value estimated at at least 20 million U.S. dollars.[703] The Italian report mentions Belarus as the seller of the weapons, the Ukrainian GTI as the company which made the deal, and Nigeria as the final destination. However, Nigerian authorities declared that the documents were fraudulent.[704] It soon turned out that the weapons were meant for Croatia, which was preparing a decisive offensive to destroy the Serbian para-state on its territory.

After investigation, Italian prosecutors established that there had been violations of U.N. resolutions. Among those they accused were Dmitry Streshinsky, Konstantin Dafermos, Yevgeny Marchuk, former chief of the Ukrainian Secret Service and a former Prime Minister, and Alexander Borisovich Zhukov, a Russian oligarch and a leader of the "Odessa mafia".[705] Zhukov and Streshinsky had business connections through GTI and companies controlling the oil industry.

The Italian authorities were able to put only some of them behind bars. When in 2001, Dafermos was arrested in Turkey on an Interpol warrant, Croatia intervened with the Turkish authorities to obtain his release. Only Streshinsky was convicted, and given a suspended sentence of a few months in jail and a fine of 500 euro. In 2004, a Turin court acquitted all the other defendants with an explanation that the Jadran Express was seized in international waters, where Italy did not have jurisdiction, and the ship was not heading for Italian ports.

From the testimony at the Turin trial it emerged that the "Odessa mafia", in cooperation with a network of international arms smugglers, sent eight ships loaded with weapons to Croatia. The value of the cargoes in the 802 containers on board was estimated at 200 million dollars.[706] The first two of the eight shipments reached Koper on October 23 and November 6, 1992, on a ship named Island. Altogether, this ship brought 96 containers of weapons loaded at the

---

703) In the containers, they discovered 30,000 AK-47 Kalashnikovs, 30 million rounds of ammunition, 400 Fagot anti-tank missiles with 50 launchers, 5,000 Katyusha rockets and 11,000 other rockets. In 2011, these weapons were stolen from the military depot on the island Santo Stefano in the archipelago of La Maddalena near Sardinia. Source: Corriere della Sera, Milan, July 11, 2011.
704) Interpol dispatch no. 50/2848, Kiev, April 30, 1994.
705) Alexander Borisovich Radkin Zhukov is the father of Dasha Zhukova, estranged wife of the Russian oligarch Roman Abramovich. Supposedly, he is also related to Marshal Georgy Zhukov, a famous Red Army general in World War II.
706) , Milano, April

Ukrainian Black Sea port of Mykolaiv. In this port, they also loaded a separate shipment of advanced Russian anti-aircraft and anti-tank weapons, which arrived in Koper on the ship Sabine.

\* \* \*

At the end of summer 1992, a message from desperate defenders of besieged Sarajevo reached Slovenia. President Alija Izetbegović and the Army of Bosnia-Herzegovina were asking for weapons and ammunition to break through the siege of Serbian forces who were shelling the city and killing its residents.

Arms traders from Slovenia, Croatia and Bosnia-Herzegovina swung into action. However, all three republics of former Yugoslavia had become members of the United Nations and were therefore obliged to adhere to the arms embargo against former Yugoslavia set in September 1991. So everybody had to be extra careful.

Dafermos and his aides decided to buy weapons in Albania, the poorest and most isolated European country, which was just freeing itself from communist dictatorship, and possessed enormous quantities of weapons, ammunition and military equipment.

This time, Dafermos' middleman was Nikša Župa, a well-mannered Croatian national, director of the theatre in Split, and later of the Slovenian Ballet. On the Bosnia-Herzegovina side, Hasan Čengić, President Izetbegović's right-hand man, was charged with importing Albanian ammunition. Perhaps the main operational person in this undertaking was Slovenia's Ludvik Zvonar, nicknamed Turn.[707]

On September 11, 1992, Zvonar boarded the ferry Hornbeam in Koper with two aides and 16 empty trucks, and headed for the Albanian port of Durres.[708] Five days later, they returned to Slovenia with the trucks loaded with ammunition sold to Zvonar by the Albanian state company Meico. In the official shipping documents, they wrote that the trucks were transporting 1,065 barrels of bitumen.[709]

The shipment, worth 11.3 million U.S. dollars, still had to be brought

---

707) *Turn* is a colloquial expression deriving from German and meaning a steeple.
708) The aides, Daniel Anžič and Matjaž Prinčič, are co-owners of the company Slotin, which

across Slovenia and Croatia to the Bosnia-Herzegovina Army, defenders of the territorial integrity of the newly formed state.[710] When a certificate confirming that the Bosnia-Herzegovina authorities owned the shipment arrived in Ljubljana, Defence Minister Janša signed a permit for its transit to that country. The permit indicated that the shipment, which transited through Slovenia on September 16, 1992, consisted of three million rounds of ammunition for automatic rifles and 5,608 mortar shells.[711]

There was a hitch at the Obrežje border crossing, because Slovenian anti-terrorism police there counted only 13 trucks rather than 16. After feverish phone calls among those involved, the convoy was allowed to proceed. Later, Zvonar explained that none of the trucks disappeared: "Three of them were empty, so they stayed in Slovenia."[712] Or someone higher up told the anti-terrorism squad to lay off.

\* \* \*

When the ship Island, sailing under the flag of Cyprus, unloaded its 96 containers of weapons and ammunition in October and November 1992, Ministers Janša and Bavčar were supposed to act in coordination with their Croatian counterparts to enable transit over Slovenian territory. However things did not go smoothly.

In autumn 1992, the poorly equipped Army of Bosnia-Herzegovina was not only fighting an unequal battle with the Serbian military, which inherited heavy weapons from the Yugoslav Army, but the relationship between Bosnians and Croatians in divided Bosnia-Herzegovina was getting worse. Armed conflict broke out between Izetbegović's Army and ethnic Croatian irregulars supported by their kin in Croatia.

Hasan Čengić, the main procurer of weapons for the Bosnia-Herzegovina Army said: "They wanted to prevent the Army of Bosnia-Herzegovina from getting the weapons transported over Slovenia and Croatia, and to make sure instead that the Croatian military in Bosnia-Herzegovina would get as many weapons as possible."[713]

---

710) The overview and assessment of activities of certain individuals in the arms trade, VIS, Ljubljana, 1992.
711) Transit permit no. 801/4891, Ljubljana, September 16, 1992.

General Ivan Čermak, the Croatian Deputy Defence Minister, and his assistant, Colonel Vladimir Zagorec, made decisions about the fate of every individual convoy bearing military equipment for Bosnia-Herzegovina. They ordered seizures of shipments, sanctioned ill-treatment of truck drivers, and charged extortionate fees for transits across Croatian territory into Bosnia-Herzegovina. Croatian military analyst Fran Višnar is convinced that they took as much as a quarter or even a third of the value of weapons which crossed Croatia and was meant for the Army of Bosnia-Herzegovina.[714]

His statement was corroborated by the VIS, which reported that Bosnians had to pay to the Croatians 20 to 30 per cent of the value of the shipments as fees. Their report stated that the Croatian Defence Minister and his assistant "privatised decisions about the arms trade with Croatia and Bosnia-Herzegovina and thus created conditions for taking commissions and bribes".[715]

Soldiers of the Bosnia-Herzegovina Army meanwhile waited impatiently for 10,700 shells and almost 10 million rounds of ammunition for Kalashnikovs, and it is possible that the delays in transit influenced the course of the war in their country.

In Slovenia, Defence Minister Janša was the only person signing permits for import, export and transit of weapons across Slovenian territory, and there was no supervision of his decisions. On a permit for the transit of a military shipment that Janša signed on November 6, 1992, he added that the consignment was being temporary stored. Anti-terrorism officials noted that "together with problems regarding the transit documents, there was also a problem with the buyers' agreement to pay for the merchandise".[716]

After about five weeks storage at the Šentvid barracks in Ljubljana, the Defence Minister signed a permit for their transit into Croatia on September 14, 1992, and next morning 28 articulated vehicles drove through the Obrežje border crossing heading south.[717]

So was Janša charging buyers from Bosnia-Herzegovina fees for transport across Slovenian territory? According to Miro Predanič,

---

714) Author's conversation with Fran Višnar, Zagreb, May 4, 2008.
715) The overview and assessment of certain individuals in arms trade, VIS, Ljubljana, 1992, page 58.
716) Information about the arms trade (Nikša Župa), without a date.

head of VIS internal protection, he was. During a home search in summer 1993, Criminal Police officers found in his safe a transit permit carrying Janša's signature and dated December 14, 1992. Predanič readily admitted that the Bosnian purchaser Čengić had to pay a fee of 30 per cent of the weapons' value: "Janša received these 30 per cent in merchandise or cash. The Croatians demanded the same percentage while they were still allowing transits. Čengić had huge costs trying to get merchandise purchased abroad into Bosnia-Herzegovina."[718]

On the Slovenian side, Nikša Župa was a much-respected man of culture. After acting as screenwriter and producer of ceremonies at the 1987 Summer Universiade in Zagreb, he became the artistic director of the Ljubljana's Ballet. But this was not his only career. He had influential connections in politics and business circles, as well as in the criminal underworlds of Croatia, Slovenia and beyond. Nikša was another middleman for supplying arms into Croatia and Bosnia-Herzegovina. At the same time, he was the agent of Dafermos for Slovenia.

In the summer of 1992, he boasted that with the help of Dafermos, he procured for Croatia weapons worth 150 million U.S. dollars. Advisors from Janša's and Bavčar's Ministries were also coming to him to take possession of weapons, as established in the report by VIS, which described Nikša Župa as "a crook who was engaged in the weapons trade only for his personal enrichment. At the same time, he did not possess the qualities necessary for such an undertaking".[719]

In autumn 1994, the Slovenian Criminal Police established that Župa financed the purchase and transport of 32 containers brought to Koper in the second Island shipment meant for Bosnia-Herzegovina. He paid with money collected by the Third World Relief Agency, which was supporting Alija Izetbegović and the Army of Bosnia-Herzegovina. With these weapons and with the shipment from the ferry Hornbeam, he "illegally obtained a large personal gain".[720] However, he did not enjoy this gain for long.

---
718) Official note of the conversation of Criminal Police officer Drago Kos with Miro Predanič, Ljubljana, August 25, 1993.
719) Overview and assessment of certain individuals in the arms trade, VIS, Ljubljana, 1992, page 20.

On August 1, 1994, he was returning home to Pula Airport from Cazin in north-east Bosnia-Herzegovina on a Ukrainian Antonov-26 transport plane, register number UR-26207, piloted by a resident of Bosanska Krajina and with five Ukrainian crew members on board. Soon after take-off it was hit by a rocket and crashed, killing all those on board. The plane came down in the dense forest of Čorkova uvala near the Plitvice Lakes in Croatia, an area of great natural beauty under the control of rebel Krajina Serbs. As the Antonov-26 was on a secret mission, officially there was no flight, and nothing could be found out about it.

In December 1994, a representative of the Krajina Serbs handed over to the Ukrainian Consulate in Belgrade a sealed coffin with remains of the Ukrainian crew members, which were probably taken to Ukraine. In 1996, a special commission of the Croatian military conducted a detailed investigation, but the findings are still classified.[721]

As for Nikša Župa, he had disappeared. His colleagues say he made a mistake by traveling together with the weapons he was in charge of. Experienced arms merchants do not do that, because they never know whether customs or police officers will be waiting for them with an arrest warrant.

Nikša the artist had been transporting weapons to Cazin, an area under the control of the Bosnia-Herzegovina Army's 5th Corps under the command of Atif Dudaković. On behalf of the central government, they were defending the enclave of Bihać from rebel Serbs and forces of the renegade Muslim leader Fikret Abdić. So who was Župa selling weapons to? The Security and Information Service (VIS) says he had a good relationship with President Izetbegović and his confidants, but also with the Interior Minister of Bosnia-Herzegovina, who was favourably inclined towards the "autonomist" Abdić.

Three months after Župa's death, the Slovenian Criminal Police searched his apartment in Ljubljana, and seized documents detailing technical characteristics of Russian MiG fighter jets and Scud ballistic missiles.

---

721) Forensic documentation regarding examination of the crash site of the Antonov-26 airplane,

* * *

So much for the mysterious demise of Nikša Župa, but what about Croatian General Vladimir Zagorec, who had so much say about weapons transiting from Slovenia through Croatia to Bosnia-Herzegovina?

In the Balkan wars, it would be hard to find anybody promoted faster than Zagorec. He started as an assistant geodetic technician and by the time he turned 30, he was already Assistant Defence Minister and Director of the state-owned company RH Alan, which bought and sold arms. Zagorec had great authority and almost free access to the state's money. Croatian journalist and analyst Fran Višnar described him as "Tuđman's right-hand man".[722]

The Slovenian Ludvik Zvonar brought Zagorec into the world of international arms trading: "On Janša's order, I introduced Zagorec to our foreign suppliers. I'm convinced that Janša was influenced by the Croatian Defence Minister Šušak and his assistant Čermak. When I first met Zagorec, he gave me the impression of a helpless young man who couldn't even afford a cup of coffee."[723]

Though he never smelled the black powder of a battlefield, Zagorec became the most decorated Croatian General. He also turned into one of the richest Croatians and moved to Vienna, where in 2000 he became Dafermos' business partner and advisor to his Scorpion International Services Company.[724]

There his shady past caught up with him, when criminals kidnapped his underage son. During a subsequent trial against the kidnappers, it turned out that one was a former associate and friend of Zagorec, Hrvoje Petrač, a leader of the Croatian underground. During the trial in which he was convicted, Petrač accused Zagorec of stealing jewels allegedly worth five million U.S. dollars from a safe in the Defence Ministry. So General Zagorec found himself under investigation, and in autumn 2008 Austrian authorities extradited him to Croatia, where he was sentenced to seven years in prison and ordered to return money equivalent to the value of the jewels.

---

722) Author's conversation with Fran Višnar, Zagreb, May 4, 2008.

Fran Višnar said Zagorec scrupulously exercised the "mental discipline" typical of the mafia, and did not rat on anybody during his trial. Višnar is convinced that the greatest enemies of weapons merchants are – other weapons merchants. "The competition in this line of business is fierce."[725]

\* \* \*

Ludvik Zvonar becomes upset when mafia and international criminals are mentioned. He goes to great lengths to present the arms purchases for the emerging Slovenian military as a very hard but legal work, where almost at every step "you were risking your life".

"It should be understood that those were deals among countries, and each country had its own task. The one that contributed the money did not sell weapons itself. For that, others were tasked, and they were procuring weapons from third ones ... Americans were monitoring the transports, and usually there was also an agent of the Israeli Mossad. Ships full of weapons are not toys you can put in your pocket." He reckons that the ships' captains were closely connected with western intelligence services. "As soon as a ship arrived in Koper, its captain would take off across the border, usually to Trieste in Italy."

Germany gave loans, and the weapons were Russian. Were they arms that the Red Army left in Eastern Europe when it retreated to Russia?

"Don't ask me about that. Our supplier was Dafermos and his middleman Oman. I don't really care where they got the weapons from." Zvonar met Dafermos in Vienna several times. He described him as a polite man with a dark complexion which gave away his Mediterranean origin.[726]

When interrogated at the Vienna office of Interpol in 1995, Konstantin Dafermos stated: "I no longer have contacts with Kosi and Oman. Now I deal in real estate, and the Scorpion Company has been terminated."[727] However the Panama registry of companies shows that in March 1994

---

725) Author's conversation with Fran Višnar, Zagreb, May 4, 2008.
726) Author's conversation with Ludvik Zvonar, Radovljica, June 28, 2010.

only the company Scorpion Navigation ceased to exist.⁷²⁸ There is no information about a liquidation of Scorpion International Services.

The Scorpion I.S. web page offers equipment for civil defence, firefighters and electronic warfare. The company is the exclusive representative of military manufacturers such as the Russian State Monopoly Rosoboronexport and Irkut Corporation (manufacturer of Sukhoi jet fighter), and Indra Sistemas, a Spanish company dealing in informatics technology and defence systems.⁷²⁹

\* \* \*

On conviction, Vladimir Zagorec lost his General's rank and medals; however, he did not remain idle in prison. He announced lawsuits against witnesses, whom he accused of landing him in prison through false testimonies. He also took university classes and received a master's degree with a paper entitled: "Helicopter Interventions in Accidents". After serving two-thirds of his sentence, he was released on parole – he received the highest grade for good behaviour – and immediately traveled to Vienna to see his friend and protector, Konstantin Dafermos.

He seemingly had a great future ahead of him: no more trading with death, but instead working with new technologies and real estate. However in December 2014, the Austrian State Prosecutor's Office filed an indictment against him.

Zagorec and three others, among them Günter Striedinger, former number two executive at the Hypo Alpe Adria Bank, were accused of embezzling almost 20 million euro from that bank, which is located in Celovec, Austria. Four of Zagorec's companies, registered in Liechtenstein and Switzerland and all without any equity, purchased several hundred thousand square metres of land on the Croatian coast. However, the prices that these companies actually paid for the land were significantly lower than the inflated prices shown in documents of the Austrian bank, which financed these business "undertakings". The difference in prices – according to the Austrian Prosecutor's office – was pocketed by Zagorec.⁷³⁰

This relatively small Austrian bank entered the Croatian market in 1994 before the war in Croatia ended. Today, its name is associated with many suspicious dealings in the Balkans. Hypo Alpe Adria Bank – with a blue hippopotamus as its trademark – became synonymous with money laundering, corruption and abuse of power – in other words, organised crime.

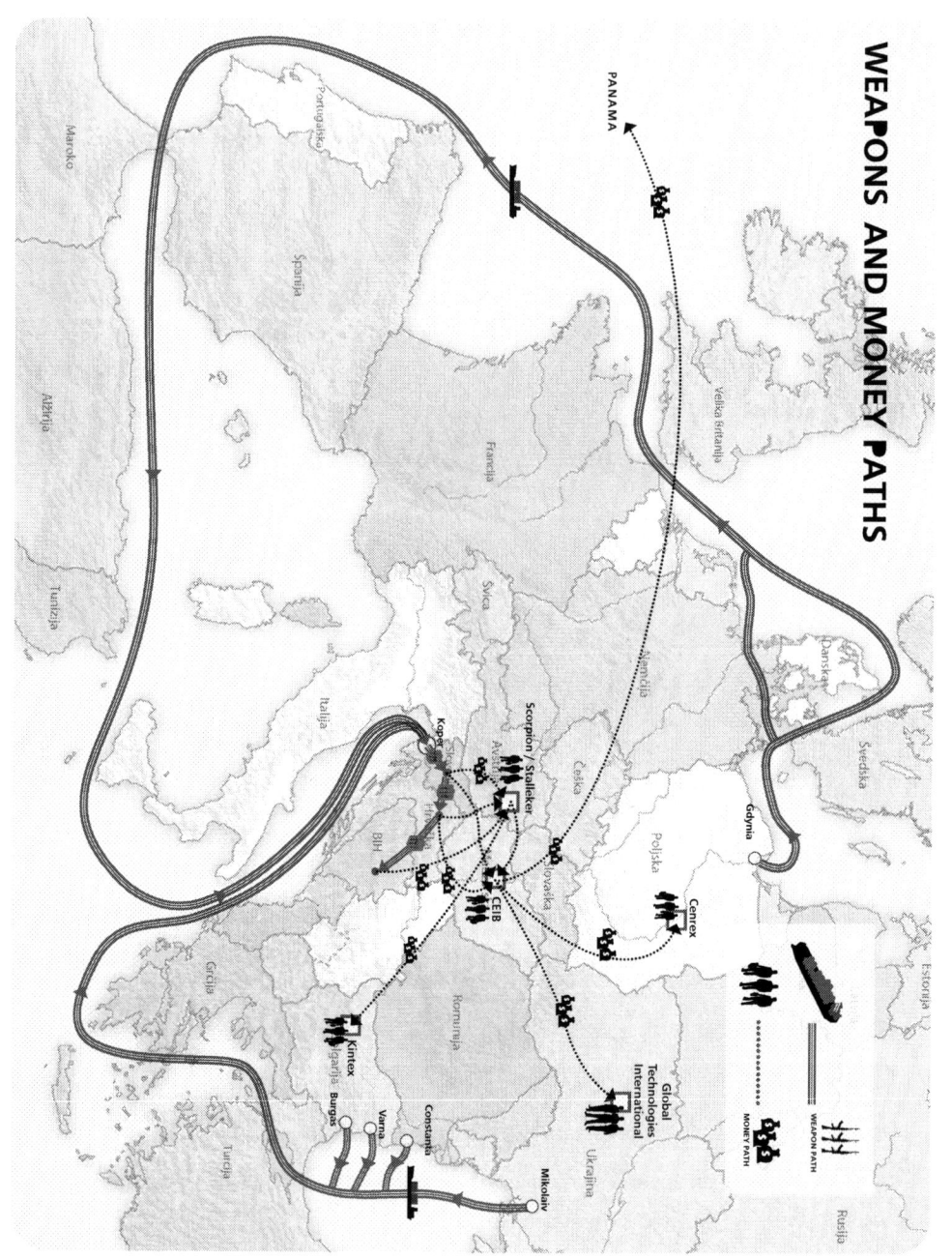

- *Chapter 18* -

# ARMS DEALER OMAN, JANŠA AND THE MYSTERY OF FUNDS IN AUSTRALIA

In the 1990s, when Dafermos and Slovenian businessman Nikolaj[731] Oman were supposedly "big fish" in the arms trade, Oman contributed greatly to making newly-independent Slovenia known in Europe, but unfortunately not so much in political and diplomatic circles as in the world of crime.

Oman was born during the World War II as the youngest child of a large workers' family in the idyllic mountain village of Podkoren, close to an intersection of the borders of Slovenia, Austria and Italy. At the age of only 20, he already had a police record: for a burglary he was given a suspended sentence of four months, and for breaking into an apartment eight months in jail.[732] He did not serve the time because he managed to leave for Australia, where he took up residence in Mordialloc, a suburb of Melbourne. There, he was arrested and convicted of making an armed attack. Police also found illegal substances in his residence.

Later he became a pilot and traveled frequently to Africa, and for a few years he worked on the archipelago of Tonga in the South Pacific. Before Oman met the new leaders of Slovenia, he was in touch with Licio Gelli, a former fascist who became Grand Master of the *Propaganda Due* (P2) Freemason lodge, a mysterious right-wing network of rich industrialists, media tycoons and influential politicians. Gelli was implicated in a scandal involving the disappearance of one billion dollars from the Vatican's Banco Ambrosiano. He was also connected to the financing of terrorists and a lethal bomb explosion at Bologna train station in 1980. He was sentenced several times, but was more often under house arrest or on the run.

---

731) Nikolaj is his birth name, but outside Slovenia he is known as Nicholas or Nicholas Alexander,

Oman confirmed that he visited Gelli three times as a plenipotentiary of Tonga's government, which was looking to attract foreign investments. In a meeting in Arezzo, Oman stated that Gelli showed most interest in creating a rubbish dump on Tonga in return for permission for him to retreat to one of the Tonga islands and receive immunity from prosecution. However, the deal fell through.[733]

Oman kept in close contact with his homeland, and when he returned to Slovenia, he was already rich. His aide and former friend Jernej Čepin said Oman told him that he cooperated with Iran during its war with Iraq in the 1980s. "At that time, he allegedly made 35 million U.S. dollars. With that money, he went into the arms trade business," said Čepin, and added that in 1990 Oman traveled to Iran again.[734]

According to information obtained by the Slovenian Security and Information Service (VIS), Oman was allegedly asked by the U.S. government to help establish ties between the United States and Iran regarding oil purchases. While looking for a middleman, he was negotiating with the Slovenian oil company Petrol but to no avail. Slovenian intelligence officers found out that for his services, Oman received from an Iranian bank a large sum of money, which he deposited into an American Bank in New York. According to VIS, Oman was also selling steel to Iran and made contacts with Yugoslav steel mills; however, the deals fell through when it turned out that he was using forged bank guarantees.[735] In a later interview, Oman stated that he made his money in Australia in real estate. He also earned money investing 20 million U.S. dollars in securities and letters of credit in Gorenjska banka and Jugobanka before the breakup of Yugoslavia.

\* \* \*

Oman returned to Slovenia a few weeks before its declaration of independence. He was staying in the former residence of Tito in Brdo pri Kranju, and in August 1991, he moved to an apartment above a police station in the Ljubljana district of Bežigrad.

---

733) Police interrogation of Nikolaj Oman in the investigation case Cheque to Cheque, Vico Equense, Italy, November 7, 1996.
734) Jernej Čepin's interrogation in the investigation case Cheque to Cheque

He boasted that he was an advisor to the King of Saudi Arabia, and in 1990, before the outbreak of the first Gulf war, he acquired a large amount of gas masks for Saudi Arabia, which refused however to accept them.[736] Ludvik Zvonar remembers this episode differently: "The Federal Directorate for Commerce, which was a Yugoslav state institution, bought the masks in Ukraine, planning to sell them to Iraq, but the Iraqi authorities refused to accept them, so they ended up in Slovenia."[737] Thus, in August 1991, two ships unloaded containers filled with 874,720 Soviet-made GP-5 gas masks in Koper, and Oman requested just over nine million U.S. dollars in payment for them.[738]

Oman tied the purchase of the gas masks to any future Slovenian acquisition of weapons. Zvonar had no other choice but to sign the order for them and was aghast when he saw the huge quantity of masks in the port. One purpose of Oman's shipment was to test the security of navigating in the Adriatic with sensitive cargo. He was concerned that the Italians or international monitors might seize ships or even sink them.[739]

So the gas masks got stuck in Koper. The official supplier, Dafermos's Scorpion, expected payment from the Slovenian Defence Ministry. However, the masks were mostly unusable. People who saw them said rubber parts had deteriorated and dust was coming out of them, as the filters were well past their expiry date.

A few years later, at the beginning of 1994, Minister Janša signed a contract with the Director of the Port of Koper to transfer ownership of the masks to the Defence Ministry and settle a bill of some 20 million Slovenian tolars for storage of the masks.[740]

Even though the masks were no good, they were for many years a point of contention between Oman and the Slovenian authorities. In 1998, the matter was taken up by a Parliamentary commission headed

---

736) Letter from Oman's colleague Lorenzo Mazzega to the British journalist Paul Holden, London, June 22, 1996.
737) Author's conversation with Ludvik Zvonar, Radovljica, December 29, 2010.
738) The ship Madget-H arrived in Koper from the Bulgarian port of Varna, and on August 9, 1991 unloaded 467,480 gas masks; on August 11, the ship Herman C. Boye brought 407,240 gas masks. The official importer was the Podkoren branch office of the Orbital Marketing Services Company. The branch office's owner was Nikolaj's brother, Friderik Oman.
739) "Nikolaj Oman, portrait of the week", *Delo Saturday edition*, Ljubljana, June 29, 1996.
740) A contract to acquire gas masks, Luka Koper, January 11, 1994. 20 million Slovenian tolars =

by Marjan Podobnik but without much success. A small quantity of the masks was later sold to Bosnia-Herzegovina, but most of them are to this day decaying in the Šentvid military depot in Ljubljana.

In the meantime, Nikolaj Oman was buying real estate in Slovenia and living the high life. He moved to the town of Bled and bought the small Grimšče chateau nearby and then a mansion named after Swiss hydropathist Arnold Rikli, who pioneered spa-tourism in Bled. Oman then rented the prestigious castle overlooking Lake Bled, where he hosted receptions and feasts. Among the guests were Slovenian ministers, businessmen, weapons traders, traffickers, forgers and mafia types from all over the world.

One of the participants of those VIP gatherings, the Italian Fulvio Leonardi, remembers that in the castle he met Georgian-born General Vladimir Kuzin, who introduced himself as a former president of the Soviet aircraft-maker Sukhoi. He admitted that he was not only selling military and civilian aircraft, but also weapons from the former Red Army. He wanted to sell several thousand tanks, which were positioned before 1990 on the Russian-Polish border. According to Leonardi, General Kuzin openly presented himself as a high-level arms trader.[741]

Spared from the devastation of the Yugoslav war, Slovenia offered Oman an ideal springboard to supply arms to other Balkan states. In autumn 1991, he was selling anti-aircraft weapons to Croatia. The invoice that Oman's company, Orbal Marketing Services, sent to the Croatian Defence Ministry shows that Oman wanted 15.6 million U.S. dollars for 80 rocket launchers and 400 missiles the British Blowpipe Mobile System.[742]

Middleman Josip Vukina had to buy these anti-aircraft systems as a condition for purchasing Kalashnikovs and ammunition which arrived in Koper on the ship Scotia, since the Blowpipes were part of a tied selling arrangement between Oman, the Slovenian Defence Ministry of the Croatian buyers.

However, in the following three weeks Oman had not received payment for these weapons from the Croatians, so on November 23,

---

741) Fulvio Leonardi's statement to the British journalist Paul Holden, London, June 5, 1996.
742) For MK40 A2 rockets, he charged US$ 28,000 apiece and for launchers, US$ 55,000 apiece.

1991 he sent to the Croatian Ministry of Defence another invoice with the same quantity and the same prices for the Blowpipe anti-aircraft system. This time, he added a charge of 113.96 million dollars for 1,400 Igla SA 16-E anti-aircraft missiles and 200 launchers.

His inflated prices were 74,800 dollars apiece for the missiles and and 36,200 dollars for the launchers.[743] Only two months earlier, Zvonar was buying the same Igla missiles for 48,000 dollars and launchers for 28,000 dollars.[744]

\* \* \*

While Croatia was paying dearly for Oman's weapons, Oman himself was embarking on a new venture concerning Liberia, an African country where in 1989 one of the most brutal wars in the history of Africa erupted.

As a result of fighting among different military factions and ethnic groups, Charles Taylor, who used to live in the United States, was elected President of Liberia in 1997. Taylor became rich by selling so-called blood diamonds and, due to the violent nature of his regime, was responsible for the outbreak of a second civil war in 1999.[745] Taylor's murderous path ended in 2012 at the Hague Tribunal, where he was sentenced to 50 years in prison for terror, murder and other atrocities.

In the summer of 1992, the Liberian regime informed the Slovenian government that it wanted Nikolaj Oman to represent Liberian interests in Slovenia. Ljubljana agreed and the Slovenian-Australian businessman became honorary consul of Liberia, receiving a diplomatic passport and diplomatic immunity.

The authors of the book The Shadow World wrote that Oman received his status in return for helping Liberian elites sell diamonds and buy weapons – despite a U.N. embargo against Liberia put in place in 1992.[746] In July 2009, a 10-member commission for truth

---

743) Orbal Marketing Services, invoice no. 91717, November 23, 1991.
744) 240 Igla SA 16-E rockets and 24 launchers arrived with the ship Ardal on September 21, 1991.
745) The first civil war in Liberia lasted from 1989 to 1997, and the second one from 1999 to 2003. In 20 years, more than 250,000 people were killed, more than a million fled the country and the economy was devastated.

and reconciliation formed by the Liberian parliament established that Nikolaj Oman was involved in illegal trading of arms and that he helped other individuals from Europe do the same.[747]

\* \* \*

Then this Slovenian-Australian merchant of death started trading with the leader of the Bosnian Serbs, Radovan Karadžić, an extremist from the town of Pale, which Bosnian Serbs made capital of their parastate known as Republika srpska during the war in Bosnia-Herzegovina. Karadžić, a psychiatrist, poet, politician and *guslar* (gusle player),[748] subsequently became one of the most hunted Serbian war leaders. He was arrested in Belgrade in 2008 and is currently defending himself at The Hague Tribunal on charges of war crimes, including the killing of more than 8,000 Bosnian Muslim men and boys from the Bosnian town of Srebrenica in 1995.

Karadžić dreamed of possessing a miniature nuclear bomb which could destroy his enemies once and for all, and sure enough Oman turned up in Pale and offered him exactly that. Lorenzo Mazzega, who participated in the meeting between the two men, remembers the discussion thus: "The leader of the Bosnian Serbs showed us oil fields in the area that was under his control. He was convinced that the Republic of Srpska would become another Switzerland and asked us to deliver a request for the recognition of the Republic of Srpska to Rome and the Vatican."[749]

Nikolaj Oman, who introduced himself as the Liberian General Consul in Slovenia, gained Karadžić's trust and early in 1995, they reached agreement that Oman would deliver to the Bosnian Serbs a "vacuum bomb" and several Russian-made BM-30 Smerch multiple rocket launchers. The value of the deal was 60 million U.S. dollars, of which Oman received 10 per cent. Mazzega brought the dollar bills to Slovenia in his car.

As a guarantee for the rest of the payment, Orbal Marketing Services received a lien on land near Bosanski Brod, where a state oil

---

747) Available on http://trcofliberia.org/resources/reports/final/volume-three-3_layout-1.pdf (last

refinery was to be built. While Oman was counting his dollar bills in his Bled chateau, Karadžić, in front of the highest representatives of the Republika srpska in Pale, opened a brass box containing "raw material" for a strategic weapon weighing just two kilogrammes. However, instead of "red quicksilver", the box contained table tennis sized balls filled with a reddish jelly. The substance was a harmless gel, maybe suitable for shaving, but definitely not for the mass destruction of enemies.

The scam committed by the honorary consul would probably have stayed hidden if it were not revealed in the New York Times by Predrag Ćeranić, head of the secret service of Bosnian Serbs, who subsequently deserted Karadžić for his Bosnian-Serb rival, Biljana Plavšić.[750]

"Table-size" nuclear bombs do not really exist. Eastern and western intelligence agencies however frequently used harmless gel-like "weapons" as bait for discovering secret channels and criminal organisations. Oman's hoax therefore may corroborate gossip that he was working for intelligence services of certain other countries. When Karadžić realised that he had been cheated, he angrily demanded Oman's head and the return of the pre-payment. For that task, he hired a member of the Serbian underground, Branislav Lainović, called *Dugi* (Lanky).

In the meantime, the Slovenian crook sent his friend Mazzega and a Russian to Bosnia, to persuade the Bosnia-Serbian leadership in Pale pay the outstanding 54 million dollars. However, they encountered *Dugi*: "I found them in a hotel in Jahorina, above Sarajevo. I took their passports away and threatened that they would not be going anywhere until Oman returned the money, but at that very time, American aircraft started to bomb Serbian positions, so they took advantage of the confusion and escaped."[751]

Branislav Lainović traveled to Slovenia to see Oman, who told him that he had already given six million dollars to the Russians to supply weapons. Oman offered 1.3 million dollars to his pursuer to spare his life and suggested to tell Karadžić that he had escaped from Europe without a trace. *Dugi* took the money and followed Oman's

---

750) "An old tale of swindle resurfaces in Bosnia",

advice, which turned out to be a fateful decision, because soon after he was dead. At the time, almost every major police organisation in the world was tracking Oman. Even though the Security and Information Service (VIS) had been eavesdropping on Oman since October 1991, Slovenian prosecutors in this case too chose not to act.

In the meantime, prosecutors from the Italian city Naples launched a vast investigation called Cheque to Cheque. It concerned a criminal enterprise, whose members were Oman, a Swiss attorney, a notary from Vienna and a Canadian businessman, suspected of trafficking arms, radioactive substances and precious stones, as well as money-laundering and illegal operations with cheques and other securities. In June 1996, Italian police opened a safe in a bank near Venice and found glass ampoules with osmium – a strategic substance to make detonators for nuclear bombs – as well as a large sum of cash, reportedly 30 million German marks, and a large quantity of documents on arms trading and money transactions. The content of the safe was registered in Oman's name. Despite this evidence of Oman's criminal activities, Slovenian law enforcement was hard put to lay hands on him. He was protected by his diplomatic immunity and had received a Slovenian medal for his contribution to the Slovenia's independence. Perhaps they were also concerned that sifting through Oman's suspicious dealings would expose Slovenian politicians.

\* \* \*

Nikolaj Oman bragged to his friends that he possessed documents about the sale of weapons to Slovenia amounting to 50 million dollars. Jernej Čepin, in his letter to the government headed then by Janez Drnovšek, stated: "I estimate that in just over a year, Oman made at least 10 million dollars from arms trafficking at the expense of the Slovenian and Croatian taxpayers."[752]

Oman told his interrogators at the police station in Vico Equense near Naples that he had no dealings with Čepin. He denied everything, even his contacts with Karadžić. For him, all the accusations about illegal dealings were only a conspiracy of the Slovenian politicians. He only admitted that he was buying weapons for Slovenia "just before the declaration of Slovenian independence" and asserted that Slovenian politicians were "as their private undertaking, selling surplus weapons to other republics of former Yugoslavia at prices that were three times higher, or even more".[753]

Slovenian authorities finally woke up. In August 1996, they took away Oman's exequatur consular recognition, the Liberian flag was no longer hoisted over the chateau of Grimšče, and they removed a plaque saying: Honorary Consulate of the Republic of Liberia in the Republic of Slovenia. In November the same year, Slovenian police who searched the chateau on suspicion of money laundering and alcohol smuggling found a large quantity of cash, gold bars and precious stones.

Oman had to leave the chateau because he stopped paying the rent, while an attempt by the municipality of Bled to take back the Rikli villa was unsuccessful, and the building collapsed. A month later, Oman was expelled from Slovenia.

In the meantime, Interpol issued a warrant for the arrest of Oman and his accomplices; however nobody involved in the Italian prosecutors' extensive Cheque to Cheque investigation was ever charged. At most, they inconvenienced members of criminal organisations by obliging them to spend time in detention during interrogations.

In February 1998, Slovenian police issued another international warrant for Oman's arrest, this time on suspicion of tax evasion, embezzlement and fraud, but the warrant was soon canceled, supposedly for lack of evidence. When Oman returned to Australia, Slovenia requested his extradition, but to no avail. In November 2006, a news report broke that Oman was on trial in Melbourne, charged with paedophilia. Judges saw film footage showing abhorrent abuse of children during his stays in Liberia and Thailand from October 1999 to December 2001. One of the victims was only five years old. Oman was sentenced to six years in prison and will be registered as a sex offender until his death.

However, Oman had one more hoax up his sleeve to play on Slovenia. Soon after he was released from prison in 2013, he informed Igor Lukšič, President of Social-Democrats (SD) in the government coalition that he would like to donate to Slovenia a large piece of land in Bosnia-Herzegovina supposedly rich in oil deposits.

Lukšič spoke about Oman's offer to Milan Balažic, Slovenian ambassador in Australia, who was a friend and colleague from school. Balažic gained the impression that the Slovenian leadership had entrusted him with an important task, and in a few days, he received a bunch of papers in which the convicted paedophile tried to prove the seriousness of his altruistic offer.

Balažic deduced from the documents that Oman had "lent" the government of Republika srpska 60 million U.S. dollars several years previously, but the country was not able to return the money, so Oman was repaid with land, supposedly abundant with oil and including a Russian oil refinery.[754] In reality, Oman did not lend any money and he was also never the owner of that land in Bosnia-Herzegovina. A refinery is nearby, but not on the property, and it is not known whether there are oil deposits underground.

It is true that the according to the contract with Karadžić, Oman received "oil land" near Bosanski Brod as collateral. However, when he delivered glycerine balls instead of materials for a lethal bomb, the collateral was taken away from him. Today, Karadžić's former aides confirm that the land was valid as a guaranty for payment only for a month and a half.

During his first meeting with Oman, ambassador Balažic was impressed by Oman's "persuasiveness" and "dependability". He informed Lukšič that the "benefactor" had his own wishes. As part of the "suitable attentiveness" that should be granted to Oman by the Slovenian state, the ambassador mentioned issuing Slovenian passports to him, his son and his daughter, a peaceful life in his Bled chateau and repayment of debts relating to favours that Oman did for the Republic of Slovenia.[755] He apparently had in mind nine million U.S. dollars for gas masks and 300,000 German marks that the Defence Ministry supposedly owed him.[756] Balažic and Lukšič exchanged many e-mails, trying to figure out how to "save the country".[757]

---

754) From the writing "About the background of the recall" that Milan Balažic presented to the journalists, Ljubljana, July 8, 2014.
755) From Milan Balažic's e-mail to Igor Lukšič, Canberra, June 24, 2013.
756) "Is trading with osmium illegal in our country?" asked Nikolaj Alexander Oman in surprise – interview with Nikolaj Oman, *Nedelo*, Ljubljana, June 1996.

When it was becoming more and more clear that there was something wrong with registering the land in question into the land registry in Bosnia-Herzegovina, Oman explained to the ambassador that the Bosnian Serbs were not trustworthy. He suggested that he would use a deed of donation, to transfer his Liberian company, the owner of the land, to Slovenia, which would in turn negotiate with the Russians who owned the refinery.

"The key factor is a paper that attests that Oman's company owns the land and Oman keeps this document in a safe place," reported Balažic to an excited Lukšič. The latter insisted that all activities regarding Oman's demands should be transparent, and he kept Prime Minister Alenka Bratušek abreast of the secret diplomacy, asking her to be discreet so "nothing could be derailed".[758]

In December 2013, Oman received a Slovenian passport and planned to visit Slovenia with a group of wealthy Australian businessmen owning large companies, to introduce revolutionary projects to Slovenia, including an environmentally-friendly recycling system for sewage waste. However, the plan blew up on March 7, 2014, when Nikolaj Oman showed up as an uninvited guest at the opening of the Slovenian Consulate in Melbourne. The story about the unwanted guest appeared in public and the secret diplomacy turned into loud, mutual accusations.

Slovenian Foreign Minister Karel Erjavec came to the conclusion that Milan Balažic violated the law through his contacts with the convicted paedophile, because he did not inform his superiors about it. He recommended that Balažic be recalled from his position as Slovenian ambassador to Australia, and this was approved by the Government. The Prime Minister Alenka Bratušek denied its involvement in the secret diplomacy, while Lukšič, who was not a member of the government, kept quiet. Milan Balažic was left hanging without a job.

However, more than the ageing swindler Oman, Slovenia should be interested in finding out who is guarding the treasure from the sale of weapons, which may well be hidden in Sydney and Melbourne banks.

\* \* \*

Janez Janša visited Australia at the end of 2006. At first glance, his visit did not represent anything unusual, but at the time, he was Prime Minister of Slovenia and his visit was not officially recorded anywhere.

The Slovenian Ministry of Foreign Affairs reported that Janša, during his term in office between 2004 and 2008, did not make an official or working visit to Australia, but it did not want to comment on any of his private trips or meetings.[759]

During his trip to Australia, Janša met with Australian businessman of Slovenian origin Dušan S. Lajovic. Slovenian Prime Minister was accompanied by his bodyguard, and at one instance, he told the bodyguard that he did not need him because he was going to take care of some personal business. However, the conscientious bodyguard did not relinquish his obligation and secretly followed Janša. He saw that the Prime Minister was visiting banks, but why remains a secret.

In January 2015 Dušan Lajovic turned 90. He comes from a wealthy family that owned the Tuba Company in Ljubljana, and in 1950, he emigrated to Australia, where he continues the family tradition of making tubes for toothpaste. As a supporter of Janša's SDS party, Lajovic made a large donation to support SDS journals, which Janša acknowledged on March 2, 1999. However, it looks as if this money from Australia did not go into a safe of the company that was publishing the journals.[760]

Dušan Lajovic was 51 per cent owner of the publishing house *Nova Obzorja* (New Horizons), which publishes the SDS newspaper and publications, the other 49 per cent being owned by the SDS.[761]

Lajovic introduced Slovenian businessman Viktor Baraga to the Australian business world, and Baraga responded by introducing Lajovic to his childhood friend, Janez Janša. Today Viktor Baraga is honorary consul of Australia in Slovenia. According to U.S. diplomatic cables revealed by WikiLeaks, Baraga admitted to the

---

759) E-mail of the Ministry of Foreign Affairs, Ljubljana, January 13, 2013.
760) The proof for that claim is in letters to Janša and Lajovic. In order to protect the individual who provided this proof, the author does not want to reveal the individual's identity.
761) Source: *Agencija za javnopravne evidence in storitve* – AJPES, (the Agency for public records and services), Ljubljana, January 14, 2013. In June 2017, Lajovic

U.S. ambassador in Slovenia that he became the president of the supervisory boards of the Slovenian companies HIT and Petrol, largely because of his longtime friendship with Janša.[762]

In spring 1992, Baraga was middleman for the purchase of weapons and ammunition from Chartered Industries of Singapore (CIS).[763] It concerned procurement of automatic rifles, machine guns, grenade launchers, Armbrust rocket launchers and ammunition, and their sale to Bolivia.[764] Proceeds from the deal were 759,630 U.S. dollars, which Bolivian authorities transferred to the account of VOMO Director Andrej Lovšin in Celovec's Bank für Kärnten und Steiermark. The money was later re-wired to the account of the company Euro Trading, with which Viktor Baraga was connected, at the Commonwealth Bank of Australia in Melbourne.[765]

Janša, who was Defence Minister at the time, wrote to Baraga with "elements of the request for an offer from the CIS", containing a list of weapons and instructions on implementation of the deal.[766]

Even though all these happened in 1992, it is certain that Prime Minister Janša did not come to Australia at the end of 2006 only to buy chocolates.

Australia is an orderly western democracy, with a sophisticated legal system that protects financial information. Despite that, foreign governments are in principle entitled to receive help from Australian authorities to gain access to information about Australian bank customers. They only have to submit persuasive arguments to the Australian authorities to be able to check the relevant financial transactions.

Will Slovenian law enforcement be capable of investigating illegally gained wealth with traces leading to Australian banks?

---

762) As seen in a cable of the U.S. Embassy in Ljubljana, Source: "Friends", *Mladina*, Ljubljana, September 9, 2011.
763) Back in 1990, Slovenia purchased from the Singapore company CIS SAR-80 rifles, which were of bad quality.
764) The receipt no. 30021, dated March 31, 1992 issued by the Ministry of the Republic of Bolivia for Internal Affairs, Migration, Justice and Defence lists the following types and quantities of weapons, worth US$ 759,630: 70 automatic rifles, 70 machine guns, 50 hand-held Armbrust rocket launchers, 510 grenade launchers and 1.25 million rounds of ammunition of various calibres.
765) Euro Trading Pty, Ltd., 3 Dalmore Drive, Scoresby, Victoria, 3179 Australia, Bank account number at the Commonwealth Bank of Australia (367 Collins Street, Melbourne, VIC 3000), is 1135-600966.

*- Chapter 19 -*

# DID SREBRENICA DIE FOR THE FREEDOM OF CROATIA?

In a referendum in March 1992, more than 62 per cent of the voters of Bosnia-Herzegovina voted for independence.[767] On April 6, 1992, the European Community recognised the independence of Bosnia-Herzegovina, followed two days later by the United States, which also recognised the independence of Slovenia and Croatia. But already the bloodiest war in Europe after World War II had erupted in Bosnia–Herzegovina. At the beginning of April, Arkan's butchers invaded the city of Bijelina in north-east Bosnia-Herzegovina wearing uniforms of the Serbian Volunteer Guard and in front of TV cameras started to kill non-Serbian residents of the city.

The constituent nations of Bosnia-Herzegovina – Serbs, Bosniaks (Muslims) and Croatians – were mostly organised in three national parties which could not find common ground. Bosnian Serbs, as well as Croatians in Herzegovina and the Posavina area alongside the Sava River, were convinced that in an independent country they would prevail over the more numerous Bosniaks.

Armed Serbs, helped by Belgrade, started to use force to seize control over Bosnian towns, villages, and institutions. They encircled Sarajevo and erected barricades. On April 5, a sharpshooter fired a shot during a peace rally and killed Suada Dilberović, a 24-year-old medical student. Suada is one of the first victims of the war in Bosnia-Herzegovina.

The country's President, Alija Izetbegović, was deeply devoted to Islam, but at the same time insisted that some kind of a "healthy" federation should grow up on the ruins of the old Yugoslavia.

At the outbreak of the war, he did not have many friends at home or abroad. However, he kept friendly contacts with his Slovenian counterpart, Milan Kučan, and frequently asked Slovenia for help. The

---

767) 63.04 per cent of the inhabitants of Bosnia–Herzegovina participated at the referendum. 62.68

Slovenian leadership answered positively to his requests, especially once the independent and internationally recognised Bosnia-Herzegovina came under attack and therefore had the right to defend itself.

However, in summer 1992 Slovenia was the only country in the neighbouring area that closed its borders to refugees from Bosnia–Herzegovina, on the excuse that it could not accept more than 70,000 Bosnian refugees, even though there were never more than 30,000 on its territory. The question whether to take in refugees and give military assistance to this ex-Yugoslav republic became the object of vicious score-settling in Slovenian politics.

Documents still exist about two sessions of the Defence Council of the Presidency of the Republic of Slovenia, in which the country's leadership talked about helping Bosnia-Herzegovina. On June 6, 1992, after a discussion with the Vice-Premier of Bosnia-Herzegovina, Defence Minister Janez Janša asked the Defence Council to advise whether to provide assistance, and the Council unanimously decided that a portion of Slovenia's weapons and military equipment be made available to Bosnia-Herzegovina. Weapons and ammunition were to be sold and used military equipment should be donated.[768]

Two weeks later, Hasan Čengić, head of logistics for the Army of Bosnia-Herzegovina and a close Izetbegović's friend, asked Slovenia for a large quantity of mortars, tank and artillery grenades and rockets for multi-barrel rocket launchers, which the Bosnian defenders intended to use to break through the Serbian encirclement of Sarajevo.

So on June 23, 1992, Milan Kučan chaired another meeting of the Defence Council in which the participants concluded that Slovenia did not possess all the requested types of ammunition. However, the participants unanimously decided to make available to Bosnia–Herzegovina the types and quantities of ammunition and equipment which they possessed and could do without. At the same time, they decided that these operations must be kept secret, and that a report should be given to the Defence Council after completion.[769]

---

768) Minutes of the 3rd session of the Defence Council at the Slovenian Presidency, no. Z-800-01-14/92, Ljubljana, June 9, 1992. The session was chaired by President Milan Kučan, and the other participants were the President of Parliament, France Bučar; Prime Minister Janez Drnovšek and the Ministers of Defence, Interior and Foreign Affairs, Janez Janša, Igor Bavčar and Dimitrij Rupel and

The Slovenian public never learned how much money Slovenia received from selling weapons and ammunition to Bosnia-Herzegovina, and this put the entire state leadership in a bad light. Soon after December 1992 elections the Defence Council was abolished, following a ruling by the Constitutional Court. "With that, the last control body which could look into the operations of the Defence Ministry, was removed," noted Milan Kučan.[770]

On August 18, 1992, Defence Minister Janša signed an agreement with a member of Bosnia-Herzegovina's Presidency, Fikret Abdić, to help the attacked state. It stipulated that the Slovenian Defence Ministry would train and arm about 1,000 Bosniak volunteers. Training took place at Svetli Potok deep inside the Kočevje's forest. However scarcely had it begun than an order came in that the training should end immediately, the training facilities should be closed and the foreign trainees sent across the border. After the last Bosniak crossed the rickety bridge over the Kolpa River into Croatia at Slavski Laz, the head of the Slovenian Territorial Defence was able truthfully to tell the public that no foreign troops were being trained in Slovenia (after reports about the training had appeared in the Austrian media).

At the beginning of October 1992, the government in Sarajevo prepared a draft agreement on friendship and cooperation between Slovenia and Bosnia-Herzegovina, which among other things called for comprehensive military cooperation in manufacturing and procuring weapons and military equipment, and in exchange of intelligence.[771] However, President Kučan was cautious and did not sign the agreement.

Despite this last-minute reticence, the Slovenian leadership continued to help the government of Bosnia-Herzegovina defend against the militarily stronger Serbs. The leadership even accepted an initiative to set up a logistics base in Slovenia for transporting foreign humanitarian aid and military equipment to the Republic's army.

That role was performed by the airport of the Slovenian city of Maribor, where on the morning of September 21, 1992, a huge Ukrainian

---

1992. The session was chaired by President Milan Kučan, and participants were Prime Minister Janez Drnovšek, Deputy Prime Minister Jože Pučnik and the Ministers of Defence and Foreign Affairs, Janez Janša and Dimitrij Rupel.
770) Author's conversation with Milan Kučan, Ljubljana, July 8, 2014.

aircraft landed after a flight from Hungary. It unloaded three large containers weighing in total 36 tons. After they were stored in airport hangars, the plane took off for Africa at 1:30 pm. This information was contained in a report to the Maribor Police Headquarters by a Maribor border police officer, who also wrote a side note that Interior Minister Bavčar was informed about the arrival of the plane.[772]

So it was strange that, when President Kučan on January 6, 1993 asked Janša and Bavčar if any shipment of weapons for Bosnia-Herzegovina had crossed Slovenian territory, neither Minister replied in the affirmative, not even Bavčar. Subsequently it became clear they were misleading the President.

That day President Kučan set up a meeting in his office with Herman Rigelnik, the newly-elected President of Parliament, Janez Drnovšek, who had just received a mandate to form a new government, and three existing Ministers of Defence, Interior and Foreign Affairs – Janša, Bavčar, and Rupel respectively.

In the political interim, Kučan wanted to form a unified standpoint on protecting Slovenia's interests from pressures of neighbouring countries and to correct numerous shortcomings in efforts to establish the rule of law. This followed numerous reports about weapons scandals published by the Slovenian magazine *Mladina*.[773]

During the meeting, President Kučan suggested that Slovenia should stop sending military assistance to Croatia and Bosnia-Herzegovina, and inform authorities of the two countries accordingly. Behind closed doors he also recommended a public statement that, after being admitted into the United Nations, Slovenia was not helping either republic with weapons or training of soldiers. "This does not meet the truth because there are two truths and one of them is for the public ...," stated Kučan.[774]

---

772) Report of the head of the Maribor border police to the Maribor Police Headquarters, Maribor, September 21, 1992.
773) At that time, the weekly magazine *Mladina* published a series of articles entitled "VIS versus VOMO" about frictions between the military and the civilian secret services. The reason for the articles was an attempt to kidnap a Cyprus weapons merchant, to whom officials at the Slovenian Defence Ministry paid almost one million German marks, but never received weapons they had ordered. A group of Slovenian adventurers under the auspices of VOMO tried to kidnap him in Celovec, but they failed because they were caught by the Austrian police. In the meantime, VIS made sure that the

This cancelled the decision of the Slovenian leadership in summer 1991 to send weapons to the attacked republics, with the Defence Ministry in charge and VIS responsible for opening borders. President Kučan doubtless sensed that there were numerous abuses around the state weapons deals, but who beside the perpetrators themselves had an inkling at the time how much money certain individuals had already made for themselves or their political party?

Later developments soon showed that weapons trade represented a vast "primary capital accumulation" in newly-independent Slovenia. Those involved thereafter sought to protect their gained capital with all their might. That is why almost all the scandals after 1991 were connected with weapons dealings.

About six months later, on July 21, 1993, Janša with great fanfare "discovered" 120 tons of weapons at Maribor Airport, in order to divert Criminal Police officers who were on his trail. The shipment of weapons, mostly made in China, came from Sudan via Hungary.[775] It was meant for the Army of Bosnia-Herzegovina, but it got stuck in Slovenia because of frictions between Bosniaks and Croats which soon deteriorated into fighting.[776] This prompted Croatian President Tuđman and his Defence Minister Gojko Šušak to try to weaken the Bosniak side's armed forces.

Money for helping the Sarajevo government that was coming from the Muslim countries – mostly from Saudi Arabia – was being deposited in the account of a humanitarian organisation in Vienna,[777] while an Austrian national[778] was in charge of weapons transportation to the Maribor Airport.

---

Ljubljana, January 6, 1993. The whole transcription is published in the book by Matjaž Frangež, *Kaj nam pa morete!* (What Can You Do to Us!), Part 2, self-publishing, Radenci, 2008, pages 230-248.

775) According to Čengić, Slovenian authorities at Maribor Airport seized 10,900 Kalashnikov automatic rifles with 758,000 rounds of ammunition, 40 calibre 82-mm mortars with 994 mortar shells, 25 RPG-7 rocket launchers with 210 rockets. The value of all weapons was estimated to be seven to eight million DEM.

776) Clashes between the Army of Bosnia-Herzegovina and the Croatian Defence Council (HVO) lasted from October 1992 to March 1994, when a Washington Agreement was signed, whereby Bosniaks and Croatians formed a Federation of Bosnia-Herzegovina.

777) The name of the humanitarian organisation was Third World Relief Agency (TWRA). Among its founders was a Sudanese national, El Fatih Al Hassanein, who studied in Yugoslavia. He operated in conjunction with the Sudanese President at the time, Hassan al Turabi, and also with the terrorist Osama bin Laden, who was then residing in Sudan.

778) The Austrian national was Dieter Hoffmann, the owner of a company called Flying Tigers.

Janša arrived at the airport in a helicopter, accompanied by the Interior Minister and the head of the Criminal Police. A day before, he informed the head of the Customs office of his forthcoming visit, and law enforcement officers were sent to the "crime site". Everything was ready for a big spectacle, only a "military brass band" was missing.

The soldiers arrived at the airport a day earlier, and armed VOMO members under the command of Lovšin blocked the roads to prevent possible removal of weapons.

In front of the cameras of journalists, Janša telephoned Prime Minister Drnovšek, who gave him the green light to demonstrate *urbi et orbi* the efficiency of the Slovenian authorities in combating "illegal weapons traffickers". Janša branded the chief of the Maribor VIS office, Silvo Komar, as the main "smuggler", alleging he was the owner of weapons and an influential former member of the League of Communists, ideologically close to President Kučan. Janša hinted that the weapons were to be made available for a possible Communist *coup d'état*.

However, who would the Communists want to overthrow? Themselves, maybe? One way or another, the country was kept busy for a whole year with the "Maribor affair", and this exactly suited the greatest arms merchant.

Instead of Slovenia helping the legally elected government of Bosnia-Herzegovina in its fight against Karadžić, General Mladić and their armed men, the Bosnia-Herzegovina authorities were called upon to help Slovenia find out the truth about the "Maribor weapons".

Hasan Čengić, who was tasked with bringing weapons over Slovenian territory to Bosnia-Herzegovina's defenders as plenipotentiary of the Presidency in Sarajevo and head of the Army's logistics, testified to an Parliamentary investigative commission of the Slovenian parliament at the request of Izetbegović in response to a plea from Kučan.[779] He explained to the quarreling members of parliament that it was an agreed inter-state operation undertaken within the framework of helping a country that had been attacked.

---

779) In 1993, the Slovenian parliament set up an investigation commission on the "Maribor weapons". However, when the commission started to look into other weapons issues, its president Zoran Madon n meetings. The last meeting took place on April

"The affair cost Slovenia several hundred million dollars, maybe even a billion in terms of lost business opportunities. Authorities in Saudi Arabia, Turkey, Pakistan, Iran and elsewhere were asking me whether it was possible at all to establish a serious relationship with Slovenia," said Čengić.[780]

He wonders even today: "What kind of a person is Janša, who accused his own country of violating embargo, while he himself was doing the same on a large scale. Even before the Maribor affair, he was selling weapons to Croatia, as well as to me, the representative of the government of Bosnia-Herzegovina."[781]

The weapons from Maribor Airport never arrived in Bosnia-Herzegovina.

The Bosniaks were faced with devastation, and Muslim countries reacted with growing anger. In the west, only the United States took action. When President Bill Clinton entered the White House at the beginning of 1993, he closed both eyes to the international arms embargo that covered the whole territory of former Yugoslavia.[782] The transport of weapons for the Army of Bosnia-Herzegovina from the Sudanese capital of Khartoum to Maribor via Budapest took place with the knowledge, and probably the approval, of the U.S. administration.[783]

One can only conjecture how the American intelligence service, the CIA, considered somebody who interrupted its operations because of an unseemly desire to settle domestic political scores.

The war in Bosnia-Herzegovina gradually turned into hellish anarchy and violence – with everybody fighting against everybody else. In autumn 1993, even Bosniaks started to fight amongst other: Fikret Abdić, a member of the Republic's collective Presidency and a cunning businessman, established his own "mini-state" in western Bosnia.[784]

---

780) Testimony of Hasan Čengić in the Slovenian parliament, July 4, 1994. The whole transcription is published in the book of Matjaž Frangež, *Kaj nam pa morete!* (What Can You Do to Us!), Part 2, self-publishing, Radenci, 2008, pages 442-490.
781) Author's conversation with Hasan Čengić, Sarajevo, May 20, 2008.
782) In spring 1994, President Clinton was criticised in Congress for allowing transport of weapons from Iran to Bosnian Muslims. Source: a report of Republican senator Larry E. Craig, Washington DC, January 16, 1997.
783) Hasan Čengić pointed out this possibility during his conversation with the author on May 20, 2008 in Sarajevo.
784) In July 2002, a Croatian court sentenced Fikret Abdić to 15 years in prison for war crimes in

Foreign interventions were ridiculous, demeaning and fruitless. European countries failed in their efforts to solve the Bosnian crisis, while the U.N. peacekeeping forces, the UNPROFOR in Bosnia-Herzegovina, were useless from the beginning. When NATO planes finally bombed Serbian positions, U.N. blue helmets were only in the way.

In July 1995, the Bosniak enclave of Srebrenica fell. Serbian troops led by General Mladić killed more than 8,000 Bosniak men and boys in only a few days and threw their bodies into mass graves in eastern Bosnia. Nobody needed Srebrenica, apparently not even the government in Sarajevo. The desperate defenders of Srebrenica, and the other Bosniak enclaves of Žepa and Goražde, were only small change in the political poker game played by the warring sides and the international mediators.

The UNPROFOR commander in former Yugoslavia, French General Bernard Javier, told his officers: "Gentlemen, don't you understand? We have to get rid of those enclaves!" After the massacre of Srebrenica's males, U.N. Secretary General Boutros Boutros Gali cynically answered a journalist's question whether this massacre was the greatest defeat of the United Nations, saying: "No, I don't think it was a defeat. We have to judge whether the glass is half empty or half full. We always help refugees. And we were able to limit clashes to the territory of former Yugoslavia."[785]

The Dutch blue helmets battalion, which was tasked with protecting Srebrenica against the Serbian attackers, did not fire a single shot.

In the weeks that followed, Clinton's emissary Richard Holbrooke forced Milošević, who was trying to save his own skin, to negotiate in the name of all Serbs, even those outside Serbia proper, including Bosnia.

Americans shut the leaders of the fighting Balkan's tribes – the participants included presidents Izetbegović and Tuđman – up in the Dayton, Ohio U.S. Air Force base until they came up with a common formula for ending the war and for the territorial demarcation of Bosnia-Herzegovina.

---

785) Both quotes are from the Jože Pirjevec's book, , (Yugoslav Wars

On December 14, 1995, the three leaders signed a peace agreement in Paris. "It's time for all the nations on this territory to look towards economic recovery, development, reconstruction and mutual cooperation," declared Milošević in a soaring speech at the signing ceremony. He returned to Serbia as a statesman who had saved the Republika srpska within the territory of Bosnia-Herzegovina, while at the same time riding Serbia – then still connected with Montenegro – of economic sanctions. However, official Belgrade did not talk about who was responsible for sanctions in the first place: the *Vožd* declared himself a "key agent of peace in the Balkans", which to a large extent was also the perception of the rest of the world.

Until the next war. In Kosovo, in the spring of 1999.

That is how Dayton's Bosnia-Herzegovina was born – an ill-defined political creation which the locals dismissively called *okruglo pa na ćošak* (the square footage of a circle). The years of Serbian siege of Sarajevo killed the spirit of the Olympic winter games held there in 1984, and the war and degenerate ethnic cleansing had a devastating effect on its residents. Today the state that was put into a strait-jacket by the Dayton Accords continues to die: only Bosniak, Serbian and Croatian nationalisms are still alive.

Slovenia continues to behave in an exploitative way towards Bosnia-Herzegovina, with thousands of Bosnian workers who came to work for Slovenian companies still waiting to be paid. Humiliated and despairing, they have been cheated by Slovenians who made their first profits thanks to the arms trade. Even though Izetbegović and Kučan remained friends after the "discovery" of weapons at Maribor Airport, the alliance and trust between the two countries have weakened.

"Alija was a good man, however fairly unlucky and also naïve. During the war, he went on a tour of Muslim countries and after his return home he said: – You know Milan, now I received help from my brothers in Allah –. I answered to him: – Leave off that, my brother, this is not good for Bosnia, not good for the democracy. After all, we live in Europe –."[786]

Since the end of the war, Bosnia-Herzegovina has been faced with the problem of Islamic fundamentalists. Islamic regimes, including the Saudi Arabia, are investing billions of dollars there: for supporting Islamic media, education, and the building of opulent mosques, so that the Bosniaks can learn about Wahhabism and other strict forms of Islam. Despite being outlawed by the Bosnian authorities, they are spreading around the country like a cancer.

In the 1970s, there were supposedly about a hundred religious extremists, gathered around Izetbegović, but they were mainly religious-political philosophers looking for inspiration in utopian pan-Islamism.

According to some studies, nowadays several hundred thousand people subscribe to the ideology of the self-proclaimed Islamic State.[787] As a result, Bosnia-Herzegovina is becoming a cesspool of jihadists. More and more youths are being attracted by suicide bombers, who are condoned by religious extremists such as Nedžad Balkan, Nusret Imamović, Bilal Bosnić and Muhamed Porča.

Most Bosnian Muslim citizens still resist Islamic fundamentalist indoctrination, but for how much longer?

\* \* \*

Twenty years after the genocide in Srebrenica, American diplomat Peter Galbraith – whom President Clinton entrusted with the most sensitive diplomatic tasks in Croatia – admitted that Washington allowed the Croatian military to penetrate into Bosnia-Herzegovina in response to the outrage, in order to strike against Serb forces.[788]

On July 22, 1995, with a help of Germany, Turkey and Iran, U.S. strengthened the weak alliance between Bosniaks and Croats. The joint offensive of the Croatian Army and the Army of Bosnia-Herzegovina which followed this agreement soon bore fruit. During the ensuing battle for the Bosnian town of Bihać, Mladić's units were no longer as brave as they were during the

---

787) One such study was published by Gordon N. Bardos, president of the company SEERECON. More on http://seerecon.com/wp-content/themes/seerecon/images/seerecon-report-web.pdf (last access: November 24, 2014).
788) *"Zbog Srebrenice nismo bili protiv operacije Oluja i ulaska HV u BiH"* (Because of Srebrenica we were not against operation Storm and the entry of Croatian military into Bosnia-Herzegovina),

earlier destruction of Bosnian enclaves. In just a few days, Bosnian Serbs lost a third of their territory in Bosnia-Herzegovina.

Then, at the beginning of August 1995, combined Croatian military and police forces launched a large-scale offensive on the territory of the self-proclaimed Republika srpska Krajina inside Croatia. In operation *Oluja* (Storm), they chased about 200,000 Krajina Serbs out into Serbia.

Several hundred residents of Serbian nationality, mostly old men and women who wanted to stay in their homes in Croatia, were shot by the "liberators" after the end of the military operation. They were considered "collateral damage", and numerous Serbian houses were looted and set on fire.

In the weeks after the massacre in Srebrenica and operation Storm, the Croatian Army captured towns in central Bosnia with ease. Mladić's army was falling apart, and panicked Serbian residents fled northwards to Banja Luka, which in September 1995 looked like Saigon right before the American withdrawal from Vietnam. The Americans did not allow Croatia to capture this largest Serbian-populated city in Bosnia-Herzegovina, but the Croatians had reason to be satisfied: they liberated their country, they gained a greater influence in Bosnia-Herzegovina and ensured a favourable negotiating position in Dayton. Did the men and boys of Srebrenica thus have to die for the freedom of Croatia?

In every war, there are winners and losers. The former write history and dispense justice. Two Croatian Generals accused of war crimes against the civilians during operation Storm went on trial at the Hague International Criminal Court.[789] They were sentenced, put behind bars, but later they were both acquitted on appeal.

It is hard to know why. But maybe Washington became concerned that the process would serve as a precedent for other courts around the world to put American and Israeli officers on trial ...[790]

The United States offered Croatia logistics support and intelligence, and as in Vietnam, Afghanistan and Iraq, they equipped and trained

---

789) They are Generals Ante Gotovina and Mladen Markač.
790) Attention to this was drawn by the Danish judge at the Hague Tribunal, Frederic Harhoff, in an e-mail to friends. His e-mail was first published on June 13, 2013 by the Danish newspaper BT

Croatian security forces.[791] Americans supported Croatian plans to recapture occupied territories, but they did not expect that the Croatians to "cleanse" the Serbian population so thoroughly.

Croatians consider themselves undisputable victors in the Balkan wars of the 1990s, which is true in the military sense. But in the moral sense they are losers, because of their blind vengefulness against the Serbs after the operation Storm. This triumphalism by Croatian politicians does little to encourage good neighbourly relations.

How much did Slovenian politics contribute towards bad relations with former Yugoslav republics? As it underwent the final acts of war in former Yugoslavia, Slovenia was mentally perturbed, caught up in disputes and deeply immersed in affairs and scandals. It was preoccupied with itself, many times dealing with imagined problems. Many foreign statesmen who expected that Slovenia would be more actively engaged in solving the Balkan conflict left Ljubljana disappointed.

Quarrels between "hawks" and "doves" started even before the final battle between the Territorial Defence and the Yugoslav Army, and then after international recognition, Slovenian politicians busied themselves with devouring each other.

---

791) The former Yugoslav Army training range of Šepurine near Zadar was at the end of 1993 made into a centre for special training of Croatian non-commissioned officers. Among instructors were French and American officers. Šepurine also hides a darker secret: the military smuggled large quantities of cocaine and heroin there. The drugs were later distributed to Dalmatia and Bosnia-

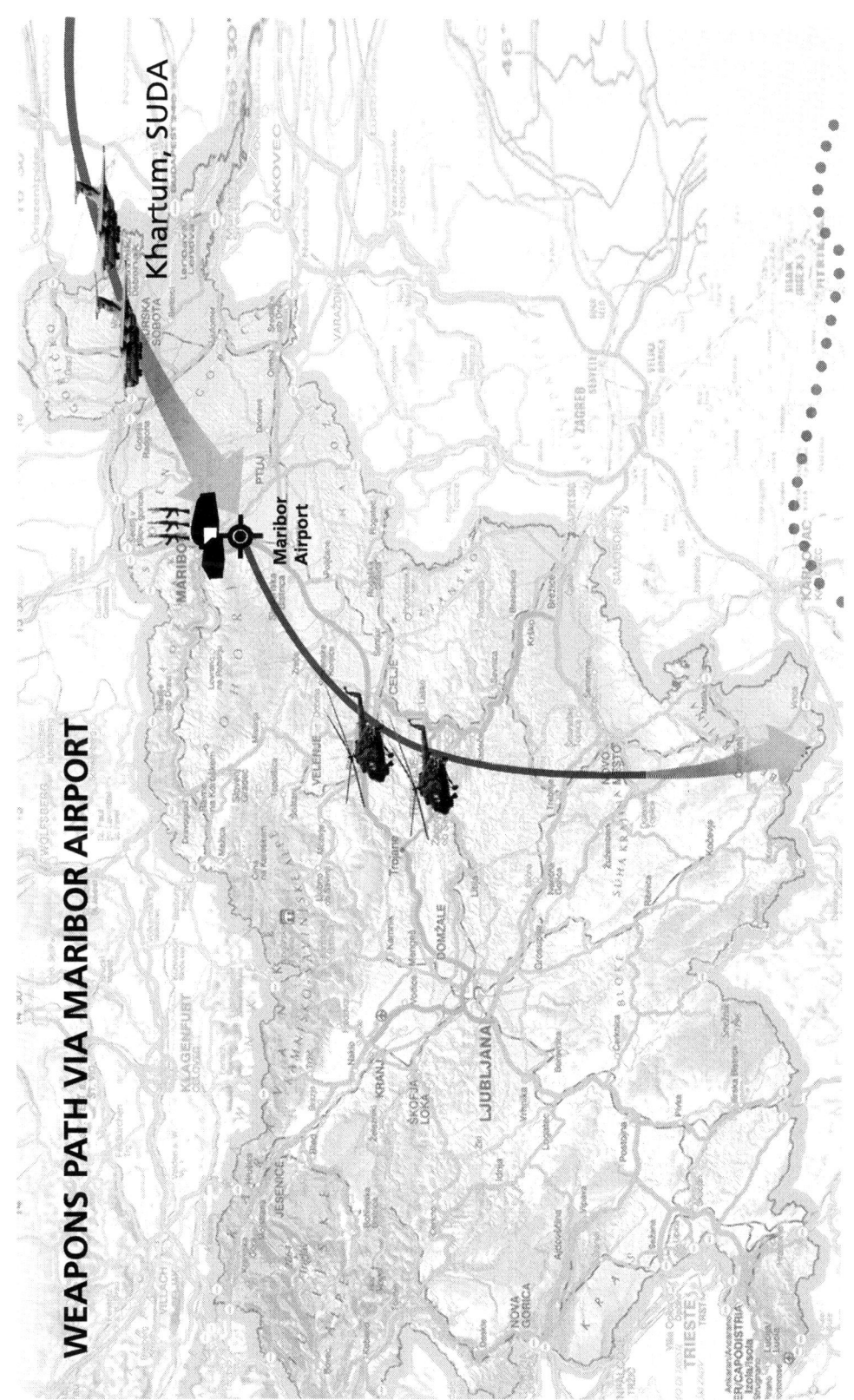

- Chapter 20 -

# DUELING SWORD VERSUS MANURE PITCHFORK

During an enlarged session of the Slovenian Presidency in early 1992, the atmosphere was relaxed because Slovenian leaders had just learned the good news that the European Community recognised Slovenia.

This is how a former member of the presidency, Ciril Zlobec, remembers the event: "In this atmosphere, Minister Janša, half jokingly, stated that in the seized archives of a Yugoslav Army eavesdropping centre, they found many interesting things, among them, a register with names of suspicious people. Then he turned to me and said: – Mr Zlobec, for you, it wouldn't be hard to become an Italian agent, had things for Slovenia gone wrong at the time, because you speak great Italian! – I responded with a logical and completely innocent question: – May I see that document? – to which he responded: – No, you may not, this is a state secret! –"[792]

With this dismissive attitude towards a member of the Presidency, which is the Supreme Commander of Slovenian Armed Forces, the Defence Minister showed that he cared little for the law. If anybody in Slovenia was authorised to read secret reports, it was the Presidency.

However, nobody criticised Janša for his autocratic attitude, except for Zlobec, who in October 1992, in an open letter complained that Janša only gives the Presidency information of his own choosing: "For example, the questions about how the weapons are purchased, and how the money for Defence is spent, have still not been on the agenda."[793]

Janša continued his private war against the poet by provocatively asking Zlobec in public how he would name the person who in June 1991 gave away the state secrets about the date of independence and the content of constitutional laws.

---

792) Author's telephone conversation with Ciril Zlobec, Ljubljana, June 6, 2014.

Janša claims that Zlobec disclosed the content of the constitutional law and the actual date of declaring independence, which was June 25, a day earlier than what was publicly announced. "That was the key element of a strategy to seize the power. Had the Yugoslav Prime Minister Marković reacted ahead of us, we would have declared our independence with the YA tanks already standing at our borders and we would have to capture them by force. On the operational-strategic level, this was a key move with which we made sure that the world did not brand us as a secessionist republic, using force, but a country which peacefully declared its independence and was then attacked afterwards."[794]

However, somebody "from the top Slovenian leadership blabbered away information about these key issues", and the YA sent tanks to border crossings a day earlier. With that Janša accused Ciril Zlobec, a poet, a Slovenian patriot and a former Partisan, effectively of causing the occupation.

The story was deliberately false, but what did Janša base this on? As a member of Slovenian Presidency, Zlobec was in charge of international affairs. In this role, he had regular contacts with foreign diplomats, among them, Fabio Cristiani, the Italian Consul-General in Slovenia.

Janša obviously got possession of the recording of a telephone conversation between Zlobec and the Italian diplomat, which was the basis for his claim that Zlobec divulged the date of actual independence – June 25 – which, according to Janša, was the cause that the YA almost won the war in Slovenia.

However Zlobec called the Italian Consul-General again and asked him which day he reported to his government as the day of Slovenian independence. "He took the necessary time to check his records and called me back within two hours. June 26 is the only date that was reported to my government, and it was taken from the interview with Janša that was published in one of the Slovenian newspapers on May 23. There was no other date that either Rome or any other Western European capital knew about," wrote Zlobec in his book.[795]

---

794) "Members of Parliament should investigate the eavesdropping affair", a interwiev with Janez Janša, *Slovenec*, Ljubljana, September 26, 1992.
795) Ciril Zlobec, (It's Nice to Be a Slovene But It's Not Easy),

Today, Zlobec does not like to talk about this. At over 90 he now mostly writes love poems. Because of his weak heart, doctors advised him not to get upset, and when the affair over the independence date was mentioned, he gave a bitter smile and said: "At that time, the Italian diplomat asked me if Slovenia was going to collect customs duties, to which I answered: – Yes, we will collect customs duties. – I also mentioned to him that we had to fulfill our referendum obligation and declare independence by June 26."[796]

For some time speculation lingered that Zlobec had acted too openly with the Italian diplomat, or even naïvely. One way or another, communication between Zlobec and Cristiani did not cause Slovenia damage. Their friendly contacts may rather have had a beneficial effect, taking into account the negative attitude towards Slovenia of Italian Foreign Minister Gianni De Michelis, who was close to Belgrade diplomats.

Ciril Zlobec resides in the Slovenian coastal region of Primorska. Because of his tireless engagement in the struggle for the Slovenian cause, he was expelled from school and was detained at the beginning of the World War II. He was a Partisan for two years, stirring the fighting spirit at Partisan meetings with his poetry, and risking his life in dangerous sabotage operations. In the 1980s, he rebelled against the "Yugoslavisation" of Slovenian culture, and he was one of the first openly to oppose the dangerous nationalism of the Serbian leader Milošević.

This was the man whom Janša went after with his "manure pitchfork". Ciril Zlobec defended himself with a dueling sword, but for dueling there are rules, and Janša did not obey them. If Janša had laid hands on an eavesdropped telephone conversation, the question arose: were they eavesdropping the Italian Consul-General and at the same time also recording a member of the Presidency, or were they eavesdropping on the Slovenian Presidency and also heard the Italian diplomat? The so-called "eavesdropping affair" caused a new round of recriminations and made Slovenia look like an immature country barely worthy of its newly-gained independence.

The case was taken up by the Parliamentary commission for monitoring the legality of the Security and Information Service (VIS) headed by Peter Bekeš. On November 9, 1992, the commission presented a report in which it determined that VIS acted illegally because it eavesdropped without a court order. On November 26, Parliament was supposed to approve the commission's final report recommending that the government change the legislation, but the session did not have a quorum.[797]

Some members of Bekeš's commission wanted to publish minutes of their meetings, but Interior Minister Bavčar objected and asked his party colleague France Bučar, President of the Parliament, to prohibit publication, to which he agreed.[798]

So VIS Director Brejc and Interior Minister Bavčar remained "clean" in the eyes of the public, even though only they could have enabled Janša to get possession of the eavesdropping records. It was later confirmed that the Interior Minister – without higher authorisation – approved the phone-tap and therefore knew the content of the discussion between Zlobec and the Italian diplomat. Bavčar accused Janša of exaggerating Zlobec's "over-trustfulness" into "blabbering and national treason".[799] The affair ended with a bitter conclusion that, under the influence of some of the "heroes of the independence", parliament was being reduced to a circus of political monkeys, who jump and scream a lot, but do nothing.

No action was taken against the the Interior Minister and the VIS Director for overstepping their authority in the eavesdropping, but Zlobec's reputation continued to suffer. One politician, Ingo Paš, filed a criminal complaint against him, and although Ljubljana prosecutor Tomaž Miklavčič threw the complaint out in February 1993, Zlobec was cold-shouldered by many of his political "friends", and also by all but one of his colleagues in the collective Presidency.

Even President Milan Kučan kept his distance. Zlobec wrote later: "I felt hurt, but I did not complain to him. I tried to understand him:

---

797) From the article: "The last parliament session was dealing with the eavesdropping affair", *Delo*, Ljubljana, November 27, 1992.
798) From the article: "Bučar withheld the public presentation", *Delo*, Ljubljana, December 24, 1992.
799) From the interview with Ciril Zlobec: "I was a metaphor for other battles,"

he's pursuing his policies, effective and at the same time, kind. As for Janša's accusations, he knew where the truth was. But still, I was being torn apart from the inside – I should have said something; when they asked Kučan in public about his opinion regarding this case, he resorted to a laconic answer that a commission would be established to see what the whole thing was about. Or something like that. Our longtime friendship was torn apart."[800]

This sinister silence was broken by Tone Pavček, a longtime friend of Zlobec, and fellow poet:[801] "How is it possible that with this action – inaction, all the defenders of the human thought, beliefs, liberty, all the supporters of human rights petitions, all the societies that were protesting the unjust and unproven sentences, who cried loudly at the announcement of sentences against the four accused by the Yugoslav military in 1989, how is it possible that during this spring, all of a sudden, they went silent and retreated into their calm comfort and their hypocritical reconciliation? No voice? No words? Are they really convinced that we are already a country with a rule of law that prevents violence against anybody? Did the voice of a person crying in the desert, an ordinary human voice, die away? And is there no child who would say that the emperor is wearing no clothes? I don't believe it. I don't want to believe it. That's why I said what I said. And that's why I'm telling you: if anybody wants to join my voice, I won't stop him! My friend remains – my friend. Stone me too!"[802]

The pain of the political campaign against him did not take away Ciril Zlobec's artistic inspiration. On the contrary, it stimulated him to write poems, including three sonnets in which he expressed his feelings about the Slovenian version of a lawless civilisation.

He ironically dedicated his Ode to Freedom to his slanderers and betrayers:

---

800) Notes from the personal archive of Ciril Zlobec, Ljubljana, July 17, 2011. Published with the author's approval.
801) Tone Pavček (1928 – 2011) was a Slovenian poet, essayist, translator and editor. In 1953, together with Kajetan Kovič, Janez Menart and Ciril Zlobec, he published a milestone book of poems titled, *Pesmi štirih* (Poems of the Four). In May, 1989, at a protest rally in Ljubljana against the YPA's sentencing of Janez Janša, he read the so-called May Declaration, in which the Slovenian opposition demanded a sovereign country for the Slovenian nation.
802) "Traitors are among us", the response of Tone Pavček in

*Earthquake, flood, hail, drought, cold,*
*No calamity sweeps everything away,*
*But in a moment, a lie washes from the memory,*
*Whatever goodness time has strewn in it.*

*Sanctimoniously it joins us at the table,*
*And forces on us its accusing memory,*
*In attack, it's versed in the art of battle,*
*And always strikes home, to the core of the word.*

*You say it's for the nation that you all sweat,*
*Now you drag me before the seat of judgment,*
*You thieves. Hold the thief, you scream.*

*But don't forget the rope, gentlemen,*
*We know: the fewer free spirits we have,*
*The tamer our freedom will be.*[803]

December 6, 1992 was the day of the first elections in independent Slovenia. Despite the cold and grey December day, the political atmosphere was heated. Zlobec announced already in October that his name would not be on the ballot. However in an election convincingly won by Janez Drnovšek's Liberal Democrats, Janez Janša's Social-Democratic Party of Slovenia suffered a humiliating defeat. At first, it looked as if it would not cross the threshold to enter parliament, but it just squeaked in.[804] Janša was invited into the new coalition government and remained Defence Minister.

Hoping that he would not have to deal with Janša anymore, Zlobec wrote: "Time heals all wounds, but the scars stay. When in a good mood, I like to joke that in the era of knights, battle scars – even those received in a vain battle – were considered an honour for whoever bore them. Well, in my verbal duel with Janša, he did not possess any knight's virtues. However on a certain level our

---

803) A poem from Ciril Zlobec's personal archive. He wrote it on December 2, 1992. The poem is published by the author's permission.
804) SDSS received 3.3 per cent of the votes, while the Democratic Party with Igor Bavčar, Dimitrij

incompatible natures brought us closer: We both have the worst possible opinion about each other."[805]

In the last few years, Janša has been spending much time in courts, both as defendant and plaintiff. When at the beginning of 2013, he could not explain the source of his assets to the Commission for Prevention of Corruption, the Special State Prosecutor's Office, which has powers to check bank accounts abroad, ordered a financial investigation against him. It also checked assets of his brother, wife and wife's cousin, who was managing assets of Janša and his spouse. At the end of April 2015, the Prosecutor's office finished the investigation and prepared a final report, but it is still not known whether it will bring charges against Janša and demand forfeiture of his illegally gained property.

\* \* \*

In March 1994 the so-called Depala vas scandal broke out. On the road between Ljubljana and Domžale, near the small village of Depala vas, Krkovič's military men clad in black shirts violently beat up a civilian who was an associate of the police.[806] After the incident, a tense atmosphere spread around the country, and there was even a possibility of armed clashes between the police and the military.

Despite organised protests by Janša's supporters in front of the parliament, Prime Minister Drnovšek finally listened to legal experts, who concluded that the military interfered in the civilian sphere, and thereupon relieved Janez Janša of his duties as Defence Minister.

Jelko Kacin, his replacement as Defence Minister, immediately started an internal investigation of the arms trade. The Ministry's Intelligence and Security Service. (OVS) set up a working group,

---

805) Statement from Ciril Zlobec's personal archive. He wrote it on December 2, 1992. The statement is published by the author's permission.
806) Police charged four members of the Slovenian military, who at the same time were associates of the illegal secret structure "Paravomo". The accused were Simon Krejan, Darko Njavro, Andrej Podbregar and Robert Suhadolnik. Prosecutor Barbara Brezigar held on to the police charges for two years. Later, all four individuals were found guilty of illegal deprivation of freedom of a civilian, but the

which buried itself in documentation and questioned Defence Ministry officials, Territorial Defence officers and others involved.[807]

In autumn 1994, the working group issued a report which said: "For the part of the trade which took place without the knowledge or approval of authorised and competent institutions of the Republic of Slovenia, there are reasons to suspect that individual top leaders of the Ministry of Defence were, in the name of the Ministry, involved in the arms trade and were using its money, and at the same time, were trying to gain illegal assets for themselves or their trade partners."

The report states that documents about the most sensitive deals were stolen or destroyed. However, from the available documentation, it is clear that the Defence Ministry sold to Croatia and Bosnia-Herzegovina more weapons than were recorded in receipts of handover and acceptance of weapons. Also, besides arms seized from the former Yugoslav Army, the Ministry was also selling weapons bought exclusively to strengthen the defence capability of Slovenia and financed from the budget. The money from these sales was never returned to the taxpayers.

The investigators described transits of weapons, official deliveries of weapons for Slovenia through the Port of Koper, and the compensation deals with Croatia called Weapons for Oil. They established indisputably that in parallel with this compensation deals, there was also selling of weapons for cash.

In collaboration with Criminal Police officers and SOVA, they also "obtained information that individual employees at the Ministry of Defence had several accounts in foreign banks and companies, where payment transactions were taking place".[808]

In the meantime, the Logistics' Administration at the Defence Ministry prepared a detailed reconstruction of the sales of weapons. "According to the Logistics estimate, the weapons that Slovenia sold to Croatia and Bosnia-Herzegovina were in total worth more than 140 million German marks.[809] This estimate was made by some very

---

807) The working group was headed by the OVS Director Marjan Miklavčič, and the members were Brane Lukač, Damjan Režek, Stane Smolej in Andrej Osterman. The operational processing of the case was titled *Trgovina* (Trade).
808) Report of Miklavčič's working group, no. 881/600-7088, Ljubljana, October 3, 1994.
809) The estimate is insufficient for several reasons. There were no detailed and collection records of weapons, and they were using prices from Janša's report, while in reality prices were much higher. The

qualified people, because I needed the estimate to start my mandate on the right track," explained Kacin later.[810]

While uncovering many shocking details about war profiteering at the Defence Ministry, which at the time was led by Janez Janša, investigators from the Interior and Defence Ministries were simultaneously drafting criminal complaints and constantly updating them. They prepared 16 criminal complaints against 21 individuals suspected of committing 25 criminal acts.

However, all of those criminal complaints got stuck in the prosecutor's office, where Barbara Brezigar became head of the Prosecutor's Group for Special Affairs in February 1996. "She's the main reason that practically none of our criminal complaints reached the courts. She stayed deaf to all appeals from us Criminal Police officers to ask the investigative judge to issue warrants to obtain information from two foreign bank accounts, the existence of which we established in our investigation. There are many examples how she wrecked our efforts," said Drago Kos.[811]

Why did Prime Minister Drnovšek refrain from uncovering the illegal trade in weapons? Marjan Miklavčič, at the time OVS Director, stated: "According to our findings, Drnovšek pulled the hand brake. Supposedly, he didn't want to aggravate the situation, and expose Slovenia abroad. The fact is that he and Janša were not quarreling at the time. Because of the Depala vas scandal, he had to get rid of Janša, but on the other hand, he also performed a purge in the Criminal Police Office."[812]

Drnovšek, who was Prime Minister four times, maybe thought endless reheating of scandals would harm the image of Slovenia and undermine its chances of joining the European Union and NATO. And when he got to know the enormous size of the arms business led by influential members of his government, he tried even harder to sweep the whole matter under the carpet.

During the third Drnovšek government, in spring 1999, Parliament set up another investigative commission which dealt with weapons.

---

810) Jelko Kacin, during questioning by the Parliamentary investigation commission, Ljubljana, March 9, 2000.
811) Author's conversation with Drago Kos, Ljubljana, December 22, 2015.

It was headed by Rudolf Moge, a member of Drnovšek's Liberal Democratic Party. The commission came to conclusions similar to those of the OVS group in 1994. However, it caused deep anger among Janša's followers, while the Prime Minister was grumbling that the whole investigation should end as soon as possible.

Members of Parliament did not have enough courage to bring the whole thing to the end, and the commission finding stayed as a non-paper in some locked book case in Parliament. Moge received death threats which came to nothing, but that summer two assailants from Czech Republic broke his spine when they drove over him in the a boat near the island of Pag.[813]

In the meantime, Drnovšek's government fell, and the Prime Minister's position went to the right-wing politician Andrej Bajuk, a descendant of Slovenian emigrants to Argentina, and Janez Janša again became Defence Minister. The government was in place only from June to November 2000. In that time, the Defence Ministry prepared another weapons report, which only mentioned compensation deals: with Croatia for 23.8 million German marks[814] and with Bosnia-Hercegovina for 3.4 million marks.[815] The total amount of 27.2 million marks is only a fifth of the proceeds mentioned by Kacin, and the report does not mention trading of weapons for cash at all.[816] After 2003, any criminal aspects of the weapon trading exceeded the statute of limitation.

During all this Janša presented himself as a perpetual victim persecuted by the old communist nomenclature. Drnovšek, who had done nothing to call Janša to account, finally revealed what he thought of him deep down. At the end of his mandate as the President of Slovenia – when he was already seriously ill – he branded the former Defence Minister *Princ teme* (Prince of Darkness).[817]

---

813) The Czech culprits were only fined by Croatian Police, and Slovenian law enforcement did not launch an investigation.
814) Weapons for Oil deals took place in such a way that the Slovenian Defence Ministry sold the Croatian oil to the oil companies Petrol and Nafta Lendava, and the proceeds went to the Ministry.
815) Because of the debt to Slovenia, the Defence Ministry seized a Bell 412 helicopter owned by Bosnia-Herzegovina. The debt was around one million German marks less than the value of the helicopter, and despite appeals from Bosnia-Herzegovina, Slovenian authorities have still not returned it.
816) Report about proceeds from the sale of weapons in the years 1991 – 1993, no. 831-00/2000-6, Ljubljana, August 3, 2000.

In spring 2014, during the coalition government of Alenka Bratušek, the Slovenian Parliament agreed to work on legislation abolishing the statute of limitation for criminal acts of war profiteering. The initiators were following the example of Croatia, which passed a similar law three years earlier and used it to launch criminal proceedings against former Prime Minister Sanader. However, these representatives of the Slovenian people have so far not summoned the political will even to prepare the draft of such a law.

\* \* \*

The last part of ex-Yugoslavia to gain independence was Kosovo. After defeats in Croatia and Bosnia-Herzegovina, Serbs started ferociously to persecute Albanians who are in a large majority in Kosovo compared with the local Serbs. A western alliance in 1999 began bombing Serbs first in Kosovo and then in Serbia, until Milošević gave up and Serbian forces quit the territory.

Fifteen months later, Milošević was defeated in elections, resigned and soon afterwards was handed over to the Hague Tribunal to be tried for war crimes. He died in 2006 in a Dutch prison before the trial ended. On February 17, 2008, Kosovo formally seceded from Serbia and declared its independence, and in November of the same year, Janez Janša's party was defeated in elections and his mandate as Prime Minister came to an end.

From then onwards the former Prime Minister and leader of the Slovenian Democratic Party (SDS) developed close relations with Kosovo, though the Slovenian public knows little about it. Slovenian companies in telecommunications, banking, insurance, oil trade and brewing moved into an anarchic Kosovo economy affording plenty of opportunities for unscrupulous profiteering.

"Companies in Kosovo did not have balance sheets, and there was no supervision. They were showing phony profits and Albanian and foreign businessmen, as well as politicians, were sharing them with each other. So the Slovenian money that was legally invested in Kosovo was coming back as a so-called black fund," said one knowledgeable person.[818]

Where did the money go? Maybe to political parties, parallel institutions of civil society, private universities, and private accounts in tax havens?

Soldiers from 28 countries including Slovenia participate in an international military force in Kosovo known as KFOR.[819] Some of them have been using their duties as a front for their personal dealings. Nobody is checking the soldiers and the regulations applying to them are different from those for ordinary Kosovo residents.

The KFOR Intelligence Service found out that at least three Slovenian officers belonging to KFOR were closely associated with certain Kosovo politicians and Albanian businessmen who were laundering money with Slovenian investors.[820]

Janša tried hard to get his confidants into the intelligence and security structures of the international peacekeeping forces, but was only partially successful. "Between 2010 and 2011, he frequently flew to Kosovo. As a reason his visits, he cited invitations from Kosovo universities in Priština and Prizren," said a KFOR intelligence officer.

In that period, the Slovenian media reported only one such visit. In February 2011, he gave a speech at the Priština Victory College and met his close acquaintance, Kosovo Prime Minister Hashim Thaçi.[821] In October 2015, the Universum College from Priština granted Janša an honorary doctorate.

---

819) Kosovo Forces or KFOR operate in Kosovo under the command of NATO.
820) One of the people who worked in Kosovo is the current OVS Director Franc Trbovšek, who

# EPILOGUE

## - PRINCE OF DARKNESS -

In the end, let's go back to 1990. Before the referendum for independence, Slovenian politicians realised that the Slovenian people should first unite among themselves, because in that way it would be easier to separate from Yugoslavia.

A few years earlier, as communist restraints on freedom of expression crumbled, stories surfaced how communist victors secretly slaughtered some 14,000 Home Guard collaborators and civilians at the end of World War II.[822] Because of the long cover-up, the revelations prompted an intense reaction among the people. Until then, they had been fed over-simplified and self-serving accounts of glorious, spontaneous resistance by the Slovenian people against nazi and fascist occupation under the leadership of the communist Partisans. Now they had to deal with the bloody stain of vengeful mass-murder on their history.

On the one side of this tragedy dividing the Slovenian people were reform communists who wanted to end their Party's repressive regimentation of society. On the other side was the Catholic Church, to which most of those slaughtered belonged. The two parties decided to seek reconciliation at a solemn ceremony in the forested hills of Kočevski Rog, where many of the soldiers were shot in the neck and buried in pits. For the families of the Home Guards, the ceremony would represent a symbolic Christian burial and an end to vilification. For the communists, an opportunity to win over the people to their new spirit of openness and cooperation.

The sociologist, philosopher and publicist Spomenka Hribar, who had for several years fought to bring the matter out into the open, and had often been roundly criticised for her pains, acted as an informal link between the two sides. She suggested to Archbishop Alojzij Šuštar, head of the Catholic Church in Slovenia, that he perform a religious service at the hidden graves in Kočevski Rog. She urged

---

822) These killings, which took place after hostilities ceased, represented 15 per cent of all World War

President Milan Kučan, a reform communist, to make a gesture as the highest representative of the Republic, because the state was responsible for the killings.

Šuštar and Kučan, both of them thoughtful and unflustered men, accepted their parts in the ceremony. Before it, they even exchanged their own reconciliation speeches, so each of them could object to parts of the other speech he did not like. That was evidence of great mutual trust. The ceremony on July 8, 1990 in Kočevski Rog took place without complications. Kučan expressed sorrow for the killings and called for reconciliation.

Spomenka Hribar remembers it as one of the brightest and most touching events in her life: "Archbishop Šuštar added the killed Slovenians to the Christian community and President Kučan to the Slovenian community. Before the ceremony, these people were in neither the book of living nor the book of the dead."[823]

Slovenians could then walk into their independence with clear hearts, but this lasted only a short time. Quickly Slovenian politics became overwhelmed with more destructive passions and calculations. Ideological divisions began to flourish again.

In 25 years of independence, nobody has come close to the symbolic reconciliation made explicit in the speeches made by Archbishop Šuštar and President Kučan, standing next to the open caves of Kočevski Rog, the final resting place of the murdered Home Guards.

Has everything happened too quickly? Or too late? Or are the people not mature enough? Spomenka Hribar believes both sides need to acknowledge more frankly what they did wrong during and after World War II: "The anti-communist right-wing, personified by the Catholic Church, was directly involved in collaboration with the occupiers during World War II, but it never apologised for it. It did not condemn the collaboration with the occupiers, nor the treason of the counter-revolution. The same can be said for the communist side … It did not apologise for the revolutionary violence during the war, or the killings, or the Communist Party dictatorship after the war. If we want to achieve real reconciliation, then each side has to assume its part of the responsibility. However, the reality is that the right side blames everything on the revolution, and the left side blames the collaborationists."[824]

So the small Slovenian nation is even today split in two regarding the past. In the meantime, new mass graves are discovered, and many people killed after the war remain unburied. This does dishonour to the principle that a proper burial is a moral obligation of a civilised nation.

\* \* \*

Janez Janša chose politics as his profession. Once a loyal communist, he transformed himself into a right-wing conservative, because he counted on the votes of country people and the support of the Catholic Church and political emigrants abroad, especially their funding. After the only victory of his Slovenian Democratic Party (SDS) in 2004, he did everything he could to take over the state and viewed any opposition as attempts by ex-communists to kidnap the state.

He turned his SDS into a stronghold of hotheads with the same mindset, ready to do everything for their leader and foment hatred against those who disagree with him. It seems that other political parties in Slovenia are afraid to get into a quarrel with Janša, and prefer to submit to his political arbitrariness.

When a successful Slovenian scientist or a businessman returns to Slovenia, when an athlete steps from the winner's podium or a pop star finishes a performance – there is always a polite SDS invitation waiting for them. The SDS party has its own media and its own people in other Slovenian media. The same goes for politics, the military, police, the economy, state institutions, civil society… In almost every social stratum Janša has implanted his followers, who in return for privileges inform the party what is going on.

Janša incites intolerance and misleads people with lies and half-truths. Those who fail to verify his statements are to a large degree responsible for his success in doing so. Equipped with these dark faculties, he tries to climb to the summit of the political Mount Olympus.

However in fact he is driving backwards, endlessly harping on divisive interpretations of history. In a speech in August 2013, he accused World War II communists of trying to destroy the people's will to rebel against injustice. He needlessly fanned the embers of a quarrel that other more balanced Slovenians have sought to lay to rest.

Such is the chosen path of Slovenia's Prince of Darkness.

This account of Slovenia's recent history began with a nonsense *Butalci* story by the popular author Fran Milčinski. Now let us finish with a poem by his son, Frane Milčinski – Ježek:

> *No sooner was he born,*
> *On to the balcony he crawled,*
> *And off the fence he tumbled.*
> *On to concrete he then sprawled,*
> *And all the doctors mumbled –*
> *A massive brain disorder.*
> *So the prognosis was humble:*
> *He'll join the "reactionary" order.* [825]

---

825) Frane Milčinski-Ježek,           (first version). From the book of poems

# REVIEWS

The global trade in weapons is the deadliest and most corrupt of trades. Accounting for 40 per cent of all corruption in global trade, the arms business is characterised by the shadowy intersection of politics and business behind a veil of national-security-imposed secrecy, and operates with virtual impunity.

This crucially important, thoroughly entertaining and deeply worrying book, illustrates how Slovenia, at the most seminal moment in its history – the country's struggle for independence - fell victim to this darkest of trades.

'Patriotism for Sale' is ultimately a tragic tale of liberation heroes gone greedy. As such it reminds me of the way in which the liberation of my own country, South Africa, was undermined by a deeply corrupt arms deal. Just four years into our hard won democracy, as Nelson Mandela was leaving office, the governing African National Congress decided to spend 10 billion U.S. dollars on weapons, with 300 million dollars of bribes going to senior politicians, officials, military leaders, executives and arms dealers. Our liberation heroes effectively sold out the country for personal gain.

'Patriotism for Sale' describes, in remarkable detail and highly readable prose, the litany of corrupt weapons transactions that littered Slovenia during the conflict and beyond. Politicians, officials, businessmen and shady intermediaries, all succumbed to the prospect of making quick bucks, undermining the struggle for independence in the process. The name of transitional Defence Secretary Janez Janša looms large, in much the same way as South Africa's first Defence Minister (and former head of the liberation struggle's armed wing) Joe Modise, dominated South Africa's ill-fated arms deal.

The South African deal continues to blight the young democracy – the country's current President faced 783 counts of fraud, corruption and racketeering, charges that were controversially dropped weeks before he was elected President. Today South Africa is riven with corruption and ineptitude in governance.

Similarly, in newly independent Slovenia the weapons trade represented a vast opportunity for "primary capital accumulation", which those, involved sought to protect with all their power. Almost all scandals in Slovenia after 1991 were connected with weapons dealing that had its origins in this period.

The disarmament of Slovenia's Territorial Defences by the Yugoslav People's Army created opportunity for & enabled corrupt deals. The TD's need to reinforce their arsenals resulted in the sourcing of arms from abroad. Many of these arms were obtained through illicit channels which enabled high government officials to make significant money.

During the 10-day war, vast amounts of weapons that were seized from the YPA were sold to Croatia by state employees, with many sales taking place in cash with no paper trail.

The main war profiteer was Janez Jansa, Defence Minister at the time, who later became Prime Minister. His Ministry was not only involved in selling and re-selling weapons but also in tied trading for which it was receiving commissions. The Interior Ministry also played a role in arms dealing, but in a more regulated way and not on the same scale.

In a move redolent of the worst practices of the illicit arms trade globally, both the Defence and Interior Ministries held secret bank accounts in Die Kärntner Sparkasse in Austria.

The highly suspect weapons trading of the Ministry of Defence was fuelled in collusion with the company Orbis, part of the Gorenje Group of companies. Orbis received weapons from the YPA depots captured by the Slovenian TD and other sources, for which Croatia paid extortionate prices in German marks. Some transactions were paid for by Croatia in oil, which, as is common in arms deals, made covering up the illegal nature of the deals much easier.

The U.N. embargo on arms imports also meant that the weapons merchants in the Slovenian Defence and Interior Ministries were now international criminals. This suited them, as all weapons channels had to be concealed, increasing the number of intermediaries involved, significantly inflating prices so that all involved could increase their profits even further.

These appalling profiteers were now acting as a state within a state, answerable only to themselves and the cabal of complicit, friendly right wing parties in Croatia and Slovenia. It should not be forgotten that in addition to the huge corruption, by shipping weapons to Croatia, they were also weakening Slovenia's own defences.

In 1992, Slovenia also provided weapons and materiel to Bosnia-Herzegovina. The Slovenian public never learned how much money the country received for these weapons.

All of this arms trading was underwritten by financial skulduggery: In order to obtain the foreign exchange that was needed to pay for weapons, the Defence Ministry took out a number of trade loans. These were arranged through a Munich-based trade company called Unimercat. The loan which financed the first purchase of weapons was assigned by the Slovenian Parliament for the construction of motorway sections. The money was transferred into the bank account of Unimercat, rather than to Slovenia, thus, metamorphosing into a trade credit. But Ljubljanska banka provided a guarantee for repayment of a loan amounting to 28 million German marks to Unimercat. In October 1991, Slovenia took out a new trade loan in accordance with a decision by Prime Minister Peterle's government. Just over 15 million U.S. dollars went towards health care and 26.6 million dollars for weapons. For the arms, the story repeated itself. Through the mediation of Unimercat, 10 million German marks from Munich went to the arms supplier's bank account in Budapest.

Despite several investigations into Slovenia's illegal arms trade in the 1990s, nobody was ever found guilty. The book makes clear that responsibility for this impunity lies with the office of the prosecutor: Barbara Brezigar became head of the Prosecutor's Group for Special Affairs in February 1996. "She's the main reason that practically none of our criminal complaints reached the courts. She stayed deaf to all appeals from us Criminal Police officers to ask the investigative judge to issue warrants to obtain information from two foreign bank accounts, the existence of which we established in our investigation. There are many examples how she wrecked our efforts," said Criminal Police officer Drago Kos.

Unsurprisingly, the same actors involved in arms trading in the 1990s, were involved in the scandal of the largest weapons transaction in independent Slovenia in 2006, the Patria case.

The worst excesses of the arms trade – avarice, deception, fraud and the undermining of national defence – are always exacerbated in times of conflict. War zones are fertile fields for fraudsters and profiteers because the circumstances allow them to charge a premium for desperately needed weaponry, in conditions in which normal controls and accountability are ignored. Profiteers wrapping themselves in the flag of patriotism take advantage of these circumstances to enrich themselves and undermine the homeland they claim to love. Tragically, as 'Patriotism for Sale' so vividly illustrates, this is what happened in Slovenia.

For the sake of the strength, integrity and future resilience of Slovenia's nascent democracy, those responsible – individuals, companies at home and abroad and intermediaries – should be publicly held to account. Nothing short of a Truth and Reconciliation process in which all the sins of the false patriots are publicly aired will enable Slovenia to move on.

If this wonderful country shirks from dealing with this sordid chapter in its recent history, corrupt arms dealing will continue to engender corruption, pollute politics and the rule of law, corrode democracy and make Slovenia less safe and secure.

***Andrew Feinstein**, the author of "The Shadow World: Inside the Global Arms Trade" and Executive Director of Corruption Watch. He is a former African National Congress Member of Parliament, and served under Nelson Mandela*

\* \* \*

The author, Matej Šurc, was able to put together a complete "story" of our independence and present its dual nature as a uniform and an interdependable process: creation of a new state, while at the same time, its abuse; the manifestation of people's desire for freedom and sovereignty, while at the same time, taking advantage of it.

Behind and inside the independence process, the weapons trade was taking place, carried out by the state employees that we trusted. The author follows the weapons "story" and puts it into a context of the "story" about gaining independence. There has not yet been a study which so adequately put together these dual "stories" into a single narrative in a documented manner: the independence process, smeared with a big black mark of trading with death. The book describes several aspects of gaining independence; first from the former Yugoslavia and then affirmation on the international level; the role and management of the war for Slovenia, with special emphasis on the role of our "hawks," who clearly wanted a "real war," because only in a real war could they come out as heroes! This is the only way to explain Janša's order (without the knowledge of the Slovenian Presidency!) to the members of the Territorial Defence and the police to attack the Yugoslav Army barracks! That particular "episode" is documented in the book by eyewitness accounts. We must say that we were very lucky to be able to escape unscathed. The YA's revenge could have been terrible: the bombing of the cities, especially Ljubljana and civilian targets, as well as energy facilities, post offices, TV facilities, governmental and parliamentary buildings. Fortunately, the lower-level commanders refused to follow the controversial order, or procrastinated in its execution, until Janša himself, at the last possible moment, canceled it! But just the thought of total war is sick because it means playing with people's lives and with the whole nation.

The "upside-down world" situation started to take place: the secret services of the Ministries of Defence and Interior, which were supposed to supervise questionable weapons (re)sale deals, were themselves the most involved in it. The same goes for the Republic's Coordination and the Defence Council: their members, Janša, and Bavčar, were the main actors in the weapons trade. The highest political body, the Republic's Presidency, was not receiving pertinent reports. In fact, it was pushed on to a side-track! The Presidency's decision to help Croatia was re-interpreted as permission, or even an order, to sell and re-sell weapons. The same applied to the Defence Council – its members Janša and Bavčar were buying and selling weapons during their tenure as ministers, and were thus abusing their authority, and consequently abusing the state.

They were selling weapons at predatory prices for cash, which never went to the state coffers. Even today, nobody knows where the money from (re)selling of weapons disappeared to, and suspect foreign bank accounts are not accessible. Budgets of the Defence and Interior ministries were outside the state budget, so that financial inspectors could not have oversight of their operations. The Slovenian weapons mafia was a "state within a state".

The misuse of the state institutions is (was) like cancer: it soon engulfed all segments of the society. The Prosecutor's office (Tomaž Miklavčič, Barbara Brezigar) was constantly rejecting indictments that were submitted by the Criminal Police, as well as by the regular police. The independence and confidence of the judiciary were mortally wounded – by, amongst other things, threats to prosecutors and judges. The weapons merchants also threatened their own colleagues! As one of them stated, "I don't know anybody who talked about the weapons dealings and survived!" This word, "survived" should be understood figuratively and also literally. There were some cases of deaths under suspicious circumstances of individuals who "didn't listen". However, the Prosecutor's office and courts did not border to look into those cases! A state with a non-functioning judiciary, whose legality is not systematically protected, and where the truth cannot be found, has an uncertain future ahead, and its citizens are left to fend for themselves against the political manipulation. It's even worse when the political manipulation – self-evidently "justified" not by equality before the law, but on inequality – is carried out by the highest legal arbiter in the country, the Constitutional Court of the Republic of Slovenia. In the case of ordering the release of Janez Janša from prison, the Constitutional Court stated: "The appellant (Janez Janša – author's annotation) is a member of Parliament and at the same time, President of the largest opposition party in Parliament. A well-functioning opposition to the executive branch is one of the foundations of democracy." It is true that opposition is one of the pillars of democracy; however one of the foundations of democracy is also equality before the law! Privileged judges are a sign of an authoritarian government or one of its branches. The Constitutional Court has reached a new low with its explanation: it was biased, ideologically-driven and subjective. Where the judiciary does not function autonomously on all levels, the state is not functioning. All its systems are falling apart, and the people are engulfed in apathy and despair. That is why it is important that we present, analyse and accept findings about the importance of the weapons (re)selling affair in the 1990s! At the time, a Slovenian state not yet fully built started to crumble: systematically, legally, politically and morally.

**Systematically**: by disrespecting the state institutions, by disregarding the Presidency as the Supreme Commander, etc.

**Legally**: demolishing the laws is not only disobeying the laws, it is also arbitrarily amending them. The fact is that Janša and Bavčar called a meeting of the coordination group of the Ministries of Defence and Interior, which, ignoring Parliament and Parliamentary procedure, amended several articles of the Law on monitoring the state border in such a way that the Ministry of Foreign Affairs was excluded from making decisions about the weapons transit across the border.

**Politically**: Janez Janša wrote his political "platform" in a text called The interpretation of the December 1992 elections and a few words about the future. In it, he rhetorically posed a question "whether all political forces deserve the same credit for gaining the independence, or do some of them deserve it a bit more"? There is no doubt that some did a little bit more. But what does all this mean (for Janša)? The answer: "From a standpoint of an individual, this can be unimportant – the results are what really count – but from the standpoint of politics the results are important because they can be sold, as in commerce sweaters of a textile company are sold." (*Okopi*, pages 147 and 148). To equate the independence or forming of a state with the sale of sweaters is equating a fundamental historic event with something profane, ordinary. Therefore, we are dealing with instrumentalisation of independence: independence is only worth the amount of profit it brings me. The state is reduced to an asset – not for achieving the common good but to benefit a certain subject, who, following this logic abuses it to influence voters and the media, in other words, for Power. To achieve it, all means are allowed, like the sale of former Yugoslavia's dinars to the south of the former Yugoslavia for foreign exchange, after Slovenia introduced its own currency (coupons), or destroying the archives/documents about the weapons trade, manipulation of history for reinforcing its own Truth. Or for causing fear among co-workers so they would not talk … Or strange, fatal accidents and suicides, even threats to children, in order to intimidate their parents …

Equating independence with commerce ("sale of sweaters") nullifies the value of the political credibility and also of politics as a profession which works to fulfill the national interests, which are: the preservation of the nation and its identity, its freedom and its culture. Which means: care of the environment and ecology, self-supply of food to the greatest extent possible, energy economics, control over transportation infrastructure, etc. When independence, therefore, your own state, is equated with selling of an article, then without a second thought everything is for sale – as in our country, the national interest becomes an object of ridicule! First, they sarcastically equated the national interest with the sale of the Slovenian brewery Union, so they could make fun of it. All the achievements of our predecessors are disappearing into foreign hands.

**Morally**: instead of people living a free life in their own country, a new atmosphere of fear is permeating. An investigator who asked why even today there still no clear picture of the weapons affair received the answer: we were afraid, we were afraid! We were afraid of consequences! Therefore, questions about suspicious fatal accidents are not completely irrational. It seems that a "political party moral" became prevalent: you believe and support the individual who suits you ideologically. In such a society, the truth is suffering, and the people are suffering even more. Public and personal morals are being affected by the absence of cooperation and reconciliation and, at the same time, by the presence of sectarianism jealousy and quarrels. You can lie, falsify history, steal, practise hate speech – everything is "normal" and acceptable. The feeling for the truth and justice and the capacity for piety are fading away and becoming non-mandatory.

I am not saying that Janša, or the weapons affair, are responsible for everything that's bad, but I am emphasising that all of that set a new paradigm of behaviour towards one's own state, even before it became one! It's not about a few hundred million dollars that they put in their pockets, it's about the way of life which they introduced into our society. The weapons affair is a paradigm of today's attitude towards the state and the state towards us, the citizens. They hijacked our state before we were able to enjoy it as a system of freedom and maintaining of our identity.

We cannot think separately about independence and the arms trade – not the trade with which we armed our soldiers and policemen, but the trade in which weapons were bought abroad with the intent to be resold (several times more expensively) to the Balkans.

You cannot separate the independence struggle from the arms trade! But it is critically important to distinguish between the two: independence as a great event in our history when we got our own state, and the arms trade as a depraved deed that should have had an epilogue in a court of law. Because this has not yet happened, we at least need an accurate and honest account of it and a moral condemnation – for the history!

That is why it is so important that the author, Matej Šurc, combined the two interdepending developments into one: our independence story and the big black mark of moral depravity of the weapons affair!

The crooked fighters for independence are not helped by the patriotism they claim! The patriotism was negated or even replaced by self-interest! The August Bebel statement is correct: "The greatest patriots make the biggest profits during war."

Addendum: in a structural sense we can draw a parallel between the World War II struggle for freedom and the struggle for independence: the WWII struggle was something positive, which had a national significance, as did the struggle for independence. Karl Marx wrote that history first happens as a tragedy and then as a farce. The same applies to us; the revolution which was going on simultaneously with the WWII struggle for the national liberation was a tragedy, while the arms trade affair, going on simultaneously with the independence struggle, was a farce. We citizens were and still are pawns on the chessboard of history, and the objects of criminals without consciences.

**Dr. Spomenka Hribar**, *Sociologist and Philosopher, ret.*

# ACKNOWLEDGEMENTS

I'd like to express special gratitude to Draga Potočnjak, my friend, author and a playwright. I never could have written the book as it is without her insightful comments and unselfish and unobtrusive advice.

The same goes for my friend, Matjaž Frangež, who gave me instructions and advice, using his healthy and just appropriate dosage of subversion. His persistence, lucidity, and legal expertise, that enabled him to get valuable documents from the government bureaucracy, are astounding and priceless.

I'd also want to thank my friend and a special educator, who wishes to stay anonymous. Without her unique capabilities and kind help, I might have been overburdened by writing the book.

Thank you to the Slovenian journalist, Tomaž Bukovec and the British investigative reporter, Paul Holden, who gave me access to part of their journalistic material.

The person, most deserving for the foreign edition of the book, Patriotism for Sale, is my dear friend and a fellow journalist, Vasilij Volarič from the United States, who offered to translate the book into English. He considered translating the book to be a unique challenge and, as a true professional, he undertook the assignment with utmost diligence.

The same gratitude goes to the British journalist and author, Marcus Ferrar who edited the foreign edition.

I'm also thankful to the photographers Joco Žnidaršič, Igor Mali, Tomaž Skale, Borut Krajnc and Luka Cjuha who kindly donated their photographs from their archives.

Special thanks also to all named and unnamed people that, without ill-intent, helped me with my journalist endeavor.

Last but not least, I'd like to point out that the publishing of the book *Patriotism for Sale* took place in very precarious circumstances for the Slovenian publishers and it would not be possible without the support, enthusiasm and understanding of Sanje Publishing Ltd and their staff to whom I'd like to express my utmost gratitude.

Here's to our supporters on IndieGoGo. Thank you.

Avgust Cvetežar, Barbara Furjan, Miljenko Horvat, IMI d. o. o., Doc Kosinski, Irina Kosinski, Tomaž Mrlak, Volker Plassmann and Jože Volarič.

# BIBLIOGRAPHY

Biserko, Sonja: *Vukovarska tragedija 1991* (Vukovar Tragedy 1991), Helsinški odbor za ljudska prava u Srbiji, Beograd, 2007.

Boljkovac, Josip: *Istina mora izaći van* (The Truth Must Get Out), Golden marketing – Tehnička knjiga, Zagreb, 2009.

Čelik, Pavle: *Izza barikad*, (Behind the Barricades), Delo, Ljubljana, 1992.

Drnovšek, Janez: *Moja resnica* (My Truth), Mladinska knjiga, Ljubljana, 1996.

Đikić, Ivica, K. D, P. B.: *Gotovina, stvarnost in mit* (Gotovina, Reality and Myth), Sanje, Ljubljana, 2014.

Feinstein, Andrew; H. P., P. B.: *The Shadow World*, The Penguin Group, London, 2011.

Frangež, Matjaž: *Kaj nam pa morete!, 1. del*, (What Can You Do to Us!, Part 1), Grafika Gracer, Celje, 2015.

Frangež, Matjaž: *Kaj nam pa morete!, 2. del*, (What Can You Do to Us!, Part 2), samozaložba, Radenci, 2008.

Guštin, Damijan, P. V.: *Boj na Holmcu, 27. in 28. junij 1991 in Koroška v vojni za obrambo neodvisnosti Republike Slovenije* (The Battle For Holmec, June 27 and 28, 1991, and Koroška in a War For Defending the Independence of the Republic of Slovenia), Ljubljana, 2006.

Hedl, Drago: *Glavaš – kronika jedne destrukcije* (Glavaš - Chronical of Destruction), Novi Liber, Zagreb, 2010.

Janša, Janez: *Premiki* (Manoeuvres), Mladinska knjiga, Ljubljana, 1992.

Jović, Borisav: *Poslednji dani SFRJ* (The Last Days of the SFRY), Kompanija Politika, Beograd, 1995.

Jurčič, Josip: *Spomini na deda in druge zgodbe* (Memories About My Grandfather and Other Stories), Mladinska knjiga, Ljubljana, 1967.

Kadijević, Veljko: *Moje viđenje raspada: Vojska bez države* (My Point of View About the Breakup: Military Without the State), Politika, Beograd, 1993.

Kafka, Franz: *The Trial*, translator: David Wyllie, Posting Date: August 13, 2012 [EBook #7849].

Kolšek, Konrad: *Prvi pucnji u SFRJ* (First Shots in the SFRY), Danas, Beograd, 2005.

Kosmač, Ciril: *Pomladni dan* (A Spring Day), Mladinska knjiga, Ljubljana, 1985.

Kranjc, F. Marjan: *Balkanski vojaški poligon*, (Balkan Military Polygon), Pro-Andy, Maribor, 2008.

Lesjak, Janez: *Na robu – Ljubljanska pokrajina med osamosvajanjem* (On the Edge – Ljubljana Region During the Battle For Independence), Modrijan, Ljubljana, 2011.

Levstik, Fran: *Martin Krpan* (Martin Krpan), Mladinska knjiga, Ljubljana, 2009.

Marijan, Davor: *Slom Titove armije* (The Breakdown of Tito's Army), Golden marketing – Tehnička knjiga, Zagreb, 2008.

Meier, Viktor: *Zakaj je razpadla Jugoslavija* (Why Yugoslavia Disintegrated), Znanstveno in publicistično središče, Ljubljana, 1996.

Mikulič, Albin: *Uporniki z razlogom – Manevrska struktura narodne zaščite*, (Rebels With a Cause – Manoeuvre Structure of National Protection), Slovenska vojska, Ljubljana, 2005.

Milčinski – Ježek, Frane: *Preprosta ljubezen*, (Simple Love), Sanje, Ljubljana, 2008.

Milčinski, Fran: *Butalci* (Dim-Witted), Karantanija, Ljubljana, 2001.

Pirjevec, Jože: *Jugoslovanske vojne 1991 – 2001* (Yugoslav Wars 1991 – 2000), Cankarjeva založba, Ljubljana, 2003.

Potočnjak, Draga: *Skrito povelje* (The Secret Order), Sanje, Ljubljana, 2013.

Praznik, Brane: *Trgovci s smrtjo* (Merchants of Death), samozaložba, Ljubljana 2007.

Prešeren, France: *Poezije* (Poems), Prešernova družba, Ljubljana, 2008.

Repe, Božo: *Jutri je nov dan* (Tomorrow is Another Day), Modrijan, Ljubljana, 2002.

Repe, Božo: *Viri o demokratizaciji in osamosvojitvi Slovenije, III. del: Osamosvojitev in mednarodno priznanje* (Sources about Democratisation and Independence of Slovenia, Part III, Independence and International Recognition), Arhivsko društvo Slovenije, Ljubljana, 2004.

Stanonik, Marija in Uršič, I.: *Moški na položajih, ženske v strahu, otroci na češnjah* (Men at Their Positions, Women in Fear and Children on Cherry Trees), Inštitut za slovensko narodopisje, Ljubljana, 2011.

Špegelj, Martin: *Sjećanja vojnika* (Soldier's Memories), Znanje, Zagreb, 2010.

Šurc, Matej, Z. B.: *V imenu države: Odprodaja*, Prva knjiga (In the Name of the State: Sale, First book), Sanje, Ljubljana, 2011.

Šurc, Matej, Z. B.: *V imenu države: Preprodaja*, Druga knjiga (In the Name of the State: Re-Sell, Second book), Sanje, Ljubljana, 2011.

Šurc, Matej, Z. B.: *V imenu države: Prikrivanje*, Tretja knjiga (In the Name of the State: Cover-Up, Third book), Sanje, Ljubljana, 2012.

Švajncer, J. Janez: *Obranili domovino* (They Defended the Homeland), Viharnik, Ljubljana 1993.

Vasić, Miloš: *Atlas organizovanog kriminala na Balkanu* (Atlas of the Organised Criminal in the Balkans), Vreme, Beograd, 2005.

Several authors: *20. obletnica Manevrske strukture narodne zaščite – zbornik prispevkov in razprav*, (20th Anniversary of the Manoeuvre Structure of National Protection - a Collection of Articles and Discussions), Zveza policijskih veteranskih društev Sever, Ljubljana, april 2011.

Several authors: *Prikrita modra mreža* (Hidden Blue Net), Inštitut za novejšo zgodovino, Ljubljana, 2010.

Several authors: *Stvaranje hrvatske države i Domovinski rat* (Creating the Croatian State and the Homeland War), Zagreb, 2006.

Several authors: *Trdna mreža: Osamosvajanje Slovenije v Zgornjem Posočju* (Firm Network: Battle For Independence in the Slovenian Region Upper Posočje), Območna ZVVS, Tolmin, 2008.

Vukšić, Dragan: *JNA i raspad SFR Jugoslavije* (YPA and the Breakup of Yugoslavia), Tekomgraf, Stara Pazova, Srbija, 2006.

Zlobec, Ciril: *Lepo je biti Slovenec, ni pa lahko* (It's Nice to Be a Slovene But It's Not Easy), Mihelač, Ljubljana, 1992.

# INDEX OF PERSONS

## A

Abdić, Fikret 331, 353, 357
Abramovich, Roman 326
Adžić, Blagoje 11, 33, 74, 83, 140, 166, 167, 171, 174, 175, 189, 190, 193
Andoljšek, Ciril 203
Andrijić, Ivan 34
Antall, József 57
Anžič, Daniel 264, 265, 277, 298, 310, 327
Arhar, France 264
Atelšek, Ivan 219, 220
Avramović, Dragoslav 268
Avramović, Života 141

## B

Babič, Bojan 109, 122, 123, 243
Badinter, Robert 70
Bajić, Ljubomir 141, 166
Bajramović, Sejdo 61
Bajuk, Andrej 374
Baker, James 80, 188
Balažic, Milan 346, 347
Balent, Jože 238
Balkan, Nedžad 360
Baraga, Viktor 348, 349
Bardos, Gordon N. 360
Bavčar, Igor 10, 42, 44, 48, 49, 56, 57, 75, 76, 77, 90, 91, 98, 138, 154, 162, 167, 168, 169, 195, 232, 233, 235, 250, 261, 284, 292, 296, 297, 298, 307, 309, 311, 312, 314, 316, 318, 328, 330, 352, 354, 368, 370, 385, 387
Bećir, Ivan 69, 78, 182, 287
Bekeš, Peter 297, 301, 368
Benko, Marijan 64
Berginc, Vito 130
Beznik, Vinko 45, 47, 48, 68, 69, 79, 91, 92, 98, 108, 169, 308
Biserko, Sonja 207, 393
Bizjak, Ivan 298, 317
Bobetko, Janko 182
Bobinac, Franjo 221
Bogataj, Miran 119, 196, 250, 255, 256, 257, 259, 260

Bohinc, Rado 298
Boldin, Franc 158, 159
Boljkovac, Josip 44, 49, 56, 233, 393
Borak, Neven (Veno Karbone) 147, 149, 150, 152, 153, 155, 156, 157, 158, 160, 161, 162, 163
Borštner, Ivan 40, 199
Bosnić, Bilal 360
Brajković, Josip 240
Bratušek, Alenka 225, 347, 375
Brejc, Miha 10, 78, 216, 233, 239, 250, 292, 296, 297, 298, 299, 300, 302, 311, 313, 316, 338, 368
Brezigar, Barbara 301, 314, 315, 371, 373, 386
Brodarac, Đuro 240
van den Broek, Hans 136, 137, 188, 191, 192, 193, 253, 254, 270
Brovet, Stane 82, 141, 208, 211
Broz, Jovanka 21, 25
Broz - Tito, Josip 19, 20, 21, 25, 34, 35, 59, 93, 125, 159, 173, 221, 229, 237, 274, 278
Brvar, Bogomil 250, 316, 317
Bučar, France 10, 49, 56, 169, 191, 205, 258, 271, 352, 368
Bulatović, Momir 70
Bush, George H. W. 80, 209, 210
Bush, George W. 101
Butara, Miha 47, 148, 149, 150, 151, 159, 160

## C

Carrington, Peter 254, 269
Cekuta, Jure Jurček 221, 222
Cerar, Miro 228
Churchill, Winston 157
Cimerman, Franci 203, 240, 241, 242, 243
Clinton, Bill 357, 358, 360
Craig, Larry E. 357
Cristiani, Fabio 366, 367
de Cuéllar, Pérez 269
Cvetežar, Avgust 151, 152, 153, 155, 156, 157, 158, 159, 160, 161, 162, 163
Cvetković, Dragan 112, 113

## Č

Čad, Marjan 82, 83, 84, 89
Čelik, Pavle 138, 393

Čengić, Hasan 11, 298, 301, 310, 327, 328, 330, 352, 355, 356, 357
Čepin, Jernej 265, 309, 310, 338, 344
Čerin, Janez 123
Čermak, Ivan 298, 329
Červ, Zdravko 182
Črnigoj, Dušan 279
Črnkovič, Ivan 221, 222, 225, 227, 228

## Ć

Ćeranić, Predrag 343

## D

Dafermos, Konstantin 11, 215, 219, 252, 277, 284, 288, 298, 307, 310, 314, 321, 322, 323, 324, 325, 326, 327, 330, 332, 333, 334, 337, 339
Dedaković - Jastreb, Mile 204
Dedić, Ibrahim 264
Delić, Mićo 72, 73, 74, 75, 77, 165
Dembowski, Jerzy 11, 323, 324, 325
Dilberović, Suada 351
Dolanc, Stane 24, 25, 26, 29
Dolenc, Pavle 300, 301
Domadenik, Milan 250
Dovžak, Marjan 131
Dragić, Dragan 110
Drašček, Vid 123
Drašković, Vuk 59
Draušbaher, Ivan 10, 63, 122, 215, 216, 217, 218, 219, 220, 221, 298
Drnovšek, Dušan 161
Drnovšek, Janez 10, 26, 32, 60, 61, 140, 162, 191, 194, 208, 244, 248, 254, 298, 299, 302, 315, 316, 344, 352, 353, 354, 356, 370, 371, 373, 374, 393
Drobnič, Anton 301, 315
Dudaković, Atif 331

## E

Eagleburger, Lawrence 209, 210
Erčulj, Karmen 104
Erjavec, Karl 222, 225, 347

## F

Fanelli, Aldo (Chiesa, Gino; Ciccarelli, Luigi) 265
Feinstein, Andrew 69, 341, 393
Ferlinc, Alojz 73, 74, 75, 77, 166
Ferš, Drago 294
Fila, Toma 20, 21
Frangež, Matjaž 102, 122, 186, 249, 355, 357, 393

## Đ

Đikić, Ivica 267, 393
Đozo, Fadil 266

## G

Gaddafi, Moamer 278
Galbraith, Peter 360
Gantar, Damjan 314
Gaspari, Mitja 280
Gašpić-Kljaković, Marinko 100, 101, 102
Gelli, Licio 337, 338
Genscher, Hans-Dietrich 167, 254
Glavaš, Branimir 238, 239, 393
Gligorov, Kiro 71
Gorbachev, Mikhail 17, 21, 59, 60, 272
Gorenak, Vinko 225
Gorinšek, Karlo 190
Gotovina, Ante 361
Gračanin, Petar 46
Grošelj, Andrej 157, 158
Grošelj, Boštjan 152, 158
Grubelič, Sandi 54, 55, 56
Grujović, Dragomir 144, 145
Guštin, Damijan 111, 112, 393

## H

Hadžić, Goran 57
Hari, Danilo 116
Ali Hassanein, El Fatih 355
Hedl, Drago 239, 393
Henigman, Žarko 136, 247
Hitler, Adolf 157, 200
Hočevar, Ivan 23, 24, 25, 27, 28, 29, 31, 35, 37, 38, 39, 45, 46
Hočevar, Marjan 134
Hofer, Manfred 66
Hoffmann, Dieter 355
Holbrooke, Richard 358
Holden, Paul 339, 340, 341, 342, 343
Horaček, Božidar 130
Horvat, Ivan 310
Horvat, Zlatko 239
Hribar, Spomenka 377, 378, 389
Hribar, Tine 275
Hulkkonen, Heikki 227
Hussein, Saddam 69, 101, 278

## I

Ilc Mortin, Nada 242

Ivić, Pavle 207
Ivković, Ante 215
Izetbegović, Alija 11, 71, 168, 300, 327, 328, 330, 331, 351, 356, 358, 359, 360

## J

Jamnik, Alojz 277
Jamnik, Janez 277
Janša, Janez Ivan 10, 13, 14, 27, 30, 31, 37, 38, 40, 42, 44, 46, 47, 48, 49, 52, 54, 55, 56, 57, 61, 63, 66, 67, 69, 72, 75, 78, 79, 84, 86, 90, 91, 95, 96, 97, 98, 99, 100, 101, 102, 103, 107, 117, 121, 124, 133, 134, 149, 151, 152, 153, 154, 155, 157, 158, 159, 161, 162, 163, 167, 168, 169, 173, 179, 183, 184, 192, 194, 195, 200, 201, 202, 203, 204, 205, 206, 210, 212, 216, 218, 220, 221, 224, 225, 226, 227, 228, 230, 232, 233, 234, 235, 238, 239, 240, 241, 243, 244, 245, 248, 249, 250, 252
Janša, Rajko 55
Jeglič, Roman 298, 299
Jelinčič, Zmago 199, 296
John Paul II. 271, 278
Jokić, Ljubivoj 259, 260
Jovanović, Ljubo 218, 220, 313
Jović, Borisav 11, 26, 27, 32, 33, 34, 59, 72, 83, 187, 189, 191, 194, 208
Jurčič, Josip 228, 229, 393
Jurjević, Zvonko 139, 140

## K

Kacin, Jelko 10, 67, 68, 96, 99, 100, 101, 103, 124, 220, 232, 246, 248, 264, 265, 280, 282, 312, 370, 371, 373, 374
Kadijević, Veljko 11, 32, 33, 34, 35, 39, 40, 58, 59, 60, 61, 83, 87, 139, 140, 141, 169, 187, 189, 190, 208, 211, 393
Kafka, Franz 105, 393
Kalan, Jože 93, 94, 99, 257
Karadžić, Radovan 342, 343, 344, 346, 356
Kardelj - Krištof, Edvard 44, 121
Al Kassar, Monzer 323, 324
Klavora, Mitja 64, 68, 226, 243, 244, 312
Klinar, Rina 141
Kmecl, Matjaž 10, 25, 38, 232, 235
Kodrič, Danilo 299
Kohl, Helmut 188, 270
Kolšek, Konrad 74, 82, 85, 89, 90, 114, 139, 140, 141, 209, 393
Komar, Silvo 72, 73, 77, 78, 356

Kos, Drago 14, 15, 238, 280, 295, 314, 315, 317, 330, 373
Kos, Franc 207
Kosi, Franc 284, 298, 307, 308, 309, 310, 311, 312, 313, 314, 315, 317, 321, 333
Kosmač, Ciril 143, 394
Kosmač, Drago 125, 126, 127, 128
Koščak, Anton 298
Košir, Franc 292
Kovačič, Danilo 299
Kovačič, Edvard 125, 126
Kovič, Kajetan 369
Kozinc, Miha 298
Kraljević, Blaž 205
Kranjc, Marjan F. 394
Kranjc, Viktor 132, 133, 134
Krejan, Simon 371
Krile, Davor 267, 361, 393
Krivic, Matevž 261
Krkovič, Anton 10, 30, 31, 42, 43, 44, 47, 49, 85, 95, 97, 98, 99, 104, 120, 121, 122, 169, 203, 221, 222, 225, 227, 228, 241, 371
Kučan, Milan 10, 25, 26, 27, 31, 32, 33, 34, 37, 38, 39, 40, 42, 49, 56, 61, 66, 70, 82, 85, 98, 137, 139, 149, 155, 162, 167, 169, 170, 187, 191, 192, 232, 234, 235, 251, 254, 263, 272, 275, 300, 351, 352, 353, 354, 355, 356, 359, 368, 369, 378
Kušar, Janez 158
Kuzin, Vladimir 340
Kuzma, Danijel 75

## L

Lacković, Milan 245
bin Laden, Osama 355
Lainović, Branislav 343
Lajovic, Dušan S. 348
Laubič, Franc 323
Leban, Tonček 129
Leonardi, Fulvio 340
Lesjak, Janez 151, 152, 153, 154, 155, 156, 157, 158, 159, 160, 162, 163, 394
Letica, Sveto 102
Levpušček, Julij 127, 128
Levstik, Fran 273, 394
Lipič, Ladislav 134
Lisjak, Srečko 128, 169
Lovšin, Andrej 10, 69, 99, 100, 101, 102, 103, 150, 151, 200, 201, 202, 203, 216, 233, 238, 239, 240, 241, 242, 243, 244, 245, 246, 247, 248, 249, 250,

Lukač, Brane 201, 372
Lukman, Miklavž 294
Lukšič, Igor 345, 346, 347

# M

Madon, Zoran 314, 316, 356
Maksimovič, Dragan 127
Makuc, Darko 299
Mamula, Branko 173
Marijan, Davor 211, 248, 394
Markač, Mladen 361
Marković, Ante 11, 52, 71, 80, 81, 83, 139, 169, 171, 191, 193, 208, 209, 210, 366
Martinović, Jozo 267
Mate, Dragutin 243
Mazzega, Lorenzo 339, 342, 343
Medeot, Marino 120
Medvedev, Dmitry 190
Medved, Roman 130
Meh, Rajko 48
Meier, Viktor 34, 36, 57, 61, 71, 167, 394
Menart, Janez 369
Mesić, Stipe 11, 36, 49, 59, 72, 136, 137, 169, 171, 191, 195, 208, 230
Mesić, Tomislav 133
De Michelis, Gianni 136, 137, 168, 187, 191, 367
Miklavčič, Franc 299
Miklavčič, Marjan 68, 372, 373
Miklavčič, Tomaž 104, 299, 300, 301, 368, 386
Mikulič, Albin 394
Milanović, Dafina 267
Milčinski, Fran 19, 81, 86, 87, 283, 380, 394
Milčinski - Ježek, Frane 380, 394
Miloševič, Vladimir 47, 72, 73, 74, 75, 76, 77, 78, 79, 165, 179, 181, 186, 196, 199, 244, 245, 248, 262
Milošević, Slobodan 11, 21, 50, 51, 52, 58, 59, 61, 70, 72, 79, 83, 187, 189, 190, 191, 193, 202, 211, 253, 254, 267, 268, 270, 306, 358, 359, 367, 375
Mladić, Ratko 36, 37, 207, 215, 356, 358, 360, 361
Mlakar, Peter 133, 134
Moge, Rudolf 100, 102, 103, 181, 195, 241, 242, 243, 245, 356, 374
Mohorko, Rafael 181, 298
Moljk, Dušan 84, 292
Mrlak, Emilija 95, 103, 104, 105
Mrlak, Toni 92, 93, 94, 95, 96, 98, 99, 100, 103, 104, 105, 117

# N

Niittynen, Reijo 227
Njavro, Darko 120, 131, 371

# O

Oman, Friderik 339
Oman, Ivan 10, 25, 38, 61, 232, 235
Oman, Nikolaj 9, 11, 264, 265, 277, 280, 281, 284, 286, 288, 295, 298, 307, 308, 309, 310, 314, 324, 333, 337, 338, 339, 340, 341, 342, 343, 344, 345, 346, 347
Orgolič, Anton 115
Osterman, Andrej 372
Ožbolt, Drago 24

# P

Pace, Barnaby 341
Paradžik, Ante 200, 204
Paraga, Dobroslav 11, 200, 201, 202, 203, 204, 205
Paš, Ingo 368
Pavček, Tone 369
Pavelić, Ante 200, 204
Pavelić, Boris 267, 361, 393
Peinkiher, Anton 123, 131, 203, 231, 232, 241, 242, 245, 276
Peperko, Edvard 151
Perko, Anton 313
Perko, Jože 121
Perković, Josip 298
Peša, Mladen 322
Petan, Branko 76, 186
Peterle, Lojze 10, 25, 27, 53, 79, 85, 139, 191, 205, 232, 233, 261, 278
Petkovšek, Jože 123
Petrač, Hrvoje 267, 332
Petrič, Ernest 252
de Pinheiro, João 191
Pirjevec, Jože 358, 394
Pirkovič, Ljubo 238, 292, 313
Plavšić, Biljana 343
Plut, Dušan 10, 25, 32, 33, 34, 37, 55, 232, 235, 236, 258
Podbregar, Andrej 371
Podgoršek, Boštjan 279
Podobnik, Marjan 340
Polovič, Jože 29, 30
Poos, Jacques 136, 137, 187, 188
Popov, Berislav 115, 116
Porča, Muhamed 360

Požar, Dušan 300, 301
Prah, Roman 313
Pravdič, Vladimir 75, 76
Praznik, Brane 202, 221, 242, 287, 394
Prebilič, Vladimir 111, 112
Predanič, Miro 298, 301, 316, 329, 330
Pregl, Živko 82
Premk, Martin 148, 150, 160
Prešeren, France 92, 93, 142, 394
Prešeren, Marko 207
Prinčič, Matjaž 327
Prtenjača, Šime 216, 217
Pučnik, Jože 25, 200, 274, 275, 299, 353
Putnik, Radomir 259

# R

Rabus, Dieter 219
Radobuljac, Frane 200, 201
Rašeta, Andrija 11, 119, 132, 135, 160, 196, 255, 256, 269, 306
Ražnatović - Arkan, Željko 35, 57
Repe, Božo 57, 59, 98, 188, 190, 191, 211, 252, 394
Režek, Damjan 102, 201, 262, 372
Ribarič, Miha 80
Riedl, Hans Wolfgang 11, 222, 223, 224, 225, 226, 230
Rijavec, Elo 98, 99, 122, 284, 285, 287, 288, 289
Rikli, Arnold 340, 345
Rozman - Stane, Franc 151
Rožič, Marjan 113, 114, 139, 140, 141
Rupel, Dimitrij 10, 49, 82, 137, 190, 191, 232, 352, 353, 354, 370

# S

Sanader, Ivo 375
Sandevski, Ljubomir 94
Sapunxhiu, Riza 61
Sibinovski, Bojanče 92, 94, 99, 104
Simčič, Darij 127
Sirithaporn, Apichat 223, 224
Sirše, Janez 298
Slapar, Janez 10, 41, 42, 46, 47, 98, 120, 140, 148, 149, 150, 152, 154, 155, 161, 248, 250, 352
Slukan, Mirko 291, 292, 294
Smolej, Stane 201, 372
Stajčić, Nikola 257, 259
Stajčić, Slobodan 175
Stalin, Josip Visarijonovič 20, 157
Stanič, Jože 220

Stanovnik, Janez 82, 162
Stipetić, Petar 175
Stojanović, Mirko 74, 75
Stojanović, Petar 323
Stojanović, Slaviša 74
Strajnić, Mile 127
Strehar, Marjan 13, 177, 178, 180, 181, 182, 184, 185, 186
Streshinsky, Dmitry 11, 325, 326
Striedinger, Günter 334
Stušek, Janko S. 37
Suhadolnik, Robert 371

# Š

Šarinić, Hrvoje 317
Šeks, Vladimir 183
Šešok, Dušan 183, 279, 280
Šiljak, Ante 201, 202
Šimčik, Josef 77
Šipčić, Tomislav 159
Špegelj, Martin 11, 44, 49, 54, 56, 57, 58, 86, 133, 183, 194, 233, 237, 395
Štajner, Alojz 75, 76
Šter, Andrej 293, 294, 295, 315
Šuligoj, Bojan 25
Šuligoj, Slavko 127
Šumandl, Franc 127
Šušak, Gojko 71, 102, 293, 306, 332, 355
Šuštar, Alojzij 377
Šuštar, Bogomir 93, 99, 257, 378
Švajncer, Janez J. 127, 395

# T

Tadić, Drago 239
Tasić, David 40
Taylor, Charles 341
Thaçi, Hashim 376
Tovšak, Hilda 279, 335
Trbovšek, Franc 139, 376
Trifunović, Vlado 116
Truman, Harry 157
Tuđman, Franjo 11, 26, 36, 50, 57, 58, 64, 70, 80, 86, 123, 137, 186, 190, 191, 193, 194, 202, 204, 211, 212, 237, 267, 332, 355, 358
Tuđman, Miroslav 237
Tupurkovski, Vasil 59, 208
Al Turabi, Hassan 355
Türk, Danilo 134

## U

Uršič, Irena 115, 394
Ušeničnik, Bojan 155

## V

Valvasor, Janez Vajkard 7
Vance, Cyrus 252, 306
Vasić, Miloš 53, 62, 395
Vasiljević, Jezdimir 267
Velikonja, Rajko 130, 131
Velišček, Jožica 242
Vičar, Stane 218
Vidmar, Marjan 93, 94, 108
Virant, Gregor 225, 302
Višnar, Fran 36, 54, 329, 332, 333
Vrabič, Rudi 159
Vučić, Borka 268
Vukina, Josip 11, 69, 135, 182, 237, 245, 246, 247, 248, 287, 288, 289, 313, 340
Vukšić, Dragan 60, 395

## W

Wickey, Paul 64
Wiitakorpi, Jorma 227
Wolf, Walter 11, 221, 222, 223, 224, 225, 226, 229, 230

## Y

Yazov, Dmitry 60

## Z

Zagorec, Vladimir 11, 298, 329, 332, 333, 334
Zagožen, Jože 221, 222, 224, 225, 226, 250
Zavrl, Franci 40
Zhukova, Dasha 326
Zhukov, Alexander Borisovich 326
Zhukov, Georgy 326
Zidar, Ivan 276, 277, 278, 279
Zimmerman, Warren 209
Zlobec, Ciril 10, 25, 38, 39, 55, 56, 84, 167, 168, 192, 232, 235, 365, 366, 367, 368, 369, 370, 371, 395
Zorc, Milovan 155
Zorko, Milan 120, 150, 151, 168, 169, 237, 240, 241, 245, 246
Zupan, Peter 69, 101, 122, 131, 196, 218, 221, 238, 244, 245
Zvonar, Ludvik 10, 30, 31, 53, 63, 66, 67, 70, 77, 216, 220, 246, 264, 265, 266, 274, 276, 277, 278, 280, 281, 282, 284, 285, 286, 288, 295, 296, 298, 309, 310, 313, 314, 327, 328, 332, 333, 339, 341

## Ž

Železnik, Anton 75
Žerjav, Radovan 225
Župa, Nikša 11, 298, 327, 329, 330, 331, 332

Made in the USA
Middletown, DE
13 December 2018